Additive Manufacturing

Artificial Intelligence and Optimization

Editors:

Vijaykumar S. Jatti

Professor, Symbiosis Skills and Professional University
Kiwale, Pune, Maharashtra, India

A. Saiyathibrahim

Department of Mechanical Engineering
Saveetha School of Engineering
Saveetha Institute of Medical and Technical Sciences
Chennai, Tamil Nadu, India

CRC Press
Taylor & Francis Group
Boca Raton London New York

CRC Press is an imprint of the
Taylor & Francis Group, an **informa** business
A SCIENCE PUBLISHERS BOOK

First edition published 2027
by CRC Press
2385 NW Executive Center Drive, Suite 320, Boca Raton FL 33431

and by CRC Press
4 Park Square, Milton Park, Abingdon, Oxon, OX14 4RN

CRC Press is an imprint of Taylor & Francis Group, LLC

Library of Congress Cataloging-in-Publication Data (applied for)

ISBN: 978-1-041-03872-6 (hbk)
ISBN: 978-1-041-06496-1 (pbk)
ISBN: 978-1-003-63567-3 (ebk)

DOI: 10.1201/9781003635673

Typeset in Times New Roman
by Prime Publishing Services

Preface

Additive Manufacturing (AM) has emerged as one of the most transformative technologies of the twenty-first century, revolutionizing the way products are designed, developed, and manufactured. By enabling layer-by-layer fabrication directly from digital models, AM offers unprecedented opportunities for complex geometries, rapid prototyping, mass customization, reduced material wastage, and decentralized manufacturing. Technologies such as Selective Laser Sintering (SLS), Fused Deposition Modeling (FDM), and Stereolithography (SLA) have significantly advanced manufacturing capabilities across industries including aerospace, automotive, healthcare, biomedical engineering, defense, and consumer products.

In recent years, substantial progress in materials science, computational modeling, artificial intelligence, and process optimization has further accelerated the evolution of additive manufacturing. Researchers and industries are increasingly focusing on improving dimensional accuracy, mechanical strength, surface quality, and production efficiency. The integration of machine learning, artificial neural networks, deep learning, and optimization techniques has opened new avenues for predictive manufacturing and intelligent process control.

Despite remarkable advancements, several challenges continue to hinder the widespread industrial adoption of additive manufacturing. Issues related to interlayer bonding, porosity, residual stresses, surface finish, process repeatability, scalability, and quality assurance remain critical concerns. Moreover, the need for standardization, sustainable material selection, and reliable process optimization continues to drive extensive research worldwide. Organizations such as ASTM International are actively contributing toward establishing standards and best practices for additive manufacturing technologies.

This book presents a comprehensive exploration of the fundamentals, applications, emerging trends, and future directions of additive manufacturing. It brings together contributions focusing on hybrid aluminum matrix composites, artificial intelligence-based prediction models, convolution neural networks, topology optimization, and advanced optimization techniques such as GRA, TOPSIS, COPRAS, and MOORA. The chapters collectively highlight the growing convergence of additive manufacturing with data-driven technologies and intelligent systems.

The editors sincerely hope that this book will serve as a valuable resource for researchers, academicians, industry professionals, and students seeking deeper insights into advanced additive manufacturing technologies and their future potential. It is our aspiration that the knowledge presented herein will inspire further innovation and contribute toward the advancement of smart, sustainable, and intelligent manufacturing systems.

Contents

List of Contributors

A. Umapathy
Department of Mechanical Engineering, SRM Institute of Science and Technology, Vadapalani, Chennai, Tamil Nadu 600026, India.

Adavayya Sangayya Chikkamath
KLE Technological University, Hubballi 580031, Karnataka, India.

Ahmet Atak
Department of Industrial Product Design, OSTİM Technical University, OSTİM, 06374 Ankara/ Türkiye.

Ajay Tanikonda
Independent Researcher, San Ramon, California, USA.

Akash Abaji Kadam
Mechanical Engineer, Servotech Inc, Chicago, IL, USA.

Hassanein I. Khalaf
Mechanical Department, Engineering College, University of Basrah, Basrah 61004, Iraq.

K. Premnath
Assistant Professor, Dhanalakshmi Srinivasan College of Engineering, Coimbatore, India.

K.G. Anbarasu
Department of Mechanical Engineering, SRM Institute of Science and Technology, Vadapalani, Chennai, Tamil Nadu 600026, India.

K.G. Ashok
Department of Mechanical Engineering, Easwari Engineering College, Ramapuram, Chennai, Tamil Nadu 600089, India.

M Arunadevi
Associate Professor, Department of Mechanical Engineering, Dayananda Sagar College of Engineering, Bangalore 560111, India.

M. Pradeep
Assistant Professor, Karpagam Institute of Technology, Coimbatore, India.

M. Satthiyaraju
Department of Mechanical Engineering, SRM Institute of Science and Technology, Vadapalani, Chennai, Tamil Nadu 600026, India.

M.G. Kulthe
Department of Metallurgy and Materials Engineering, COEP Technological University, Shivajinagar, Pune 411005, Maharashtra, India.

Murthy S.S.S. Chavali Yadav
Department of Research and Development, Dr. Vishwanath Karad MIT World Peace University, Pune 411038, Maharashtra, India.

Nilesh Kolhe
Department of Research and Development, Dr. Vishwanath Karad MIT World Peace University, Paud Road, Kothrud, Pune 411038, Maharashtra, India.

P.P. Deshpande
Department of Metallurgy and Materials Engineering, COEP Technological University, Shivajinagar, Pune 411005, Maharashtra, India.

Pralhad Pesode
Department of Mechanical Engineering, Dr. Vishwanath Karad MIT-World Peace University, Pune 411038, Maharashtra, India.

R. Majusha
Department of Electrical and Electronics Engineering, St Peter's College of Engineering and Technology, Avadi, Chennai, Tamil Nadu 600054, India.

R. Venkatesh
Assistant Professor, Coimbatore Institute of Engineering and technology, Coimbatore. India.

Raheem Al-Sabur
Mechanical Department, Engineering College, University of Basrah, Basrah 61004, Iraq.

Sagar Wankhede
School of Mechatronics Engineering, Symbiosis Skills and Professional University, Pune, Maharashtra, India.

Sathiesh Veera
Solution Architect, TruStartIT LLC, Richardson, TX, USA.

Shailendra Dayane
Department of Mechanical Engineering, Dr. Vishwanath Karad MIT-World Peace University, Pune 411038, Maharashtra, India.

Siddhalingeshwar I. G.
KLE Technological University, Hubballi 580031, Karnataka, India.

Subba Rao Katragadda
Independent Researcher, Lathrop, California, USA.

V.S. Kathavate
Department of Defence System Engineering, SRM University AP, Neerukonda-Tadikonda Road, Mangalagiri Mandal, Guntur District, Neerukonda 522 502, Andhra Pradesh, India.

Vedant Utikar
Research Engineer, Fatigue and Materials Centre of Excellence (FMCE), The Automotive Research Association of India (ARAI).

Vikas R. Gulia
Symbiosis International (Deemed) University, Symbiosis Institute of Technology, Pune, India.

Vinayakumar R.
Assistant Professor, Department of Mechanical Engineering, Dayananda Sagar College of Engineering, Bangalore 560111, India.

Vinod Kumar V. Meti
KLE Technological University, Hubballi 580031, Karnataka, India.

1 Fundamentals of Additive Manufacturing

Vikas R. Gulia[1*]

1. Introduction

The fourth industrial revolution is characterized by a radical change in the manufacturing landscape, which is being driven by digitalization and automation as well as interconnected systems. One of the most significant changes has been facilitated by additive manufacturing (AM) that has transformed the way goods are designed, developed, and produced in many industries such as aerospace, automotive, medical, and consumer goods among others. AM is based on layer-by-layer printing of objects of desired shape using digital models, unlike conventional subtractive technology manufacturing methods in which the product is formed by removing material on a solid block until a desired shape is obtained, or formative technology manufacturing methods in which the final product is made by molding material or forging it into shape. AM has a fundamental impact on the economics of modern manufacturing, the design possibilities provided by technology, and the supply chain dynamics of manufacturing products [1].

The importance of AM is not limited to technological novelties. This manufacturing model allows developing complex forms and intricate geometries that are impossible to work with conventional technologies, and at the same time has a positive impact on the environment because of less material waste and energy use [2]. The capability of AM to fabricate advanced components with design freedom never experienced, the capability to incorporate various materials on a voxel basis and produce products to meet customer-specific requirements has made AM a key enabler of innovation in a wide range of application fields [3]. Since it has been used to create implants tailored to patients and lightweight aerospace parts, as well as to allow rapid prototyping and on-demand manufacturing, AM has shown to be a transformative technology that is constantly growing as the technology becomes more mature.

Nevertheless, in spite of the major improvements that have been made in the last 40 years, AM has not been able to live up to its hype as a revolution in serial manufacturing, especially of metal parts. The underlying issues include the lack of understanding of processes, material sources, cost of production, quality assurance, and the strength of value chains globally [4]. The distance between the theoretically possible AM and its practical industrial application is still very

[1] Symbiosis International (Deemed) University, Symbiosis Institute of Technology, Pune, India.
* Corresponding author: vikas.gulia@sitpune.edu.in

DOI: 10.1201/9781003635673-1

wide, so further research, development, and standardization efforts are required to overcome technical, economical, and organizational obstacles.

The chapter is a general introduction to AM, which gives the reader a well-rounded overview of the principles of the AM, its variety of technologies, limitations at present, and prospects. Being the beginning of a book, it forms the fundamental knowledge base on which later chapters are going to be constructed and provides breadth of information and depth of essential areas. The chapter also summarizes the recent academic literature published last decade, which means that the readers of the book can have an idea of the latest trends, developments, and challenges that shape the current AM landscape. Using various prisms of AM, including historical, technical, economic, and strategic, this chapter provides the reader with the conceptual foundation to comprehend the role of AM in transforming manufacturing in the 21st century and beyond.

2. Principles of Additive Manufacturing

2.1 Conceptualization and Key Terms

The term "additive manufacturing" is defined as reference to the process of assembling materials to create objects based on three-dimensional (3D) model data in accordance with the international standard ISO/ASTM 52900, that is, by joining materials, rather than by removing material, as is done in subtractive manufacturing methodologies [5]. AM is also the official, industry-wide name of what is informally referred to as "3D printing" but the latter term tends to be considered in terms of consumer-style uses unlike AM which represents both ends of the spectrum of desktop printing all the way to industrial manufacturing systems.

The main idea of all AM technologies is the conversion of digital design information into some physical objects with the help of some layer-by-layer fabrication process. This digital-to-physical process starts with a 3D computer-aided design (CAD) model which the model undergoes on the computer through special software to slice the CAD into thin cross-sectional layers. These digital slices are then used as instructions for the AM system on where, when, or how to deposit, fuse, or solidify material to recreate the original 3D geometry [3], and which, after being stacked sequentially, reconstruct the original 3D geometry. This first-hand relationship between the digital design and the act of physical fabrication has removed much of the intermediate steps, tooling needs, and geometry restrictions of traditional manufacturing processes.

One of the most important features of the AM conceptual background is that it is material-agnostic. Although certain AM operations are designed to work with certain material groups metal, polymer, ceramics, or composite materials, the concept of building up in layers is universal. This flexibility has allowed AM to grow beyond its original interest in polymeric materials to include a wide variety of materials that can be used to make functional parts that can be used in the end [1]. The development of the technology has been marked by ongoing growth of the material capabilities, both in the kind of materials that can be worked, and in the properties that can be obtained in finished components, which has seen AM being transformed into a useful means of production rather than a prototyping technology.

2.2 Historical Evolution

AM dates back its history to the 1980s, when the first commercial AM systems were developed to be used as a rapid prototyping tool. The AM industry started with the pioneering efforts of Charles Hull who developed the process of stereolithography (SLA) in 1984 and established 3D Systems Corporation in 1986 [6]. In his patent of SLA, Hull proposed a process to build 3D objects by reinforcing a single layer of ultraviolet-curable material over another, and this principle would be used to build the underlying layer-by-layer pattern through which the field would be known. There were several AM techniques such as selective laser sintering (SLS), developed by Carl Deckard at the University of Texas at Austin [7], and fused deposition modeling (FDM), developed by Scott Crump at Stratasys [8], among others which increased the variety of materials and applications, and also the development of other AM processes in the late 1980s and 1990s.

AM was used mainly in the rapid prototyping stage of the first 20 years, the physical models were produced at a very fast rate to visualize the designs, test form and fit, and help in communication. The value of the technology was based on speed and design flexibility as opposed to mechanical properties or economic factors of production. Early AM-system-produced parts were largely regarded as inappropriate to functional usage because of the constraint in material characteristics, surface finish, and dimensional control. Nevertheless, this prototyping period was essential to establish the core capabilities of AM and create the knowledge base that would be used to advance in the future [2].

The 2000s saw a slow change in the spheres of rapid prototyping to rapid manufacturing with the increased possibilities of process control, the development of material science, and machine capabilities. The production of functional parts with reasonable mechanical properties became a possibility. This change was especially obvious in industries with high value products like aerospace and medical equipment, where the capabilities of AM to produce complicated geometries, save weight, and modify items were worth the expenditure of higher expenses per part. Around the end of the 2000s and the beginning of the 2010s, the expiration of major patents in AM technology triggered a democratization of the technology, spawning an explosion of low-cost desktop 3D printers and opening up a renewal of innovation in the field [3].

The epoch, which began around 2015 up to the present, has witnessed AM becoming a production technology that was able to produce high-end products in various industries. The major progress has been made around multi-material printing, large-scale fabrication, automation of the processes, and the addition of digital manufacturing ecosystems [1]. The technology has been demonstrated to be worthwhile not only to low-volume and high-complexity but also to medium-volume production when competitive benefits can be gained through the customization, fast iteration, or supply chain benefits. The strategic value of AM was further emphasized by the COVID-19, with manufacturers using the technology to quickly make personal protection equipment, medical devices, and other essential supplies following disruptions in the supply chain [9].

2.3 Fundamental Principles and the Concept of Layer-by-Layer

The concept of the layer-by-layer construction is the key feature of all AM procedures which makes them unique in relation to the traditional manufacturing process. This principle works on the idea of constructing 3D objects by the superposition of two-dimensional (2D) cross-sections, similar to the way a stack of sheets of paper forms a 3D object. Each of the layers is a thin slice of the final part geometry, generally between 20 micrometers and a few millimeters thick as applied to a particular AM process and the specific desired resolution [3].

The layer-by-layer strategy has a number of inherent strengths that justify the transformative power of AM. To begin with, it can allow geometric complexity without any further cost or time penalty. In traditional production, specialized tooling, multiple operations, or even impossible to create are required in complex features like internal channels, undercuts, or complex surface textures. With AM, complexity is literally free: a very complicated part takes no longer or more expensive to manufacture than a plain part of the identical volume since the machine merely scans the digital instructions to the given layer irrespective of the geometric complexity [2]. This property has far-reaching implications on the optimization of designs, and engineers will be able to come up with structures that are optimized to perform and not to be manufactured.

Second, the layer-by-layer principle makes it easier to customize materials and combine different materials. Since all layers are built separately, it is possible to change the material composition, microstructure, or properties at any point in the volume of a part. This allows functional grading of materials, whereby properties continuously vary between one part to another, or multi-material components, where different materials with different properties are manufactured as one component [10]. This type of material heterogeneity is highly challenging or even impossible to realize by traditional production techniques, which creates new opportunities in terms of optimization of component performance.

Third, the mass customization and personalization made by the layer-by-layer construction is made feasible at scales that were not economical before. Because AM systems create parts based on digital models without the need to use specific tooling or fixtures, all that is needed to switch one design to another is to load a new digital file. This enables the manufacturers to make personalized products to meet the needs of individual customers without the set-up costs and time wastage of retooling the traditional manufacturing processed. This has specifically helped the medical device industry, which has created patient-specific implants, prosthetics, and surgical guides that fit specific anatomies.

Nonetheless, the layer-by-layer principle presents certain limitations and challenges as well. The discrete layering process forms anisotropic material properties, in which mechanical strength and other properties vary with the direction of measurement with reference to the build direction. The parts are usually the strongest in the plane of each layer but weak in the direction that is perpendicular to the layers where the bonding between layers can be incomplete [11]. Surface finish is also another effect of layering in that, the stair-stepping effect of the discrete layers is fabricated to create surface roughness which can lead to post-processing to obtain smooth surfaces needed in applications. Also, the layer-by-layer method may take longer to build simple geometries or high-volume manufacture than traditional techniques due to the fact that all individual layers have to be finished before any other part may start.

3. Classifications of AM Technologies

3.1 Overview

The wide variety of AM technology has led to the need to create classification systems to organize and comprehend the different processes. The best adopted classification framework is defined in the international standard ISO/ASTM 52900 which classifies AM processes into seven main technology groups based on their fundamental working principles and material deposition mechanisms. This standardized classification has offered a common language between industries, the academia, and regulatory agencies that enables an easy flow of communication and comparisons between various AM technologies.

The seven main types that are defined according to ISO/ASTM 52900 are: (1) vat photopolymerization [12], (2) material extrusion [13], (3) powder bed fusion [14], (4) material jetting [15], (5) binder jetting [16], (6) directed energy deposition [17], (7) sheet lamination [18]. All categories include several specific processes that share the same principles but can vary in details of implementation, material, or scale. This classification system has been proven robust and comprehensive, and it has kept pace with the evolving world of technology to include both established technologies as well as emerging processes.

Alternative schemes of classification have been suggested on different basis such as mode of fixing the layer (solidification, binding, or joining), type of material (polymers, metals, ceramics, or composites), state of the feedstock material (liquid, powder, solid, or wire). Although these other frameworks present useful insights in certain applications, the ISO/ASTM system has gained the universal status of the reference framework. The knowledge of this classification is necessary to work in the AM landscape, choose the right technologies to use in certain applications, and understand the unique strengths and weaknesses of various solutions.

The subsections below discuss each of the seven large categories of AM technologies individually, taking a closer look at their operational principles, common process variations, materials they use, most common applications, and their characteristic strengths and constraints. This all-in-one survey gives readers the base knowledge needed to understand the variety of technology available in AM and make informed decisions about which process to use for applications.

3.2 Powder Bed Fusion

Powder bed fusion (PBF) is one of the most commercially important and technically mature types of AM especially for metal parts. The process of PBF is used to produce parts by selective fusion of areas in a powder bed with thermal energy, usually a laser or electron beam. The basic process is to spread a thin layer of powder material on a build platform, selectively melt/sinter the powder in the location representing the cross section of the part for that layer, lower the platform by one layer thickness, spread a new layer of powder material, and repeat until the part is finished [14].

The major PBF processes are selective laser sintering (SLS) for polymers [19], selective laser melting or laser powder bed fusion (LPBF) for metal [20], direct metal laser sintering [21], electron beam melting (EBM) [22], and multi-jet fusion (MJF) [23]. Both variants use various energy sources and process parameters which are optimized to apply to material classes. SLS was developed in the 1980s which uses the CO_2 laser to sinter the thermoplastic powder like polyamide (nylon) and produce parts with good mechanical properties and no support structures required as the unfused powder supports overhanging features. The technology has found widespread application in functional prototyping, low-volume production, and manufacturing of complex assemblies. The high-value industries, especially the metal component manufacturing processes, have been transformed by metal PBF processes, especially by LPBF and EBM. LPBF involves high power laser (typically fiber or Nd:YAG) to melt metal powder particles completely producing dense components with mechanical properties that are like or surpass those of wrought materials [19]. Complex geometries such as internal channels, lattice structures, topology-optimized designs can be created through the process that cannot be produced through conventional metalworking [24]. Moreover, titanium alloys, aluminum alloys, stainless steels, nickel-based superalloys, and cobalt–chrome alloys are the common materials. On the other hand, EBM uses an electron beam as a source of energy and processes in a vacuum environment at high temperatures and has the benefit of being used for reactive materials such as titanium and reduced residual stresses compared to laser-based processes [22].

MJF is a more recent innovation of polymer PBF created by Hewlett-Packard (HP). Rather than use a laser to selectively fuse powder, MJF uses an array of inkjet printers to selectively deposit fusing and detailing agents on the powder bed, then uniformly heat the bed with infrared lamps to fuse areas of the bed. It allows reaching much higher build speeds and material uniformity than laser-based SLS. The benefits of PBF technologies are huge. The powder bed surrounding also has a built-in support capability of overhanging features, which removes or minimizes support structures. Nesting of parts can be done effectively in the build

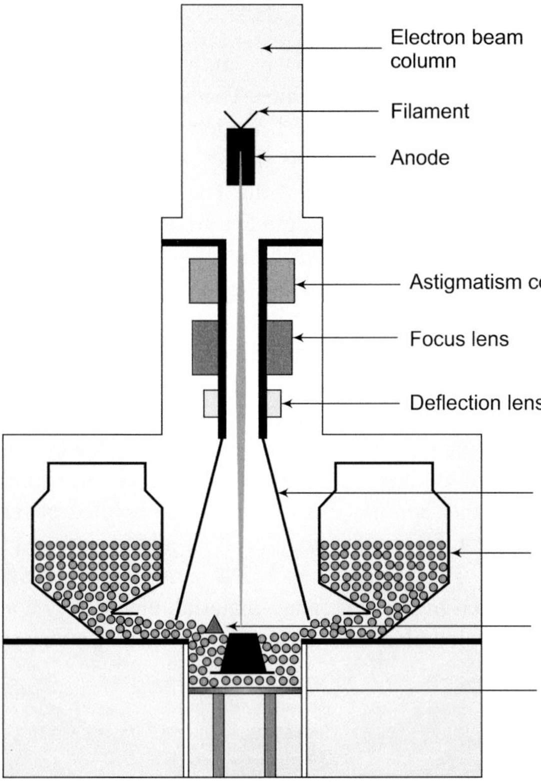

Fig. 1 EBM technology schematic [14].

volume leading to high packing density and productivity. The material properties are typically great with metal PBF components being close to full density and mechanical properties that can handle tough structural tasks. The technology supports a large variety of materials, and the process is relatively mature, and parameter sets for common alloys are established. However, PBF processes also have major limitations and challenges. The powder handling requirement poses safety issues especially when dealing with reactive metal powders that can lead to explosion or fire. Powder recycling and quality control are important issues as the characteristics of powders are degraded as they are used repeatedly, with the quality and consistency of parts affected. Surface finish is usually rough because of partially sintered powder particles adhered on part surfaces, which requires a post-processing for applications requiring smooth surfaces. Construction speeds are quite slow, especially when one is working on large pieces, since every layer must be done in series. Stresses left behind by the high heating and cooling rates may lead to distortion or cracking of parts, and it is necessary to optimally tune the process and, in many cases, post-build heat treatment. The cost of both the PBF systems and materials remains on the higher side, restricting the adoption in price sensitive [23].

3.3 Material Extrusion

Material extrusion (ME) is a group of AM technologies that uses selective dispensing of material through a nozzle or orifice. The best-known ME process is the FDM often referred to as fused filament fabrication (FFF) which has become synonymous with 3D printing for the desktop market and is the most accessible and widely deployed AM technology in the world. The basic principle is to feed a thermoplastic filament into a heated nozzle whereby it melts and deposit the molten resin in a specified pattern to form each layer. As the material comes out of the nozzle and is deposited, it cools and solidifies and bonds to the previously deposited material.

FDM/FFF printers are available in consumer level desktop printers that cost a few hundred dollars as well as industrial devices that can manufacture large parts using engineering grade materials. Common materials are commodity thermoplastics such as polylactic acid (PLA) and acrylonitrile butadiene styrene (ABS), as well as engineering polymers such as polycarbonate (PC), polyetherimide (PEI/ULTEM), polyether ether ketone (PEEK), and various filled or reinforced composites containing carbon fiber, glass fiber, or other additives to improve the mechanical properties [25]. The versatility of FDM material has grown greatly and more recently it has seen high-temperature material, electrically conductive filaments, and even metal-filled filaments which can be sintered after printing to form metal parts.

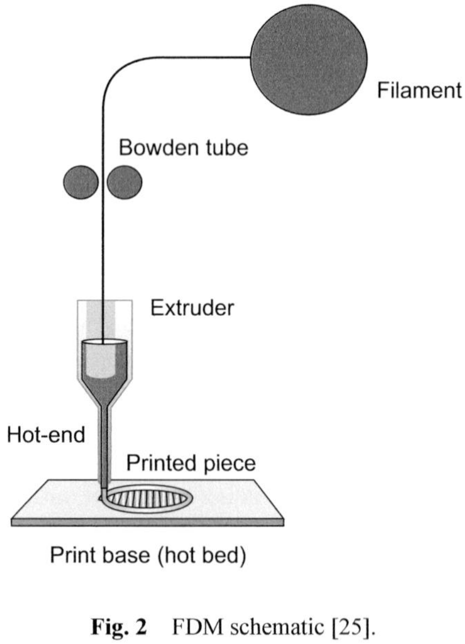

Fig. 2 FDM schematic [25].

The other significant ME variant is direct ink writing (DIW) which extrudes paste-like materials (inks) that are solidified by several different processes such as solvent evaporation, chemical reaction, or thermal gelation. DIW allows processing of materials that cannot be shaped into filaments, such as ceramics, hydrogels, and biological materials, which increases the scope of ME from thermoplastics. This has enabled DIW to be especially useful in bioprinting processes in which cell-laden hydrogels are extruded to create tissue constructs.

Newer technologies in the field of ME have been continuous 3D printing, whereby part production can occur continuously because it can add material during the build process, and multiaxial printing systems, which can deposit material on nonplanar toolpaths and thus circumvent the size limitations and anisotropy of the traditional layer-by-layer methodologies. These advances improve some of the traditional limitations of ME while adding new capabilities to enhance the production applications of ME [26].

The benefits of ME are many and account for its ubiquitous use. The technology is straightforward, nonhazardous, and not expensive relative to other processes of AM, and so, it is available to educational, hobbyist, and small businesses users. The cost of materials is usually cheaper compared with other AM technologies and the materials available are ever increasing. The process operates at atmospheric pressure and at moderate temperatures (except in the case of high-performance polymers) simplifying machine design and operation. Multi-material and multi-color printing is easily realized with the addition of multiple extruders to achieve functional integration and esthetic customization of the printed parts. However, ME also has significant limitations that limit its scope of application. The anisotropy of part performance is high, that is, the performance build-direction (perpendicular to layers) is generally 50–70% of the in-plane strength because not all links occur between layers. The surface finish is not good and layer lines and stair-stepping on curved surfaces can be seen. The nozzle diameter and layer height are the limiting factors to resolution that are typically between 100 and 400 μm. The overhanging features need support structures which must be removed after construction, and which leave surface artifacts. The build speed of large parts is comparatively low since the material is deposited in a serpentine manner with just a single nozzle (or few nozzles). During cooling, thermal stresses can cause warping and delamination especially in large parts or materials with large thermal expansion coefficients [13].

3.4 Vat Photopolymerization

Vat photopolymerization (VP) is the oldest type of AM with its roots dating back to the invention of SLA by Charles Hull in 1984. VP processes are built up by selective curing of a liquid photopolymer resin using either ultraviolet or visible light. The basic mechanism is to have a vat filled with liquid resin that has a build platform just below the surface of the resin. A light source (laser or digital projector) is used to expose the resin areas which are the cross section of the part causing photopolymerization of the exposed areas. After every layer is cured, the platform moves in a vertical direction, by the thickness of the layer, and a fresh resin flows over the cured layer and the whole process repeats [27].

These are the principal VP processes SLA [28], digital light processing (DLP) [29], and continuous liquid interface production (CLIP) [30]. A laser beam is used in SLA and is scanned over the surface of the resin one point at a time tracing the geometry of the part. DLP does utilize a digital projector to simultaneously display a full layer, greatly increasing the speed of the build when compared to laser-based SLA. CLIP, as an innovation by carbon, combines continuous projection with oxygen permeable window to enable continuous part formation as opposed to layer-by-layer which drastically increases build rates as well as surface finish and mechanical strength.

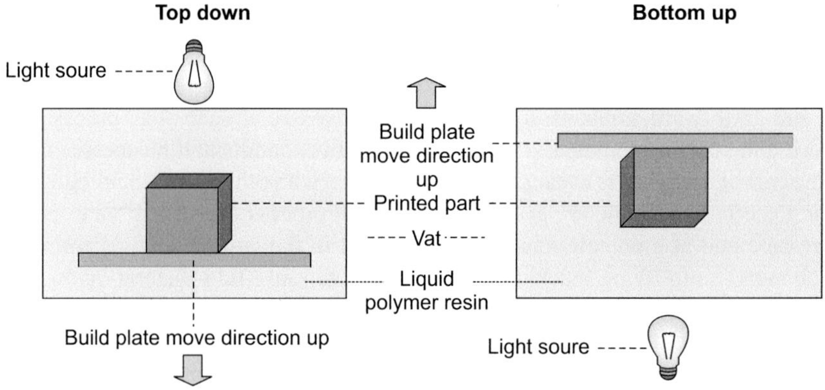

Fig. 3 Vat photopolymerization printer orientations [32].

VP technologies are good at creating parts with superior surface finish and fine feature resolution, normally with layer thicknesses of 25–100 μm and feature detail of tens of micrometers scale. This fine detail makes VP especially appropriate to work in areas that have complex details like jewelry, dental models, hearing aids, and investment casting patterns. The finished surface is often smooth and thus not much post-processing is required, but parts normally must be washed in resin to remove the uncured resin and then be post-cured by ultraviolet light in order to reach full mechanical properties. The mater palette behind VP has grown very much as compared to the early acrylates and epoxies. Modern formulations include rubber-like elastic resins, thermal-resistant resins, biocompatible resins to be used in the medical industry, and resin filled with ceramics which can be sintered after printing to produce ceramic components. The dental and medical industries have been major impetus for the development of biocompatible resins that can be meant for temporary or permanent implantation. Engineering resins with excellent mechanical properties, chemical resistance, and thermal stability has led to the expansion of VP to functional prototyping and end-use parts. The benefits of VP are good surface finish and resolution, parts which are very detailed with smooth surfaces can be constructed. Speed in building can be fast especially in DLP and CLIP processes which cure whole layers at the same time. VP allows complicated shapes, with fine detail, thin walls, and complicated detailed designs. Modern resins have mechanics performance appropriate for a lot of functional applications [31].

Limitations of VP are relatively limited range of material (mostly resins are thermoset photopolymers which cannot be remelted and recycled), brittleness and potential degradation when exposed to UV light or moisture, the need for support structures in overhanging features, which have to be removed carefully, the danger of handling liquid resin which requires proper ventilation and protective equipment, and the need for washing and post-curing. Photopolymer resins can be costly, compared to thermoplastic filaments and the volume of the build can be very limited compared to other technologies [32].

3.5 Material Jetting

In material jetting (MJ), the parts are built by deposition of micro-droplets of build material selectively, in a process like 2D inkjet printing but in 3D form. In most cases, MJ system will jet droplets of photopolymer onto a build platform and immediately cure them with UV light layer by layer. MJ systems also jet a removable support material which can be easily removed after build and thus allow complex geometries without surface artifacts of support removal as in other processes.

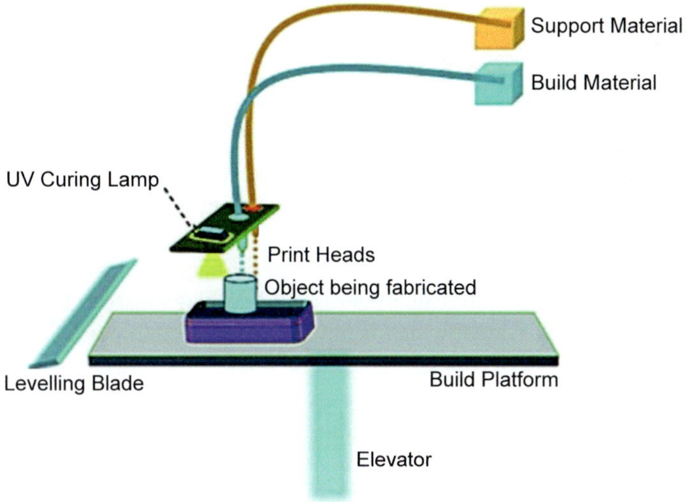

Fig. 4 Material jetting printing [33].

The most common technologies in the field of primary MJ are PolyJet introduced by Stratasys and MultiJet Printing (MJP) introduced by 3D Systems. These systems contain arrays of inkjet print heads able to print a few different materials simultaneously and are therefore able to manufacture multi-material and multi-color parts in a single build. Voxel-level material composition allows the fabrication of components with spatially differentiable properties, such as rigid and deformable areas of one component or full-color models for further visualization and communication. MJ has several unique advantages. The technology provides better surface finish and fine features resolution as compared or better than VP. The capability of multi-materiality is a distinguishing feature, in the ability to functionally integrate and creatively customize esthetically, difficult and impossible with other AM processes. Support material can be made easy to remove (typically by water or melting), minimizing after-processing and surface artifacts. The reaction is carried out at ambient temperature and atmospheric pressure making machine design and operation easier. Parts are usually ready to use immediately after build with little post-processing. Still, MJ has serious shortcomings. The material characteristics are on average worse than other AM technologies and components are prone to be fragile and to UV degradation. The material collection is dominated by photopolymers, despite the recent advances in the field and some waxes for investment casting have been introduced, along with some experimental metal jetting systems. Build speed is rather slow for large parts as material is built up one droplet at a time. Material and equipment are costly compared to ME which limits the use to cost-sensitive applications. Parts may have some anisotropy in them due to the layered deposition but usually not as much as commonly seen for ME [33].

3.6 Directed Energy Deposition

Directed energy deposition (DED), which includes processes that construct parts with the material being, melted as it is deposited, usually by a concentrated energy source, such as a laser, electron beam, or plasma arc. In contrast to PBF where powder is already laid out and then selective melting is applied, DED systems allow the input of material (powder or wire) into a melt pool formed by the energy source, creating the parts one pass at a time by being deposited. This way not only can new parts be fabricated but also old components can be repaired, and new features can be added, which is unique to AM technologies. The prominent DED processes are laser metal deposition (LMD) [34], electron beam additive manufacturing (EBAM) [35], and wire arc additive manufacturing (WAAM) [36]. LMD is based on the principles of creating a melt pool on the substrate with a laser and feeds metal powder following the nozzles with the laser beam. The process has the potential to deposit material at very high rates and is most useful in repairing expensive components like those in turbine blades or in introducing features to an existing part. EBAM uses an electron beam and wire feedstock in a vacuum environment and has the ability for very high deposition rates for large structural components. WAAM uses an electric arc as a heat source with wire feedstock which has the highest deposition rates of the DED processes and is suitable for large-scale metal fabrication.

DED technologies have several unique advantages. For the most part, deposition rates are higher than PBF, which is why DED is suitable for large parts where build time is a critical consideration. The feature of repairing and adding functionality to the existing components offers distinct values to the maintenance and remanufacturing processes. Functionally graded materials and control of compositions multi-material deposition is a possibility by changing powder or wire feedstock, which allows functionally graded materials. The process can be applied to multi-axis robotic systems, and this allows for nonplanar deposition since it overcomes some of the geometric limitations of layer-by-layer approaches. However, DED also has some major limitations. Surface finish and dimensional accuracy are also generally lower than PBF and usually need a great deal of machining to get to final dimensions and surface quality. The size of the melt pool limits the size of the features that can be resolved, and thus DED cannot resolve fine details or thin walls. The overhanging features are often in need of support structures. To protect against oxidation, a vacuum or inert atmosphere is usually required during the process which complicates the process and adds to its cost. The high thermal gradients may result in distortion or cracking through residual levels of stresses. The technology generally has a higher complexity and cost compared to other AM processes, so it is restricted to applications that offer concrete value from its unique capabilities [37].

3.7 Sheet Lamination

Sheet lamination (SL) consists of bonding pieces of material together and using cross-sectional profile of the part as a guide to cut the material. Two main SLs include laminated object manufacturing (LOM) [38] and ultrasonic additive manufacturing (UAM) [39]. LOM involves bonding adhesively in paper, plastic, or metal sheets and each sheet is cut into shape before the next sheet is bonded. UAM uses ultrasonic welding to join metal foils or tapes, with the CNC milling being used to cut each layer to the desired profile.

SL has a few benefits such as high build speeds for large parts—material addition (bonding a full sheet) is fast compared to point-by-point or line-by-line deposition. Material costs might be low, especially for paper-based LOM. The process runs at low temperatures (particularly LOM) and no thermal stresses and distortion occur. UAM allows electronics, sensors, or other elements to be embedded into metal structures, which create smart or multifunctional parts.

However, the SL also has significant limitations that have limited its adoption. Material waste can be very high as the whole sheet needs to be used even if the part only occupies a small portion of the sheet area. Internal structures and complicated geometries are hard or impossible to manufacture because the cutting process can only reach to the surface of each layer. The material can be anisotropic where the inter-lateral bonding is weak compared to the intra-lateral bonding. Surface finish post-processing is necessary, because the stratum of layers is stepped in appearance. The technology has experienced limited commercial success when compared to other AM categories with a relatively small number of systems being available in the market [38].

3.8 AM Technologies Comparative Analysis

The variety of AM technologies is in tune with the range of applications, materials, and performance demands of various industries and applications. No single AM technology is best for all applications but instead each has unique respective advantages and limitations that make them more or less suitable for different applications. These trade-offs are critical in effective technology selection and development of the application.

AM technologies are characterized by several important parameters that can be used to select the choice. Resolution and surface finish are quite different among them with VP, and MJ having the best details and smoothest surface and DED and binder jetting having rougher ones. The speed of building is dependent on the deposition process, and the geometries of the part, binder jetting, and SL are typically the fastest processes with large parts whereas the ME process is very slow. PBF and DED of metals can be effectively used with material properties and can achieve near-wrought properties, and MJ and some vat photopolymerized materials form parts with poor mechanical performance.

Material range varies considerably with ME and binder jetting supporting the greatest variation in the range of material classes and VP and MJ being mostly confined to photopolymers. The sizes of parts vary as small (VP, MJ) to very large (DED, binder jetting, SL). The needs of support structures are various, and no specific support is necessary in powder bed processes or binder jetting processes (powder gives support), whereas ME, VP, and MJ processes need removable supports. MJ and part of VP applications do not require many post-processing needs, whereas PBF metals, binder jetting, and DED do. Apart from all other above parameters, cost is one of the prominent parameters.

The cost factors include cost of capital equipment and cost of operating such as cost of materials, labor, and cost of post-processing. ME is the lowest cost and operating cost for polymers, and metal PBF and DED is the highest capital and operating cost. The scale of production is suitable with AM technologies being most cost-effective at low to medium volumes due to the cost of tooling being prohibitive, but higher-volume tooling development is increasing AM competitiveness [40].

Since AM technologies are still evolving and new processes are being created, the distinction between categories might become unclear and new hybrid techniques can be created to bring the benefits of one, two, or several technologies together. The knowledge of the core principles, abilities, and drawbacks of each of the major AM categories will be the cornerstone in negotiating this changing environment and making wise decisions regarding the technology adoption and application development.

4. Challenges and Limitations

Despite remarkable advancement for four decades, there are still significant challenges and limitations that limit its wider adoption and inhibit it from fulfilling its transformative potential. These challenges range from technical, material, economic, quality, and standardization issues from researchers, industry, and regulatory bodies to address. These restrictions should help an individual to have realistic expectations, priorities in the research, and strategies to overcome barriers to AM implementation.

4.1 Technical Challenges

Technical challenges are perhaps the most visible and discussed limitations of AM, and include process physics, part quality, and manufacturing performance issues. Surface quality is a continuing issue with most AM technologies. The layer-by-layer nature of construction provides a surface roughness outcome of the stair stepping effect with partially bonded powder particles, layer lines, or support structure attachment points, further degrading surface finish. While post-processing can be used to improve the surfaces, this introduces time, cost, and complexity into the manufacturing process. Applications that demand smooth surfaces as a performance requirement (aerodynamic), fluid flow, or esthetics may frequently demand a considerable amount of finishing work to offset certain AM time and cost benefits [41].

Dimensional accuracy and geometric precision are still a challenge, especially when the parts or materials used are large or have a large thermal expansion coefficient. Thermal stresses that occur during fabrication may result in warping, distortion, or cracking, which results in parts with dimensions that deviate from those intended. Residual stresses that are accumulated during layer-by-layer deposition can cause distortion during/after building, or even part failure during the build process. These effects are offset only by systematic process modeling and parameter optimization, but it is not yet possible to ensure consistent dimensional accuracy of different geometries and build conditions [42].

Defects and porosity are the key issues, especially for metal AM parts for safety-critical applications. Other typical defects are lack-of-fusion porosity (between scan tracks or layers) caused by unstable melt pool dynamics, keyhole porosity (caused by trapped gases or powder contamination), and cracks (caused by thermal stresses or material susceptibility). While process optimization can reduce defects, it is difficult to eliminate them and small defects are hard to detect with sophisticated inspection techniques. Formation of defects is stochastic in nature therefore even optimization processes can sometimes result in defective parts and that requires strong quality control measures [43].

Competitiveness to produce higher volumes of production is hampered by the build speed and productivity constraints of AM. Most AM processes are sequential (layer-by-layer) build processes that take anywhere from hours to days depending on the part size and complexity. Although AM removes tooling time, the fabrication time per part of a simple geometry or high volume is usually longer than that of conventional processes. The need to have faster builds by using more laser power or more lasers or by using other deposition processes must mitigate against part quality and preventing defects. The size limits the largest dimensions of parts that can be manufactured and the volumes of builds are in the tens of cubic centimeters of some desktop machines and in the several cubic meters of large-format machines. While large format AM systems are still increasing the size possibilities, producing very large structures remains difficult because of thermal management, powder handling, and machine design constraints. The solution to size constraints on assemblies can be achieved through part consolidation, that is, combining more than one component into one AM part, but not on the size of individual components.

The process of providing support structure to features that are overhanging introduces complexity, waste of materials, and subsequent post-processing. While powder bed processes enjoy the advantage of requiring no extra support structures since the powder is self-supporting, other technologies require dedicated support structures to be designed and built, in addition to being removed. Unloading support may be error-prone and time-consuming and leave artifacts or damage part surfaces. Designing parts to reduce the support requirements is an important aspect of design for AM, but there may be geometric constraints that limit this optimization [44].

4.2 Limitations Related to the Material

Material limitations are a fundamental constraint on the applicability of AM because the number of materials that AM technologies can process is still limited compared to what is available for conventional manufacturing. Limited material selection is a common theme in AM technologies to a greater or lesser extent. While the number of available AM materials has increased significantly, it is still a small portion of the thousands of alloys, polymers, ceramics and composites that are used in conventional manufacturing. Developing new AM materials requires extensive research to understand processability, optimize parameters, and characterize properties making material development a lengthy and expensive process [45].

Material properties and performance of AM parts are often different from the conventionally manufactured parts of the same nominal material composition. Microstructural differences due to rapid solidification, thermal cycling, and directional heat flow may make that the grain structure, phase distribution, and defect population are different than cast, wrought, or molded materials. While these differences sometimes lead to superior properties (e.g., fine grain structures in metal AM parts), they may also lead to inferior performance, more especially in relation to fatigue, fracture toughness, or high-temperature applications [46]. The absence of comprehensive material property databases for AM materials makes the design and certification process difficult because engineers cannot use handbook values designed for traditional materials. On the other hand, the cost and complexity of powder production, handling, and recycling are major operation challenges, especially for reactive materials such as titanium or aluminum alloys.

Material compatibility and multi-material processing are still largely underdeveloped even though there is great interest in them. While there are a few AM technologies (specifically, MJ) that can work with multiple materials in a single build, most processes are restricted to single materials or narrow ranges of compositions. A good bonding between dissimilar materials, control of various processing requirements, and forecasting of multi-material interfaces offer significant technical challenges. The vision of spatial compositional variation of printed parts has yet to be realized in production applications.

Material recyclability and sustainability are emerging issues as the use of AM has increased. The thermoplastic materials used in ME can be recycled, but generally the material properties change for the worse with each recycle iteration. Thermoset photopolymers in VP and MJ cannot be remelted and recycled. Recycling of metal powder can be done but there is a need to pay special attention to quality as to preserve powder characteristics. The environmental impact of AM material production, usage, and disposal need to be more thoroughly evaluated through life cycle analysis [47].

4.3 Economic Barriers

One of the major barriers to the extensive application of AM is economic factors, particularly those applications that the conventional manufacturing is cost-effective. Entry barriers are high since industrial AM systems are very costly in terms of capital. These high capital investments will need to be recovered on any quantity of production and therefore AM will be the most cost-effective on any of the low to medium quantities that would make the cost of tooling prohibitive. Direct costs include operating costs like materials, labor, energy, and maintenance which make huge cost in the total part costs. As mentioned above, the material costs are usually bigger than traditional feedstock. The cost of workforce can be high, particularly in those processes that require much after processing, support renovation, or maintenance. Power requirements of the process (which can be large) determine the power it consumes per process, particularly in cases where laser or electron beam systems are employed. The complexity of AM systems like laser replacement and optics cleaning and calibration incurs maintenance costs thereby driving up the operation cost [48]. Post-processing charges may make up a huge percentage of the part costs, particularly where metal AM is involved. Other operations like support removal, heat treatment, hot isostatic pressing, machining, and surface finishing may inflate the cost of the as-built part by 50–200%. The post-processing cost is reducing the time benefits of AM and adding a level of complexity to the production process. Although applications of as-built parts allow utilization with very minimal post-processing, there are numerous demanding applications that involve significant amounts of finishing in order to meet the desired properties and surface quality. The inability to compete in terms of production in volume is due to lack of economies of scale by AM. The traditional manufacturing processes like injection molding, die casting, or stamping have a high price of tooling but very low cost per part of high volume where economies of scale are good. The toolless nature of AM reduces the setup costs without providing high volume cost reduction due to the basic similarity of the amount of time and material needed to build one part, so the price will not change with the number of production units. Cost structure is the most attractive to AM in cases where the volumes are low, the parts are customized, or the design richness is charged on the cost of a part counterbalanced by extremely high parts costs. The throughput constraints and the productiveness constraints of AM limit its economic viability because of either time-sensitive or large volume usage. The hours to days per part to build a part compared to conventional processes which could build parts in seconds or minutes per part. Although it can be constructed several parts simultaneously during a single build, it remains limited by the volume and time of the build. Any attempt to boost productivity, whether in the form of increased rates, higher volumes of builds, or more machines must be counterbalanced by capital and operating cost rises [49].

5. Lacunae: Research Gaps and Unresolved Issues

Although the topic of AM is studied intensively and has been investigated in research over a period of 40 years, the gaps in knowledge and unanswered questions remain, which do not allow releasing the full potential of the technology and restrict the scope of its use. These gaps are known to determine the direction of future research work, the allocation of resources, and realistic expectations of the capabilities of AM soon. This part looks at important gaps in research in various fields and where the basic knowledge is still missing and innovations could be necessary. It is important to have basic process understanding that will be used during the project. There is a critical discontinuity in the basic knowledge of physical phenomena that govern AM processes, especially when it comes to metal systems with complicated thermal, fluid dynamic, and metallurgical interactions. The dynamics of melt pools in laser and electron beam processes include rapid melting, surface tension gradient-driven fluid flow (Marangoni convection), evaporation and solidification time and length scales of microseconds and micrometers. The relationship between the energy beam, powder particles, and melt pool presents an extremely dynamic system that is hard to experimentally measure and hard to compute computationally. Poor comprehension of melt pool behavior restricts the possibility to forecast and manage defect development, microstructure development, and material property. Complex thermal cycling is associated with thermal history and development of residual stress in layer-by-layer construction whereby addition of each new layer resets the previous deposited material and forms thermal gradients that promote the accumulation of residual stress. Although computational models could be used to simulate thermal fields and predict residual stresses, validation is not easy since it is difficult to measure temperature and stress fields in a building. The connection between process parameters, thermal history, residual stresses, and distortion of parts is not clearly understood; thus, constraining the capacity to design parts that will achieve dimensional specifications without significant optimization through trial and error. Unlike traditional processing, the evolution of microstructure in AM materials is fundamentally based on the high solidification velocity, directional heat flow, as well as repeated thermal cycles. The resultant microstructures whose characteristics are the morphology of the grain, the distribution of the phases, and the population of defects dictate the material properties but are not yet predictable comprehensively based on the process parameters. Treatment of the process conditions and solidification behavior with relation to final microstructure is an ongoing research topic, and knowledge gaps, especially about complex alloy systems, multi-material interfaces, and non-equilibrium phases, are still significant. Most AM material systems have not been fully characterized in terms of process–structure–property relationships between process parameters and microstructure, and between microstructure and the mechanical properties. Though there are empirical correlations which exist between common materials and processes, there are no predictive models of materials, which can be used to predict reliably the properties of materials based on their process parameters, without broad experimental validation. The disadvantage is that it prevents the possibility to design processes to align with certain property targets and makes material and process development more complicated [50].

5.1 Optimization and Design Engineering Tools

There are major gaps between design tools and methodologies that are specifically designed to be used in AM, which deny the engineers the full capabilities of AM. Design for additive manufacturing (DfAM) principles are still mostly descriptive and process-oriented, with no broad, physics-based models as is the case with conventional manufacturing operations. Although there are guidelines to minimize the support structure and optimize the orientation and the size of the features, they are not always actively researched but mostly rely on rules of thumb. Comprehensive DfAM frameworks, where geometric, material, and process are considered, are still an open challenge [51]. The optimization of topology of AM has been enhanced to include the possibility of producing light-weight structures that have been optimized according to loading conditions. However, optimization methods in use today tend to overlook manufacturing considerations, thermal considerations in the building process, or post-processing considerations and therefore produce designs that are both theoretically optimal but practically challenging to produce or that do not execute the performance requirements when AM-specific considerations are considered. Incorporating the manufacturing constraints, thermal simulation, and property forecasting into the topology optimization processes is a current research topic. Multi-scale design and optimization schemes, which optimize the geometry of parts, internal architecture (lattice structures), and material composition concurrently, are not developed so far [52]. Although each of these design aspects were discussed separately, there are no integrated frameworks which optimize on a variety of scales and take into consideration manufacturing constraints and process structure property relationships. These multi-scale methods may open vast opportunities of performance enhancement but need to be supported by innovations in computational strategies and their underlying conceptual knowledge. AM-specific design aid and software are not as mature as traditional CAD/CAM. Even though the large vendors in CAD have included AM-specific functionality, full-fledged design environments, where AM concepts, process simulation, build preparation, and post-processing planning are well coordinated, do not exist yet. The absence of built-in design tools prolongs the learning curve to adopt AM and reduces productivity [53].

5.2 Material Development and Characterization

The development of materials to be used in AM lags far behind the process development, and the choice of materials available is limiting the applicability of AM. The development of a new material in AM would have to be heavily researched to grasp the processability, manipulate the parameters, and define the properties, which is time-consuming and costly. It is now an open question how to accelerate the development of materials by using computational materials design, high-throughput experimentation, or machine learning. AM is restricted in the responsiveness to new application requirements by the absence of expedited material development approaches. Databases of material property of AM materials are much less comprehensive than conventional materials, making it more difficult to design and certify. Although there are databases of typical AM materials, they are usually not as comprehensive in terms of conditions (temperatures, strain rates, environments) and statistical rigor (number of specimens, variability measurement) as those required by allowable design in safety-critical applications. To create full property databases, it would take a lot of testing on a variety of machines and build conditions as well as material batches, which is a resource-intensive task that has been done on only a small portion of AM materials. A multi-material system is a largely untouched direction of AM, and the basic questions of interface bonding, property prediction, and compatibility of processes remain open. Though there are AM processes that may deposit more than one material, it has been difficult to comprehend how to make multi-material components, forecast their characteristics and reliability in performance. The opportunity to achieve the new functionalities with the use of functionally graded materials and multi-material components is relatively large, yet the achievement of this opportunity can become feasible only with the improvement of the underlying knowledge and processing [54]. Many AM systems contain material–process interactions which are not well known, especially when the material is a new material or at non-standard process conditions. The correlation between material qualities (thermal conductivity, absorptivity, viscosity, surface tension) and process behavior (stability of melt pools, adhesion of layers, formation of defects) is complicated and not completely defined. Such a discontinuity restricts the possibility of forecasting the processability of new materials and optimization of the factors without conducting large-scale experiments [55].

5.3 Multi-Material and Multi-Scale Integration

Although there is a great interest in research, the AM vision of allowing parts with spatially different composition, microstructure, and properties has hardly been fulfilled. The ability to bond dissimilar materials, to control various processing needs (melting temperatures, solidification properties, thermal expansion), and forecast properties of multi-material interfaces are the fundamental challenges of multi-material processing. Although certain steps toward this direction have been undertaken in certain material combinations, there are no general models of multi-material AM. Multi-material integration to produce parts with embedded functionality (sensors, electronics, actuation) can be seen as a new field with significant promise and little implementation. Issues such as safeguarding embedded constructs against processing environments, providing electrical connectivity, and combining dissimilar manufacture procedures exist. Hybrid AM processes, which are additive and conventional manufacturing operations, have not yet been mature enough but are still developed [56]. Multi-scale architectures multi-scale part geometry with meso-scale (lattices, cellular structures) and micro-scale material design are challenging to design, optimize, and make. This gap is caused by computational issues in the modeling of multi-scale systems, manufacturing issues in the production of features at multiple length scales, and characterization issues in the measurement of multi-scale properties. The increment in the design and production on a large scale might allow substantial performance to improve but needs breakthroughs in various fields.

5.4 Sustainability and Life Cycle Assessment

AM as an environmental subject is not fully characterized and there is great gap in the life cycle assessment (LCA) data and the knowledge of the net environmental effects of AM. AM processes have a wide range of energy consumption due to technology, part geometry, and the material concentration and there has been limited extensive comparisons with conventional manufacturing taking into consideration the entire production chain (including tooling, material production, and post-processing). Although AM can lessen the amount of waste materials, the intensity of certain processes (namely metal PBF) could balance these gains out [57]. Most AM materials have not been fully addressed in terms of material sustainability such as their recyclability, their environmental footprint in terms of feedstock production, and their end-of-life disposition. The sustainability of powder production, photo polymer synthesis, and the use of support material needs to be investigated more comprehensively. Sustainable AM material and processes development is also a valuable area of research that has not been extensively studied in comparison to performance and cost aspects [58]. The LCA systems that are directly related to AM are not developed yet, and most of the literature emphasizes the narrow scope of systems or case studies instead of the overall assessment. The absence of standardized LCA methodologies on AM makes it difficult to compare it with traditional manufacturing and restricts the capacity of making informed decisions concerning the adoption of AM on the sustainability level. The elaboration of the LCA frameworks, which consider the AM-specifics, is a significant area of research [57].

6. Future Prospect and Future Trends

AM is at an inflection point as it is no longer a niche technology that is used to prototype and in highly specific applications, but as a mainstream manufacturing technology that can alter production in wide range of industries. The technological innovation and the market demand will influence the future direction AM is going to take, focusing on the materials, processes, digital integration, and application development. This area will look at new trends and prospects that will see AM increase in terms of abilities, restrain existing hindrances, and unlock new uses.

6.1 Materials and Multi-Material Advanced Systems

The creation of novel materials that are tailored to AM is an important frontier to developing the capabilities and uses of the technology. High-performance polymers such as PEEK, PEI, and polyphenylene sulfide (PPS) have better mechanical properties, chemical resistance, and thermal stability than commodity thermoplastics, and allow AM of functional components to be used in demanding applications. More recent developments have made these materials more processable and increased their supply in a variety of AM technologies. The future advancements will probably be made in terms of extension of the scope of high-performance polymers, making them easier to process and less expensive so that they can be widely used. Sustainable and bio-based materials are a new trend guided by the environmental issue and the concept of the circular economy. Renewable biodegradable polymers, thermoplastics that are recyclable, and materials that end-of-life can recover provides opportunities to more sustainable AM. The studies of bio-based photopolymers, natural fiber-reinforced composites, and closed-loop recycling systems of materials should decrease the environmental footprint of AM, yet may not impact performance [59]. As the environmental regulations increase and the market needs sustainable products, the realization of sustainable AM materials will probably increase. Fiber reinforced polymers (carbon, glass, aramid) or particle reinforced polymers (ceramics, metals) are composite materials that allow improvement of properties and functional adaptation. Fiber-reinforced composites manufactured continuously by AM have strength-to-weight ratios close to those of aerospace-grade composites and can be applied in structural components. Enhanced mechanical, electrical, or thermal properties can be available to nanocomposites that contain carbon nanotubes, graphene, or ceramic nanoparticles. The development of future progress will be aimed at enhancing fiber–matrix bonding, regulating the orientation of fibers, and creating new composite formulations to amuse AM processing.

Functionally graded materials: The achievement of this vision demands development of process control and material compatibility as well as designing procedures. Smart and responsive substances such as shape memory alloys [60], shape memory polymers, and stimuli-responsive hydrogels can be 4D printed, that is, the printed component alters its shape, properties, or functionality upon exposure to an external stimulus, such as temperature, moisture, or electromagnetic fields. Although 4D printing is seen as relatively in the research stages, it is possible that it can be used to create deployable structures, adaptive devices, and biomedical systems. The direction ahead will be on the widening of the scope of responsive materials, enhancement of shape variation predictability and controllability, as well as creation of design tools to 4D printed structures [61].

6.2 Implementation of Artificial Intelligence and Machine Learning

The combination of artificial intelligence (AI) and machine learning (ML) and AM is a paradigm shift, which would help to overcome numerous existing shortcomings and open new opportunities. With high-dimensional parameter spaces, process optimization with AI/ML has the potential to find the best parameter combinations more effectively than the conventional design-of-experiments optimization. Prediction of part quality, sensitivities to parameters, and suggestions to make changes that can produce the desired results can be made by ML models trained on process data. Recent development has shown AI-based optimization of laser powers, scan speed, and so on to reduce defect and enhance properties [62]. In future, development will broaden these possibilities to the complex optimization goals, multi-objective optimization [63], and real-time parameter control. The AI/ML-based closed-loop control can optimize the process parameters in real-time and ensure quality and avoid defects. The development of these capabilities is expensive in terms of the amount of training data, well-developed algorithms, and interconnection with machine control systems. Without process simulation, which is expensive to implement, predictive modeling based on AI/ML is able to predict the part properties, distortion, and defect based on process parameters and part geometry. Models based on experimental data or simulation are data-driven models that can be trained quickly to give predictions used to design and plan processes. Hybrid models that integrate physics with ML have an opportunity to use the advantages of both techniques to generate physically consistent predictions with better accuracy and lower computational costs [64]. Future research is aimed at enhancing the accuracy of models, increasing their application to new materials and processes and incorporating the use of predictive models into design processes. The algorithms of design automation and AI-generative design have the potential to search through large sets of designs to determine optimal geometries with respect to performance goals. The algorithmic nature of generative design has the potential to generate completely new geometries that the human designer would not have imagined and use AM geometric freedom to overcome better performance. Topology optimization can be maximized more quickly with ML, asked to do this as part of a prediction of manufacturability, or as a design suggestion to enhance high-quality parts or cost reduction. These tools will allow more engineers to exploit the capabilities of AM without necessarily being well versed in AM specific requirements in design.

6.3 Bioprinting and Regenerative Medicine

Bioprinting, which is applying AM technologies to create biological tissues and organs, is one of the most radically promising possible uses of AM, which has potential implications in regenerative medicine, drug engineering, and basic biological research. Bio-printing of tissues seeks to produce functional tissue engineered tissue constructs toward transplantation, disease modeling, or drug testing. The modern bioprinting methods are capable of generating simple tissue models with living cells, yet more complex, vascularized models with different cell types are not yet possible [65]. The next round of developments will be to enhance cell viability during and after printing, develop vascular networks to support thick tissues, and reproduce the complicated structure of cells in native tissues. Bioprinting bioinks and biomaterials should meet several conditions such as biocompatibility, bio printability, mechanical properties like native tissues, and cell growth and cell functionality. Natural hydrogel (collagen, alginate, gelatin) is much better biocompatible but worse in mechanics and synthetic polymers have better mechanics but not always biologically active. In the development of bioinks that can meet these needs, it is an active area of research [66]. The future trends are expected to be based on composite bioinks with natural and synthetic materials, cell-responsive materials that can remodel depending on cellular activity, and materials degrading on controlled rates as native tissue is formed. The final aim of bioprinting is organ printing, which entails developing functional organs to be transplanted to the body to deal with the severe scarcity of donor organs. Although full organ printing is a long-term objective, advances toward this aim have been made in printing of tissue models to test drugs, the generation of organoids to study diseases or tissue patches to heal a wound or repair an organ. The issues involve fulfilling the complicated structure of organs, building operational vascular systems, and providing long-term survival and connectivity with the host tissues. The developments in multi-material bioprinting, vascularization concepts, and the knowledge on tissue development will play a critical role in the advancement of organ printing.

6.4 Nanoscale and 4D Printing

Nanoscale AM is also designed to produce structures apart in nanometers, which is between molecular assembly and traditional AM. Two-photon polymerization, electrohydrodynamic printing, and atomic layer deposition are techniques that are used to produce nanoscale layers used in photonics, electronics, and biotechnology. Although nanoscale AM is still mostly in its research stages, it may find extensive applications in nanoelectronics, metamaterials, and drug delivery systems. The next innovations will be to enhance the throughput, add more material choices, and combine the nanoscale components with bigger structures. The AM is expanded by 4D printing, adding time as the fourth dimension, and producing structures, which alter shape, properties, or functionality after fabrication and react to external stimuli. Programmed shape changes are realized by shape memory materials, swelling hydrogels, and bilayer structures with different rates of expansion. They can be used as deployable structures in aerospace, as adaptable medical devices, and soft robotics [67]. Further development of 4D printing will need the creation of new responsive materials, computational tools that will be used to design shape changing structures, and the relationship between the printed structure and the responsive behavior. Nanoscale integration with macro-scale components (e.g., surfaces with controlled wetting, adhesion, or optical properties) or bulk materials with improved nanoscale reinforced properties would be possible. Multi-scale hierarchical structures would be able to replicate a natural material such as bone or wood and surpass the performance by multi-scale optimization. The concept of multi-scale integration needs innovations in both nanoscale and traditional AM and the means of a consistent connection between features of different length scales [68].

6.5 Industry 4.0 Integration and Digital Manufacturing

The combination of AM and Industry 4.0 technologies such as the Internet of Things (IoT), cloud computing, big data analytics, and cyber-physical systems have the potential to make AM add more than isolated machines to connected and intelligent manufacturing systems. Simulation, optimization, and monitoring of AM processes are made possible by the creation of digital twins: virtual representations of physical AM systems that can be used to replicate the behavior of AM systems in the real world. Digital twins can forecast the quality of parts, mathematical optimization of processes, and anomalies that lead to part failure can be detected before they occur. The integration of process models, sensor data, and ML, which are part of the creation of comprehensive digital twins, is an active field of development. AM can be used to create distributed manufacturing that can enable production to be done near the location of use and incurs lower transportation costs, lead time and facilitates mass customization. The AM networks, which are linked by digital platforms, may allow the production of spare parts, custom products, or emergency supplies on demand. The COVID-19 pandemic showed that distributed AM manufacturing could be used to respond to supply chain disruption promptly. The future is based on the creation of digital platforms of distributed manufacturing, the homogeneity of quality standards across facilities, and business models of distributed production. Transformation of the supply chain by AM might change the centralized mass production and intercontinental distribution to the localized and on-demand production. This change may decrease inventory costs, be able to respond quickly to demand changes, and be more sustainable through reduced transportation. Nevertheless, to achieve this vision, AM productivity, cost cutting, quality assurance, and creating digital infrastructure in design distribution and production coordination must be developed. Quality improvement, cost reduction, and innovation can be promoted by data analytics and continuous improvement using the large amounts of data produced by AM processes. Comparing process data between builds, machines, and plants can find best practices, systematic problems, and can inform the development of processes. Data aggregation and analysis solutions on the cloud may facilitate learning across the industry and catalyze AM maturation [69].

7. Conclusion

AM has become one of the most revolutionary manufacturing innovations of the industrial age, and it has radically changed the model of typical relationships between digital design and material production. The key principle of layer-by-layer fabrication, as shown in this chapter, is the one that allows unprecedented geometric freedom, functional encapsulation, and mass customizations capabilities previously unattainable when considering traditional subtractive and AM techniques. AM has experienced its historical development, starting with the development of rapid prototyping in the 1980s to the present-day production technology that is a viable and continuous improvement in process control, material science, and digital manufacturing integration. As of today, the seven standardized varieties of AM technologies, such as PBF, ME, VP, MJ, DED, and SL, individually present a broad technological ecosystem that can support a large variety of industrial applications, such as high-precision medical implants, or massive structural components.

With these impressive breakthroughs, AM still suffers from severe technical, material, and economic limitations that restrain its further introduction on the industrial level. Recurrent problems with reliability of the process, defect origin, anisotropic material behavior, scarce material supply, and cost of production issues still limit its extensive implementation and especially its use in high-volume and safety-related cases. Moreover, a lack of process–structure–property models, standardized qualification models, and mature design methodologies also shows the necessity of more scientific knowledge and engineering innovation. The gaps in research mentioned in this chapter, such as imperfect background knowledge of the physics of the process, inadequate material databases, inadequate capabilities of integration of different materials, and the unavailability of sound sustainability evaluation systems, are some of the major obstacles that need to be overcome in order to make the most out of AM.

However, the future perspective of the AM is truly bright. The new technologies in the fields of advanced materials, process optimization with the use of AI, multi-material manufacturing, nanoscale production, bioprinting, and integration of Industry 4.0 are likely to address most of the limitations presently and considerably enlarge the range of applications. The combination of additive and digital technologies, such as digital twins, ML, distributed manufacturing systems, etc., will also allow us to realize intelligent, decentralized, and highly responsive production paradigms. Such developments cannot only increase the efficiency and product performance in manufacturing, but they can also change the entire world supply chains changing them into on-demand, localized, and sustainable manufacturing.

To sum up, AM is at the major turning point in its technology development. It still has significant challenges, but its special features and further technological development make it one of the pillars of next-generation manufacturing systems. The overall knowledge of the basics of AM, technologies, constraints, and perspectives that this chapter offers is a necessary starting point of researchers, engineers, and practitioners. Further interdisciplinary studies, technological breakthrough, and industrial partnerships will be needed to transform AM as a specialized production methodology to a full-fledged production technology, which is reliable and can be implemented in every sector to redefine the future of manufacturing.

References

[1] Kanishka, K., Acherjee, B. and Kuar, A.S. (2023). Revolutionizing manufacturing: A comprehensive overview of additive manufacturing processes, materials, developments, and challenges. *Journal of Manufacturing Processes*, *107*, 574–619.

[2] Islam, M.N., Boswell, B. and Pramanik, A. (2023). Recent advances in additive manufacturing techniques: An in-depth review. In *Additive Manufacturing: Materials, Design, and Processes* (pp. 1–42).

[3] Ahmed, N. (2025). Fundamentals of additive manufacturing. In *Advanced Manufacturing Technologies* (pp. 45–78).

[4] Bernard, A., Lohfeld, S., Soshi, M., Emmelmann, C., Mognol, P., Perry, N., Gu, D., Ahuja, B., Klahn, C. and Wegener, K. (2023). Vision on metal additive manufacturing: Developments, challenges and future trends. *CIRP Journal of Manufacturing Science and Technology*, *47*, 18–58.

[5] Martínez-García, A., Monzón, M. and Paz, R. (2021). Standards for additive manufacturing technologies: Structure and impact. In *Additive Manufacturing* (pp. 395–408). Elsevier.

[6] Bhuvaneswari, V. (2024). Brief Glimpses of Additive Manufacturing Techniques. *Additive Manufacturing with Novel Materials: Processes, Properties and Applications*, 1–22.

[7] Deckard, C.R. (1988). *Selective Laser Sintering*. The University of Texas at Austin.

[8] Crump, S.S. (1991, June). Fused deposition modeling (FDM): Putting rapid back into prototyping. In *Proceedings of the 2nd International Conference on Rapid Prototyping* (pp. 354–357).

[9] Choong, Y.Y.C. (2022). Additive manufacturing for digital transformation. In *Digital Manufacturing* (pp. 145–182). Elsevier.

[10] Nazir, A., Gokcekaya, O., Billah, K.M.M., Ertugrul, O., Jiang, J., Sun, J. and Hussain, S. (2023). Multi-material additive manufacturing: A systematic review of design, properties, applications, challenges, and 3D printing of materials and cellular metamaterials. *Materials & Design*, *226*, 111661.

[11] Sarabia-Vallejos, M.A., Rodríguez-Umanzor, F.E., Pérez-Monje, D.A., Cerda-Iglesias, F.E. and González-Henríquez, C.M. (2022). Innovation in additive manufacturing using polymers: A survey on the technological and material developments. *Polymers*, *14*(7), 1351.

[12] Davoudinejad, A. (2021). Vat photopolymerization methods in additive manufacturing. In *Additive Manufacturing* (pp. 159–181). Elsevier.

[13] Goh, G.D., Yap, Y.L., Tan, H.K.J., Sing, S.L., Goh, G.L. and Yeong, W.Y. (2020). Process–structure–properties in polymer additive manufacturing via material extrusion: A review. *Critical Reviews in Solid State and Materials Sciences*, *45*(2), 113–133.

[14] Hossain, M.S., Gonzalez, J.A., Hernandez, R.M., Shuvo, M.A.I., Mireles, J., Choudhuri, A., ... and Wicker, R.B. (2016). Fabrication of smart parts using powder bed fusion additive manufacturing technology. *Additive Manufacturing*, *10*, 58–66.

[15] Yap, Y.L., Wang, C., Sing, S.L., Dikshit, V., Yeong, W.Y. and Wei, J. (2017). Material jetting additive manufacturing: An experimental study using designed metrological benchmarks. *Precision Engineering*, *50*, 275–285.

[16] Lores, A., Azurmendi, N., Agote, I. and Zuza, E. (2019). A review on recent developments in binder jetting metal additive manufacturing: Materials and process characteristics. *Powder Metallurgy*, *62*(5), 267–296.

[17] Dass, A. and Moridi, A. (2019). State of the art in directed energy deposition: From additive manufacturing to materials design. *Coatings*, *9*(7), 418.

[18] Pilipović, A. (2022). Sheet lamination. In *Polymers for 3D Printing* (pp. 127–136). William Andrew Publishing.

[19] Han, W., Kong, L. and Xu, M. (2022). Advances in selective laser sintering of polymers. *International Journal of Extreme Manufacturing*, *4*(4), 042002.

[20] Chowdhury, S., Yadaiah, N., Prakash, C., Ramakrishna, S., Dixit, S., Gupta, L.R. and Buddhi, D. (2022). Laser powder bed fusion: A state-of-the-art review of the technology, materials, properties & defects, and numerical modelling. *Journal of Materials Research and Technology*, *20*, 2109–2172.

[21] Venkatesh, K.V. and Nandini, V.V. (2013). Direct metal laser sintering: A digitised metal casting technology. *The Journal of Indian Prosthodontic Society*, *13*(4), 389–392.

[22] Körner, C. (2016). Additive manufacturing of metallic components by selective electron beam melting—A review. *International Materials Reviews*, *61*(5), 361–377.

[23] Singh, A.P. and Pervaiz, S. (2021, November). Current status and prospects of multi-jet fusion (MJF) based 3D printing technology. In *ASME International Mechanical Engineering Congress and Exposition* (Vol. 85550, p. V02AT02A023). American Society of Mechanical Engineers.

[24] Colosimo, B.M., Grasso, M., Garghetti, F. and Rossi, B. (2022). Complex geometries in additive manufacturing: A new solution for lattice structure modeling and monitoring. *Journal of Quality Technology*, *54*(4), 392–414.

[25] Cano-Vicent, A., Tambuwala, M.M., Hassan, S.S., Barh, D., Aljabali, A.A., Birkett, M., ... and Serrano-Aroca, Á. (2021). Fused deposition modelling: Current status, methodology, applications and future prospects. *Additive Manufacturing*, *47*, 102378.

[26] Saadi, M.A.S.R., Maguire, A., Pottackal, N.T., Thakur, M.S.H., Ikram, M.M., Hart, A.J., ... and Rahman, M.M. (2022). Direct ink writing: A 3D printing technology for diverse materials. *Advanced Materials*, *34*(28), 2108855.

[27] Khan, K., Hussain, M.I., Tareen, A.K., Asghar, A., Hamza, M. and Chen, Z. (2026). Advances in vat photopolymerization 3D printing: Multifunctional materials, process innovations, and emerging applications. *Materials Science and Engineering: R: Reports*, *167*, 101120.

[28] Schmidleithner, C. and Kalaskar, D. M. (2018). *Stereolithography*. IntechOpen.
[29] Chaudhary, R., Fabbri, P., Leoni, E., Mazzanti, F., Akbari, R. and Antonini, C. (2023). Additive manufacturing by digital light processing: A review. *Progress in Additive Manufacturing*, *8*(2), 331–351.
[30] Joralmon, D., Khairnar, S. and Li, X. (2025). Rapid fabrication of multi-metal 3D objects with adjustable gradients by continuous liquid interface production (CLIP). *Manufacturing Letters*, *44*, 1064–1075.
[31] Vyas, A., Garg, V., Ghosh, S.B. and Bandyopadhyay-Ghosh, S. (2022). Photopolymerizable resin-based 3D printed biomedical composites: Factors affecting resin viscosity. *Materials Today: Proceedings*, *62*, 1435–1439.
[32] Davoudinejad, A. (2021). Vat photopolymerization methods in additive manufacturing. In *Additive Manufacturing* (pp. 159–181). Elsevier.
[33] Chokshi, S., Gangatirkar, R., Kandi, A., DeLeonibus, M., Kamel, M., Chadalavada, S., ... and Ravi, P. (2025). Medical 3D printing using material jetting: Technology overview, medical applications, and challenges. *Bioengineering*, *12*(3), 249.
[34] Mahamood, R.M. (2017). *Laser Metal Deposition Process of Metals, Alloys, and Composite Materials*. Springer.
[35] Negi, S., Nambolan, A.A., Kapil, S., Joshi, P.S., R,M., Karunakaran, K.P. and Bhargava, P. (2020). Review on electron beam based additive manufacturing. *Rapid Prototyping Journal*, *26*(3), 485–498.
[36] Wu, B., Pan, Z., Ding, D., Cuiuri, D., Li, H., Xu, J. and Norrish, J. (2018). A review of the wire arc additive manufacturing of metals: Properties, defects and quality improvement. *Journal of Manufacturing Processes*, *35*, 127–139.
[37] Gibson, I., Rosen, D., Stucker, B. and Khorasani, M. (2020). Directed energy deposition. In *Additive Manufacturing Technologies* (pp. 285–318). Cham: Springer International Publishing.
[38] Mekonnen, B.G., Bright, G. and Walker, A. (2016, January). A study on state-of-the-art technology of laminated object manufacturing (LOM). In *CAD/CAM, Robotics and Factories of the Future: Proceedings of the 28th International Conference on CARs & FoF 2016* (pp. 207–216). New Delhi: Springer India.
[39] Hehr, A. and Norfolk, M. (2020). A comprehensive review of ultrasonic additive manufacturing. *Rapid Prototyping Journal*, *26*(3), 445–458.
[40] Kadir, A.Z.A., Yusof, Y. and Wahab, M.S. (2020). Additive manufacturing cost estimation models—A classification review. *The International Journal of Advanced Manufacturing Technology*, *107*(9), 4033–4053.
[41] Abdulhameed, O., Al-Ahmari, A., Ameen, W. and Mian, S.H. (2019). Additive manufacturing: Challenges, trends, and applications. *Advances in Mechanical Engineering*, *11*(2), 1687814018822880.
[42] Lee, P.H., Chung, H., Lee, S.W., Yoo, J. and Ko, J. (2014, June). Dimensional accuracy in additive manufacturing processes. In *International Manufacturing Science and Engineering Conference* (Vol. 45806, p. V001T04A045). American Society of Mechanical Engineers.
[43] Brennan, M.C., Keist, J.S. and Palmer, T.A. (2021). Defects in metal additive manufacturing processes. *Journal of Materials Engineering and Performance*, Volume 30, Issue 7, pp. 4808–4818.
[44] Sarabia-Vallejos, M.A., Rodríguez-Umanzor, F.E., González-Henríquez, C.M. and Rodríguez-Hernández, J. (2022). Innovation in additive manufacturing using polymers: A survey on the technological and material developments. *Polymers*, *14*(7), 1351.
[45] Lee, J.Y., An, J. and Chua, C.K. (2017). Fundamentals and applications of 3D printing for novel materials. *Applied Materials Today*, *7*, 120–133.
[46] Vafadar, A., Guzzomi, F., Rassau, A. and Hayward, K. (2021). Advances in metal additive manufacturing: A review of common processes, industrial applications, and current challenges. *Applied Sciences*, *11*(3), 1213.

[47] Powell, D., Rennie, A. E., Geekie, L. and Burns, N. (2020). Understanding powder degradation in metal additive manufacturing to allow the upcycling of recycled powders. *Journal of Cleaner Production, 268*, 122077.
[48] Baumers, M., Dickens, P., Tuck, C. and Hague, R. (2016). The cost of additive manufacturing: Machine productivity, economies of scale and technology-push. *Technological Forecasting and Social Change, 102*, 193–201.
[49] Post, B.K., Lind, R.F., Lloyd, P.D., Kunc, V., Linhal, J.M. and Love, L.J. (2016). The economics of big area additive manufacturing. Solid Freeform Fabrication Symposium – An Additive Manufacturing Conference Reviewed Paper, pp.1176–1182.
[50] Pei, E., Bernard, A., Gu, D., Klahn, C., Monzón, M., Petersen, M. and Sun, T. (Eds.). (2023). *Springer Handbook of Additive Manufacturing*. Springer Nature.
[51] Salem, H., Abouchadi, H. and El Bikri, K. (2020). Design for additive manufacturing. *Journal of Theoretical and Applied Information Technology, 98*(19), 3043–3054.
[52] Ranjan, R., Samant, R. and Anand, S. (2017). Integration of design for manufacturing methods with topology optimization in additive manufacturing. *Journal of Manufacturing Science and Engineering, 139*(6), 061007.
[53] Bandyopadhyay, A. and Traxel, K.D. (2018). Invited review article: Metal-additive manufacturing—Modeling strategies for application-optimized designs. *Additive Manufacturing, 22*, 758–774.
[54] Han, D. and Lee, H. (2020). Recent advances in multi-material additive manufacturing: methods and applications. *Current Opinion in Chemical Engineering, 28*, 158–166.
[55] Jandyal, A., Chaturvedi, I., Wazir, I., Raina, A. and Haq, M.I.U. (2022). 3D printing–A review of processes, materials and applications in industry 4.0. *Sustainable Operations and Computers, 3*, 33–42.
[56] Mahmud, M.S., Mkawe, P. N. and Islam, M.T. (2023). Multi-Material Additive Manufacturing for Integrated Electromechanical Systems. *American Journal of Interdisciplinary Studies, 4*(04), 52–79.
[57] Kokare, S., Oliveira, J.P. and Godina, R. (2023). Life cycle assessment of additive manufacturing processes: A review. *Journal of Manufacturing Systems, 68*, 536–559.
[58] Agnusdei, L. and Del Prete, A. (2022). Additive manufacturing for sustainability: A systematic literature review. *Sustainable Futures, 4*, 100098.
[59] Vidakis, N., Petousis, M., Spyridaki, M., Mountakis, N., Dimitriou, E. and Michailidis, N. (2026). Ultra-and high-performance polymers for material extrusion additive manufacturing: Recent advancements, challenges, and optimization perspectives. *Materials Science and Engineering: R: Reports, 167*, 101086.
[60] Dzogbewu, T.C. and de Beer, D.J. (2024). Additive manufacturing of NiTi shape memory alloy and its industrial applications. *Heliyon, 10*(1).
[61] Megdich, A., Habibi, M. and Laperrière, L. (2023). A review on 4D printing: Material structures, stimuli and additive manufacturing techniques. *Materials Letters, 337*, 133977.
[62] Soori, M., Jough, F.K.G., Dastres, R. and Arezoo, B. (2025). Additive manufacturing modification by artificial intelligence, machine learning, and deep learning: A review. *Additive Manufacturing Frontiers, 4*(2), 200198.
[63] Wang, S., Shi, Y., Cheng, W., Fan, K., Wang, Q., Li, Y., ... and Sun, H. (2025). Multi-objective optimization of additive manufacturing process parameters of nitinol alloys for elastocaloric cooling. *Engineering Applications of Artificial Intelligence, 154*, 111012.
[64] Vashishtha, G., Chauhan, S., Zimroz, R., Yadav, N., Kumar, R. and Gupta, M.K. (2025). Current Applications of Machine Learning in Additive Manufacturing: A Review on Challenges and Future Trends: G. Vashishtha et al. *Archives of Computational Methods in Engineering, 32*(4), 2635–2668.

[65] Saini, G., Segaran, N., Mayer, J.L., Saini, A., Albadawi, H. and Oklu, R. (2021). Applications of 3D bioprinting in tissue engineering and regenerative medicine. *Journal of Clinical Medicine*, *10*(21), 4966.

[66] Iqubal, M.K., Shaikh, R., Dongala, B.P., Pansare, S.J., Thota, S.K., Charoo, N.A., ... and Rahman, Z. (2025). Additive manufacturing and criteria to fabricate bioinks. *Advanced Nanomaterials in Biomedical Implants: Processing, Structures, Properties and, Applications*, 251–306.

[67] Bora, K.M., Jena, A., Dey, U., Varshney, S.K. and SivaKumar, C. (2025). 4D Printing in Micro/Nano Scale: Technologies, Challenges, and Applications. *4D Printing Technology: Principles, Materials and Applications*, 205–243.

[68] Guo, S., Cui, H., Agarwal, T. and Zhang, L.G. (2024). Nanomaterials in 4D printing: Expanding the frontiers of advanced manufacturing. *Small*, *20*(30), 2307750.

[69] Parvanda, R. and Kala, P. (2023). Trends, opportunities, and challenges in the integration of the additive manufacturing with Industry 4.0. *Progress in Additive Manufacturing*, *8*(3), 587–614.

2 Additive Manufacturing: Progress, Applications, and Future Directions

Vikas R. Gulia[1*]

1. Introduction

Additive manufacturing (AM) is comparatively a new technology in the manufacturing industry, which allows objects of three dimensions to be created through the layer-by-layer creation of objects based on the digital model. In contrast to conventional subtractive manufacturing methods which involve cutting away material of a work piece, AM is additive and provides new levels of design freedom, material efficiency, and customization. AM has since developed over the years, beginning as a rapid prototyping system in the 1980s through to a feasible production technology to produce end-use parts with complex geometries and improved material properties. During the last decade, AM technologies have advanced significantly, as the technologies became necessitated in high-value sectors, including aerospace, biomedical, defense, and car automotive segments [1]. This change has been marked by a great change in hardware systems, process control, material development, and integration with digital manufacturing ecosystems. The technology has unique benefits over traditional manufacturing such as the capability to make low-volume, individualized products, which have complex geometries, minimized material waste, shorter time to market, and the capability to make parts on demand [2]. The drawbacks notwithstanding, AM experiences structural obstacles that are limiting its popularization. Quality control is one of the most important issues since the extremely dynamic character of AM processes often leads to the appearance of defects, such as porosity, cracks, dimensional error, and microstructural variations [3, 4]. The impossibility of repeatability and reproducibility of AM processes, as well as the absence of adequate standardization and certification systems, pose serious obstacles to industrial adoption, especially in safety-critical systems. Such material issues as powder behavior, flowability and correlation between process parameters, and eventual part characteristics all remain the subject of considerable research and development [5, 6].

The chapter covers an in-depth overview of the developments in AM between the year 2012 and 2026 in terms of technology, use in industries, quality control strategies, materials, and the future. The analysis is based on the large body of literature that comprises various AM processes such as powder bed fusion (PBF),

[1] Symbiosis International (Deemed) University, Symbiosis Institute of Technology, Pune, India.
* Corresponding author: vikas.gulia@sitpune.edu.in

DOI: 10.1201/9781003635673-2

directed energy deposition (DED), binder jetting, and other emerging technologies [7, 8]. This chapter seeks to fill in the gaps in the existing literature by synthesizing what is already known and identifying gaps in literature to establish the role of AM in the fourth wave of manufacturing and how it will transform manufacturing among various industries.

2. Progress in Additive Manufacturing

2.1 Evolution of AM Technologies

The short-term outlook of additive manufacturing since 2012 to 2026 indicates a shift in the applications of additive manufacturing to industrial-scale manufacturing capabilities. This has been brought about by the continuous advancements in hardware, software, materials, and process knowledge. The technology has evolved through various developmental phases, up to the advanced metal AM types of technology that can fabricate safety-critical parts of high demands [9]. Additive manufacturing has a wide family of technologies that are differentiated by their capabilities and areas of application. Main metal AM processes are PBF, which includes laser powder bed fusion (L-PBF), electron beam melting (EBM), and DED, which also has laser direct metal deposition (LDMD) and the wire arc additive manufacturing (WAAM) [5, 7, 10]. The technologies vary in terms of the source of energy, feedstock materials, build rates, resolution, and material compatibility, which is why they are applied in various applications and geometry of parts. High-value industries have been paying more attention to the progress of metal AM, and now the technology allows producing components directly with aerospace, automotive, biomedical, and energy applications. This has been due to increased reliability of processes, quality of parts, and economic viability to use in this industry though there are still major challenges that are facing the effort to attain consistency and certification standards required to facilitate its mass deployment [11].

2.2 Laser-Based AM

One of the most popular metal AM technologies has become L-PBF due to the possibility to fabricate parts with high resolution and high quality of surfaces. L-PBF has rapidly grown and has experienced a big development in the 1990s in the development of the laser technology, scanning, handling of powder, and control of the process. The technology works by the selective melting of thin layers of metal powder with a focused beam of laser, layer by layer to create items in an inert environment to prevent oxidation as shown in Figure 1. The major developments in the L-PBF technology are the underlying enhancement of the laser systems by which the production of more powerful lasers has led to increased build rates as well as processing of harder materials [12]. Multi-laser systems have also been implemented with the purpose of being more productive and enable more beams to operate concurrently on the build plate in separate areas. These systems have shown an increase in the printing speed with multiple nozzle applications, which has made L-PBF more economical with respect to larger production volumes [13]. Increased material palette among designers and manufacturers has been created through the creation of new metal powders and alloys that are specifically created to comply with L-PBF. Studies have concentrated on maximizing the powder properties such as the particle size distribution, morphology, and flowability to ensure greater stability in the process and quality of the part. Titanium alloys, aluminum alloys, and stainless steels were also widely tested in terms of their mechanical characteristics and possible use in L-PBF systems [6, 9].

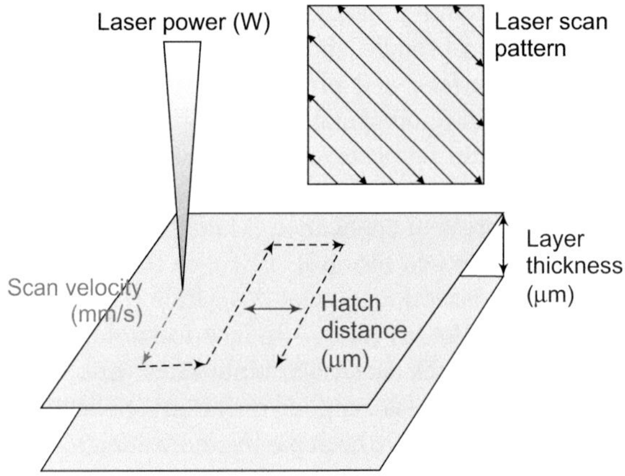

Fig. 1 Process parameters in LPBF [19].

A different form of L-PBF is selective laser melting (SLM), which has already shown the possibility of producing quality parts with good surface finish, but with difficulties in making large-size components [8]. Optimization in the process parameters, for example, reduction in the hatching distance and laser scanning speed, has been demonstrated to minimize the necks and voids resulting in high-quality surface finishes. Fraunhofer Institute developed SLM technology of metal materials, and this gave rise to commercial equipment such as the EOSINT series which has gained vast popularity in industry [10]. Another notable type of laser-based AM technology is DED, which is especially useful in the case of large-scale parts, repair, and multi-material components [14]. The LDMD can print large-size metal items, able to print large titanium alloy parts with sizes of up to 3 meters and weighing 196 kilograms [15].

2.3 *Electron Beam Technologies*

Electron beam melting (EBM) is another add-on technology to metal AM that employs a focused beam of electrons as the heat source used to melt metal powder in a vacuum chamber. The EBM technology has received much interest in its capability to manufacture complex metal components with superior mechanical properties, especially in materials that involve high temperature such as titanium alloys. EBM processes are associated with a vacuum atmosphere and high preheat temperatures that lead to less residual stresses and desirable microstructures relative to laser-based systems Fig. 2.

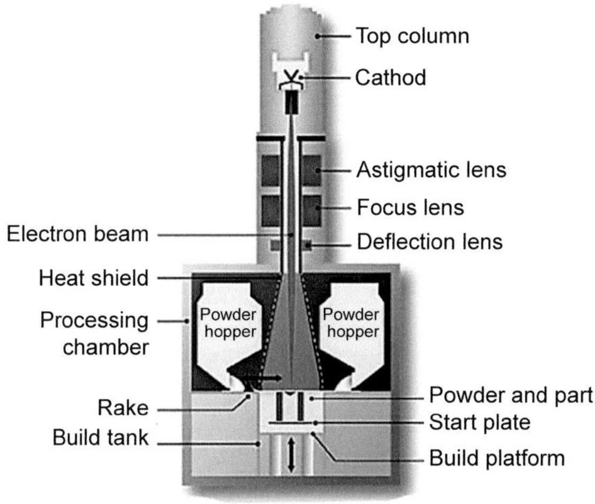

Fig. 2 Electron beam melting [16].

The existing capacity of EBM and its trends in future have been widely discussed and focus has been given on its usage in aerospace and biomedical domain. This can produce complete metallic parts directly based on the electronic information, which is beneficial in the processing of reactive materials that would otherwise oxidize in the inert gases environments in the laser-based systems. Successes have been seen by the technology in the production of titanium alloy components to aerospace uses and patient-specific implants to medical uses [16]. Special techniques of powder manufacturing and characterization of materials peculiar to EBM have been invented, the needs of powder metallurgy being given attention. Complex metal components can be produced with high performance using the technology, but the size of forming is constrained by the size of the powder bed and vacuum chamber [17]. Studies have dealt with parameter development platforms to optimize process conditions on various materials and geometries with the objective of solving the current problems and extending the scope of the technology application. EBM has managed to help in the manufacturing of vital aerospace parts such as satellite TC4 titanium alloy parts [18], which have taken a lot less time than the traditional ways of manufacturing a part. The high temperature material processing capability of the technology with low residual stress allows it to be especially applied in aerospace where component reliability and performance are of utmost importance. Nevertheless, there are still issues with the surface finish, dimensional accuracy, and the increase of the number of materials processable.

2.4 Process Improvements and Automation

Over the last two decades, there has been tremendous improvement in the efficiency, reliability, and automation of AM process. Such improvements include the advancement of the hardware architecture, software systems, process planning, and integration with the digital manufacturing ecosystems [6]. Manual intervention has been minimized and process repeatability increased by the creation of automated powder systems, build platform systems, and post-processing systems. The recent directions of automation and quality control in AM have been centered on the creation of smart manufacturing systems with the capability of real-time monitoring of the processes and adaptive control. The availability of the advanced sensing technologies in AM systems allows in situ observation of large volumes of information at different scales, one melt pool or all build layers. Such data-based framework along with machine learning (ML) models helps in real-time quality control and process improvement, which is missing in the repeatability that has been a barrier to a wide AM adoption [20]. Increment in post-processing has played a major role in attaining surface finish, dimensional accuracy and material properties needed in end-use parts. The heat treatment to eliminate residual stresses and to obtain a microstructure with optimized results, machining to reach final dimensions and surface finish, and surface treatments to enhance fatigue strength and corrosion resistance are the post-processing operations. The concept of hybrid manufacturing systems integrating additive and subtractive manufacturing processes on a single platform has also come into the perspective as a promising method to minimize the post-processing needs and enhance the quality of parts. The use of artificial intelligence (AI) and the Internet of Things (IoT) is one of the current trends in the AM process improvement [21]. The optimization of AI-based processes can be used to automatically tune the parameters, removing incipient defects and enhancing the stability, and the IoT connection can be used to monitor remotely, perform predictive maintenance, and link with enterprise resource planning systems [22]. These technologies assist in shifting AM to smarter, digital and high value manufacturing in line with the principles of Industry 4.0.

Optimization of process parameters has been given great attention in order to enhance the quality of parts and efficiency of the process. It has been shown that laser power, scanning speed, hatching distance, layer thickness, and powder properties have a great impact on the final properties of the parts. The systematic optimization methods, such as design of experiments and ML-based models, have been devised to find optimal combinations of parameters to work with certain materials and geometries [23]. These developments have helped in achieving better process knowledge and higher levels of part quality in multiple AM systems and operators. Process parameter optimization has been extensively studied to improve part quality and process efficiency. Research has demonstrated that parameters such as laser power, scanning speed, hatching distance, layer thickness, and powder characteristics significantly influence final part properties. Systematic optimization approaches, including design of experiments and ML-based methods, have been developed to identify optimal parameter combinations for specific materials and geometries. These advances have contributed to improved process understanding and more consistent part quality across different AM systems and operators [23].

3. Applications of Additive Manufacturing

3.1 Aerospace Applications

In the aerospace sector, AM has been leading AM adoption, due to the technology that allows the manufacture of lightweight, complex components whose lead times and material wastes are reduced. Aerospace uses are both in prototyping and tooling and the manufacturing of flight-critical parts of aircraft and spacecraft. The industry has been a major contributor to the AM development and commercialization due to the harsh performance standards and readiness to invest in the latest technologies in manufacturing. Additive manufacturing has brought a revolution in the making of complex, lightweight, and practically waste-free aerospace parts. The technology makes it possible to produce complex engine and structural parts, as well as auxiliary systems, which were not easy or feasible to make with the traditional processes [24]. AM coupled with topology optimization (TO) enables the designers to develop structures optimized according to the desired loading conditions leading to considerable weight savings without deteriorating or worsening mechanical performance. Certain aerospace uses are thin-walled aircraft engine parts, complex geometries, and lightweight structural components. The LEAP engine fuel nozzle, produced by AM, recorded a 25% weight saving and the nozzle was found to be 5 times more durable than the traditionally produced versions. This element represents one of the examples of AM providing performance and costs improvements in aerospace applications.

Fig. 3 Satellite solar panel deployment mechanism (left). Satellite sandwich panel (right) [32].

Studies have reported making of different aircraft components with the help of AM, such as joints, keel webs, pylon ribs, thrust beams, and wing root rings [25].

TO and SLM have already made a significant step toward minimizing the volume of the materials by 54% and a weight reduction in aircraft sample parts of 28%. These weight savings are directly converted to fuel efficiency gains and payload capacity which are very strong economic arguments in favor of AM adoption in aerospace applications, one such example as shown in Figure 3. The technology also achieves speed of production where lead times in hours or minutes are taken to produce finished goods compared to weeks and months in weeks and months, making supply chain agile and lowering inventory requirements. The technology allows repairing damaged parts and even manufacturing spare parts without storing them in large stocks and saving money and helping to increase the availability of aircraft. This has been a useful feature especially in legacy aircrafts where the original tooling might not be in existence or low volume spares where normal production would not be economical [26].

3.2 Medical and Biomedical Applications

AM has been adopted by the medical and biomedical fields due to its capability to generate customized devices, implants, and surgical equipment to an unprecedented level. The versatility of the technology in terms of geometry and materials is especially beneficial when it comes to a patient-specific approach to medical applications, where anatomy and needs of each patient are specific. AM has revolutionized the medical device production and provided innovative approaches to tissue engineering, surgical planning, prosthetic, and implantable devices. AM has also found medical use as tissue engineering scaffolds, surgical implants, prosthetics, and anatomical models to plan surgery. The technology offers the production of individualized designs that are suitable to the anatomy of the patients and enhance fit, function, and clinical results. Medical implants made with AM on a patient-specific basis have shown both better integration with the host tissue and better patient satisfaction over standard devices kept on the shelf [27] Fig. 4.

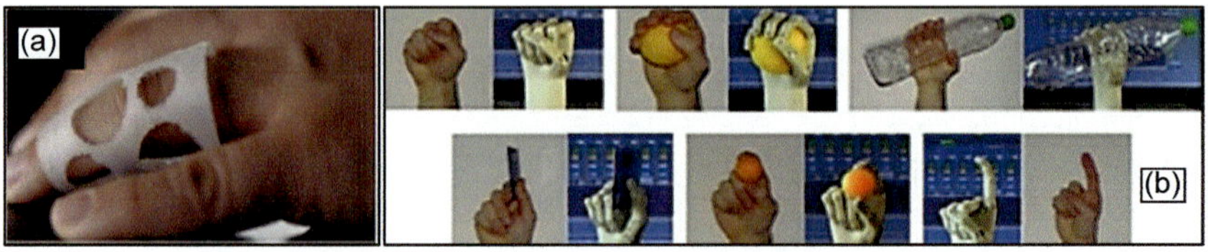

Fig. 4 (a) Personalized medical 3D-printed PHBH/CNC-based device for finger dislocation [60]. (b) Exo-prostheses: multifunctional prosthetic 3D-printed hand prototype fabricated through fiber-reinforced nylon [60].

One of the orthopedic applications is a significant domain of AM implementation in medicine, as 3D printing is applied to customize implants and repair them personally. The technology can be used to add porous structures that facilitate bone ingrowth and osseointegration leading to better long-term stability of implants.

Titanium alloys, especially Ti-6Al-4V, have been utilized widely in orthopedic implants because they exhibit good biocompatibility, corrosion resistance, and mechanical properties [28, 29]. The new application of AM technology is bioprinting, which allows printing tissue-like structures with living cells and biomaterials. Possibly, in the future, this technology will find its use in tissue engineering, drug testing, and one more time organ transplantation. Bioprinted though mostly in its infancy stage, bioprinting is a radical way to apply AM principles to the biological world and could revolutionize the field of regenerative medicine. The medical device sector has been advantaged by AM being able to make complex internal geometries including flow channels on surgical instruments and drug delivery tools. The technology allows incorporation of various functions in single elements, less assembly needs, and higher reliabilities of devices [30]. AM-produced surgical guides and patient-specific instrumentation have enhanced surgical accuracy and shorter operating durations, which have resulted in improved patient outcomes and less healthcare expenses. The dental medicine has also been used in AM, such as manufacturing crowns, bridges, dentures, and orthodontic appliances. The technology can be used to produce custom dental devices fast in a highly esthetic manner, enhancing the comfort of patients and their treatment results. Direct manufacturing of these devices out of digital scans simplifies the work process and saves time between the first consultation and the final provision of the device [28, 29].

3.3 Automotive Applications

The automotive industry has been embracing AM to prototype, tool, and manufacture the final parts of use in their end products due to the technology which has the capability of shortening the time required to develop a particular part, optimize designs, and facilitate mass customization. Although automotive AM use has historically been limited to design validation by rapid prototyping, it is now used in production of functional components, especially in high-performance and low-volume cars. AM allows the automobile sector to manufacture lightweight products to enhance fuel efficiency and vehicle performance. Weight loss is a vital goal in the design of automotive vehicles, especially in the case of electric vehicles where the lighter an object is, the farther it will move. The technology enables topology-optimized structures and combining multiple portions into single portions which minimizes the complexity of assembly and the weight of the vehicle. Auto uses are engine parts, brackets, special interior parts, and manufacturing process tooling. Complex engine parts have been created through the technology incorporating built-in cooling routes and geometries that result in enhanced performance and efficiency. Individual brackets and mounting parts can be quickly designed to suit the vehicle needs without having to require costly tools [31].

Automobile manufacturing has also taken advantage of AM to manufacture tooling, jigs, and fixtures to be used in manufacturing processes. These tools can be easily fabricated and modified to certain production needs, and it saves on lead times and cost as opposed to traditionally manufactured tooling. The automotive industry has been an early user of AM in its high-performance and motorsport segment, in which the technology has allowed the creation of optimized lightweight components to give them a competitive edge. The racing teams have adapted AM to quickly develop designs and create customized components that suit conditions of the track and the setup of the vehicle. The flexibility of the technology allows responding quickly to new needs and maintaining a high level of performance throughout a racing season. There are opportunities and challenges associated with integrating AM and automotive supply chains [31].

3.4 Other Industrial Applications

Outside the aerospace, medical, and automotive industries, AM has been used in a wide variety of industries, such as energy, construction, consumer products, electronics, and food production [33, 34]. Such applications can use the capabilities of AM peculiar to certain problems of the industry and allow designing new products and production methods. AM has been applied to manufacture parts in the oil and gas equipment, nuclear reactors, and renewable energy systems in the energy industry. The technology is utilized to manufacture complicated heat exchangers, turbine parts, and tailored equipment to energy applications. The on-demand production of parts is especially useful in maintenance and repair of detached areas where the traditional supply chain is difficult. AM has been pursued in the construction industry to create complicated architectural forms, construction elements, and even structures. The AM systems in large scales can print concrete structures minimizing waste of materials and time used in construction compared to traditional construction method. The technology facilitates making of complicated geometries and custom designs that otherwise would be hard or costly to pursue using traditional construction methods. Another potential use of AM to consumer products is the creation of custom jewelry, footwear, eyewear, and household application. The technology allows mass customization through which products can be customized to suit the personal preferences of the customers without the cost penalties normally related with customization. Direct-to-consumer AM services have been introduced, and the customers can design and place custom products online. The electronics production industry has been able to create embedded, structural, and stretchable parts such as sensors and antennas through.

Food industry has pursued AM toward customized dishes with added taste and nutrition. Technology food printing can be used to produce intricate forms and textures, can allow the accurate control of food nutritional content, and can minimize food waste. Food AM, which is still on the rise, is the new use of technology that has a potential of changing food production and consumption. AM has been embraced by fashion and apparel industries in the creation of custom clothes, accessories, and shoes that enable them to design their clothes or shoes as they have never had before. The technology allows the creation of complicated geometries and fabricated structures, which are not feasible through traditional textile manufacturing. Orthotic devices and shoes tailored to fit an individual by using AM offer better comfort and performance in comparison to regular size [34].

4. Quality Control in Additive Manufacturing

4.1 Process Monitoring Technologies

The additive manufacturing process poses a challenge to quality control, which is difficult to manage because of the layer-by-layer strategy of the manufacturing process, complex thermal cycles, and the numerous process parameters, which affect end part quality. Monitoring technologies of processes have become one of the most essential tools of the quality of parts, detection of defects, and control of processes in AM systems. Those technologies include various sensing methods capturing the process signatures at various scales down to the scale of a single melt pool to whole build layers. By introducing sophisticated sensing technologies to AM systems, the in situ monitoring of large volumes of data is possible at different scales. Photodiodes, cameras, and pyrometers are optical sensors that are commonly used to scan melt pool properties, thermal radiation, and the morphology of the layer through the build process. Such sensors give instant feedback concerning the stability of the process and can identify anomalies, which can cause defects [3]. Infrared cameras and pyrometers can be used in thermal monitoring; this allows temperature distributions in the build platform to be measured, as well as thermal history to be traced during individual layers. The temperature is a critical parameter in the study of the development of residual stress, microstructure, and the structuring of thermal-related defects including warping and cracking. Very fast thermal images can be used to measure melt pool dynamics at a frequency rate that can resolve single laser scan tracks.

The optical imaging systems such as high-resolution cameras and structured light scanners are employed to monitor the quality of the powder layers, identify the spreading defects, and quantify the layer geometry. A profilometry of phase measurement has been developed to provide in situ inspection of powder bed delivery allowing the detection of powder delivery defects that happen during the build cycle. Calibration technique used and compensation algorithms can effectively monitor powder defects in these systems using improved accuracy.

Monitoring AM processes has also been studied using acoustic emission sensors to detect the process of crack formation, delamination, and powder spatter. The acoustic signals complement the optical and thermal measurements and thus can be used to identify defects that might not be detected by other modalities of sensing. Multi-modal monitoring systems are more thorough in understanding processes and detecting defects because they combine several types of sensors that enhance the knowledge of the processes [11].

4.2 Defect Detection Methods

Defect detection in AM involves in situ inspection of AM during the building process and post-building inspection of completed parts. AM processes are often very dynamic, and as such, they often lead to different kinds of defects, such as porosity, cracks, lack of fusion, surface roughness, and dimensional error. An efficient system of defect detection is an element of the quality of parts and the opportunity to use the AM components in safety-related process. The typical metal AM defects are pores, cracks, inclusions and non-fusion defects and require sophisticated technology to control defects. Porosity may be due to entrapment of gases, incomplete melting, or key holes formed during the melting process. Cracks could be because of thermal stresses, shrinkage on solidification or brittle material. The defects of lack of fusion arise due to either the inadequate energy to completely fuse the powder or a lack of overlap between scan tracks which leaves them vacant. Machine learning systems have become effective techniques of automated defect detection in AM. These techniques can process high amounts of sensor data to find out patterns related to defect formation and categorize part quality, in real time [35]. Algorithms that are trained supervised, such as support vector machines, random forests, and neural networks, have been trained to recognize types of defects based on sensor measurements. The convolutional neural networks as deep learning methods have demonstrated potential of image data analysis in optical monitoring systems. The methods of image processing are widely applied to process optical monitoring data and identify flaws. These methods involve attribute detection, feature identification, as well as pattern recognition algorithms that can detect anomalies in powder layer quality, melt pool features, and surface morphology. In real-time image processing, it is possible to detect when there is an error in process stability, either through lack of sufficient powder, faulty supports, or damaged coaters, which affects the quality of parts. The methods of post-build inspection are either a non-destructive testing (NDT) method or a destructive one. NDT procedures like X-ray computed tomography (CT), ultrasonic testing, and eddy current inspection make it possible to recognize internal defects without causing harm to the part. X-ray CT has a three-dimensional representation of internal porosity, cracks, and dimensional deviations; hence it is more useful in intricate AM parts [36]. The standardization of non-destructive testing of AM parts has however been a challenge that needs to be tackled by developing AM specific inspection protocols and acceptance criteria.

4.3 In Situ Monitoring Systems

The in situ monitoring system is an imperative innovation to AM quality control, which makes it possible to observe the process of building in real time and identify defects early. Such systems have a series of sensors built into AM machines to record process data in real time during the build without disrupting production. Information on every deposited layer is recorded by layer-by-layer monitoring systems which can be used to identify defects in powder spreading, non-complete melting, and geometric error. These systems normally rely on optical imaging or structured light scanning to quantify layer height, identify powder anomalies, and ensure that a layer has been deposited to quality specifications before the next layer is deposited. Automated rules to detect defects may raise an anomaly to the attention of operators or cause process modifications [35].

Melt pool monitoring systems are devices that capture the molten material in the melting process by use of high-speed cameras, photodiodes, or spectrometers. The characteristics of the melt pools such as size, shape, temperature, and stability give an indication of the process quality and are used to predict the formation of defects. Examples of problems that the deviation of the melt pool behavior can reveal include inadequate powder, contaminated expectations, and inappropriate process parameters. The profilometry of powder beds has been accurately measured using in situ inspection systems, which were developed by measuring the topography of powder beds with phase measurement profilometry. To achieve these systems, the time synchronization method is used to minimize the time to capture data which is about 2 seconds per inspection and thus can be integrated into production AM systems. Surface fitting algorithm-based protocols, which are used to calibrate, eliminate the effects of phase error and random noise, and improve the accuracy of the inspections [37].

The application of process control strategies with the help of in situ monitoring data is essential to the building of the future smart AM systems. Closed-loop control systems are capable of automatically changing the process parameters to anomalies identified to enhance the stability of the process and achieve quality parts. Nevertheless, it is not an easy task to implement effective process control because there is a complex relationship among process parameters, sensor signatures, and end part properties. This requirement of the need of in-process data analytics and statistical monitoring tools that are used in automated defects detection has motivated the study in machine learning and applications of artificial intelligence in AM [37].

4.4 Standards and Inspection Techniques

The standardization and standardized inspection methods are key elements of universal implementation of AM in the industrial processes, especially on safety-related elements. Absence of standard quality assurance procedures and consistent requirements of certification has been found to be one of the major obstacles to AM adoption. Standards bodies such as ISO and ASTM have actively been working on AM-specific standards, such as terminology, test methods, process specifications, and material requirements [38]. Surface quality influences significantly the local microstructure, internal anomalies, and accuracy of products in AM parts. Influences on the quality of the surface are material properties, particle size and distribution, part orientation, layer thickness, STL file creation, surface angle, finish type, scanning parameters, and post-processing. It has been identified that AM surface texture needs, recommendations, and optimization standards are necessary to obtain part consistency. Surface imagery and AM parts examination are done using contact and non-contact profilometers. Such tools determine the roughness, the waviness of the surfaces, and form deviations, giving quantitative information to determine quality. Optical non-contact techniques such as the laser scanning and structured light systems allow the quick process of measuring complex geometries without the possibility of destroying delicate features. The dimensional inspection of AM components involves special methods because of complicated geometries and inside features that are frequently generated. The verification of parts against dimensional specifications is done using coordinate measuring machines (CMMs) and optical scanning systems [39]. Nevertheless, X-ray CT scanning might be needed to measure internal features and the intricate geometries, which contains all the geometric information that is three-dimensional in nature.

Testing and characterization of AM materials and processes are critical in qualifying their materials and processes. Mechanical tests such as tensile, fatigue, and fracture toughness tests should consider the anisotropic nature of AM components, which is usually the case. In design and certification, the development of full database of mechanical properties of AM materials is of great importance. Optical and electron microscopic microstructural characterization give information on grain structure, phase composition, and defect populations. Process qualification and part certification procedures unique to AM are currently under development. Such processes should consider the peculiarities of AM, such as the effect of build orientation, support structures, and thermal history on the properties of parts [40]. The material allowable generation requirement and the influence of AM process parameters on porosity and other flaws have been highlighted to be required in the aerospace applications.

5. Material Challenges in Additive Manufacturing

5.1 Powder Characteristics and Properties

The characteristics of powders are important factors that define the quality and reliability of the powder bed fusion AM processes. Metal powders are sensitive to their properties such as size distribution, morphology and flowability, and packing density, which affect the behavior of powder spreading, melt pool dynamics, and ultimate quality of the part. One of the challenges that have been cited regarding the technology is the necessity to enhance the quality and yield of AM powders [41]. One of the essential powder properties is particle size distribution, which has an influence on the processability and ultimately on the end of part characteristics. L-PBF processes have optimal particle size distributions of 15 to 45 μm, which is sufficient to maintain high flowability, as well as provide sufficient packing density. Finer powders have the capability to enhance surface finish, and allow thinner layers, but have poor flowability and elevated oxidation potential because of the greater surface area. Powder morphology, especially particle sphericity and satellites or irregular particles, has an impact on the flowability and packing behavior. The smooth surface of the spherical particles allows them to move more freely than the irregular ones, providing an opportunity to spread the powder, and make layers more uniformly [42]. The most usual technique of producing AM spherical metal powder is gas atomization, although even more spherical particles can be made using plasma atomization in harsh applications. The flowability of the powder is needed to maintain the continuous spreading of powder and formation of homogenous layers during the powder bed processes. Lack of flowability may cause unequal distribution of the powder, partial spreading of the powder, and final part defects. Flowability depends on particle size, morphology, surface chemistry, and moisture content. The nature of powder behavior has been described using standardized flowability tests, such as Hall flowmeter and Carney funnel tests, but these tests do not necessarily provide sufficient information about the conditions of deep flow in AM systems. The density of powder packing also influences the quantity of material to be melted and porosity of the finished component. Increased packing density usually leads to more dense (fewer voids) parts, although very high packing density can hinder the spreading of powder. The correlation between powder properties and packing density is complicated depending on the distribution of particle size, morphology, and forces between them. The reuse and recycling of powder pose an economic problem and a quality problem in AM. Materials are usually recycled many times to save on the material cost, but with consecutive recycling powder properties deteriorate as particles size distribution varies, growing oxygen levels and spatter particles deposition is built up. Powder management strategies and quality control procedures of recycled powders must be developed so that the performance of the processes is uniform [43].

5.2 Microstructure and Mechanical Properties

Contrary to traditionally produced materials, microstructure of AM pieces varies dramatically because of the peculiarities of thermal cycles during the construction of the part in layers. Control and knowledge of microstructure is critical to desired performance and the control of parts in service. AM research has also paid significant attention to the connection between process parameters, thermal history, microstructure, and mechanical properties. The AM parts are microstructurally anisotropic and heterogeneous due to directional heat flow and repetitive thermal cycling during the build process. Grain structure columnar grains with the build direction representing the leading direction of heat flow in solidification are common. It is an anisotropic microstructure that may give rise to directional mechanical properties, where the strength and ductility are different based on how it is oriented with the build direction. The high rates of solidification of AM processes, especially L-PBF and EBM, lead to non-equilibrium fine microstructures. Some AM processes have cooling rates more than 106 K/s which produces refined grain structures and long solid solubility unlike traditional processing. Such small microstructures can be used to give greater strength, but they can also lead to higher residual stresses and to crack propagation [44].

Microstructure, defects populations, and surface quality affect mechanical properties of AM parts, such as tensile strength, yield strength, ductile, and fatigue strength. The ultimate tensile strength and surface finish of different alloys have improved in liquid- and powder-based AM processes. Nevertheless, to a considerable degree, surface roughness and porosity can considerably decrease the fatigue life of AM components, especially when it comes to titanium alloy components. The quality of the surface of AM parts has a severe influence on the local microstructure and internal anomalies. To maximize the performance of the microstructure and mechanical properties of AM parts, heat treatment and post-processing are commonly required. The heat treatments used in stress relief eliminate the residual stresses that are acquired throughout the construction process, which greatly reduces the chances of distortion and cracking. The optimization of the precipitation of the strengthening phases in age-hardenable alloys may be optimized by solution treatment and aging [45].

5.3 Common Defects and Formation Mechanisms

The flaws of AM components may considerably affect the mechanical characteristics, size control, and the performance of the parts. A proper knowledge of defect formation mechanisms is the key to implementing ways of preventing or preventing defects by using process optimization and quality control. The defects are different and caused by different factors based on the type of AM process, material, and process parameters used. Porosity represents one of the most frequent flaws in metal AM, which occurs under a few different mechanisms, such as gas entrapment, the absence of fusion, and keyhole development. The gas porosity is caused by gases dissolved in the melt or released trapped in particles of the powder that are not completely liberated during melting. Porosity and lack of fusion can be caused by a lack of energy to completely melt the powder or inadequate overlap between scan tracks resulting in empty spaces. Keyhole porosity occurs when the high energy input forms deep and unstable melt pools which collapse in place entrapping gas. Cracking is a severe flaw that may be present in the process of the build or in the cooling. The solidification cracking is caused by the thermal stress and shrinkage during solidification, especially in the materials with broad solidification ranges. The liquation cracking may take place in the heat-impacted area of the earlier layers when the low-melting phases fuse and consequently break apart. This is because cold cracking can be formed after build completion because of either hydrogen embrittlement or stresses left behind [46].

The typical AM issues are surface roughness and dimensional inaccuracy. The element that affects the quality of surface is staircase effect due to layer-by-layer construction, partially melted powder particles on surfaces, and development of dross on lower surfaces. Thermal distortion, cooling shrinkage, and spreading errors or layer position error can cause dimensional inaccuracy. The necessity to enhance the quality of the dimensional accuracy and surface has prompted the research in the optimization of the processes and the methods of the post-processing. The strong thermal gradients and the repeated heating and cooling processes the AM parts undergo during the build process result in residual stresses developed in the AM parts. Such stresses may lead to distortion during the build, warping when parts are separated off the build platform, or cracking in the extreme case. AM parts also show residual stresses that influence the fatigue performance and dimensional stability of AM parts. Managing residual stresses strategies can be preheating the build platform, scan strategies optimizing, and heat treatments that reduce stress. The process-related phenomena include spatter and denudation which may also occur in powder bed AM part quality. The spatter are molten or partially melted splashes thrown off the melt pool, and may land on the powder bed, leaving defects in the next few layers [47].

5.4 Material Development and Characterization

One of the primary research directions has been the creation of new materials aimed specifically at AM, which has increased the scope of alloys and composites that can be successfully processed. The AM material development must consider the end part characteristics as well as the processability aspects of the material such as the melting characteristics, cracking tendencies, and the compatibility to the AM process requirements. Titanium alloys and especially Ti-6Al-4V have been widely used in AM applications because of their very high ratio of strength to weight, resistance to corrosion, and biocompatibility [48].

The aluminum alloy poses special problems to AM because they have a high thermal conductivity, are highly reflective to laser radiation, and prone to hot cracking. One of the areas that have been under active research is the development of AM-specific aluminum alloys with altered compositions so as to lower the phenomenon of cracking. The aluminum alloys are favored especially in aerospace and automobile where they are used because they have low density and their mechanical properties are good [49]. Various AM technologies have been used to process stainless steels such as austenitic, martensitic, and precipitation-hardening steels. These materials tend to process well and can attain comparable or even higher mechanical strength than the traditionally manufactured materials. Stainless steels can be utilized in numerous applications with aerospace components to medical equipment and industrial tooling all being examples of applications [50].

In addition to that, the nickel-based superalloys prove especially interesting in terms of high-temperature aerospace applications, such as in turbine components. The AM has become an interesting option because these materials are difficult to process in a conventional way because they are highly strong and work-hardening, which means that the conventional way cannot be applied. Nevertheless, superalloys have been found to crack in AM because of their broad solidification and the formation of high thermal stresses. There has been the development of materials characterization techniques specific to AM to gain insight into how the process parameters, microstructure, and properties are related to one another. These methods are sophisticated microscopy to define grain structure and defect populations, X-ray diffraction to detect phases, and quantify residual stresses and mechanical testing to measure properties in various orientations. Powder metallurgical needs of AM such as the production of powder and the method of characterization have been well investigated. The functional graded and multi-material materials are a new frontier in AM. These materials join distinct alloys or change composition progressively within one part, permitting the optimization of property to a particular application. However, multi-material AM is very difficult in terms of processing control, interface quality, and property prediction [51].

6. Future Prospects

6.1 Emerging Technologies and Trends

The future of additive manufacturing will be marked by a number of new technologies and trends that would further enhance the possibilities and uses of the technology. Such developments include improvement in hardware, material, process, and co-ordination with other manufacturing technologies. The further development of AM is projected to contribute to the change in manufacturing into smarter, digital, and high-value production in line with Industry 4.0 ideals. Innovations in polymer AM include high-speed sintering and continuous production of liquid interfaces, which enhance the build rates considerably. One of the main limitations of the traditional AM processes, the low production speed, is dealt with by these technologies and made AM more competitive in higher volume production. The invention of high-speed metal AM technologies, such as more than two laser systems and increased energy sources, is continuing to enhance the economic feasibility of metal AM. AM systems on a larger scale which can form parts in the meter scale are appearing and used in aerospace, construction, and infrastructure. Those systems use different technologies such as WAAM of metal components and extrusion-based systems of concrete and polymer structures. Mass AM allows creating components that could not be produced or had many limitations with traditional techniques [52].

The 4D and 5D printing technologies are future prospects which go beyond the traditional 3D printing. The 4D printing entails the production of structures, which alter shape or characteristics with time in response to the external stimuli like temperature, moistness, or light. 5D printing adds more degrees of freedom to the printing process, allowing the production of curved structures and non-planar surfaces, which enhance mechanical properties and reduce the use of support structures. Mixed-method manufacturing systems (also referred to as hybrid manufacturing systems) which integrate additive and subtractive manufacturing processes on a single platform are becoming more popular. Such systems allow one to manufacture parts of the geometric complexity of AM and the surface finish and dimensional accuracy of machining [53–55]. Existing parts can also be repaired and modified using hybrid systems, to add features or to restore scratches. The development of bioprinting still goes on, and the further uses of bioprinting are considered in tissue engineering, drug testing, and regenerative medicines. This is likely to be facilitated by the development of bioinks with the right mechanical properties and biocompatibility in conjunction with enhanced knowledge in tissue development and vascularization to develop more complex and functional tissue constructs. Bioprinting is a revolutionary application of AM principles, even though there are still major challenges, and it has the potential to transform the world of healthcare [56].

6.2 AI and Machine Learning Implementation

One of the most promising directions to be developed has been the integration of AI and ML with additive manufacturing. AI and ML technologies provide the opportunity to solve most of the existing issues of AM, such as optimization of the processes, quality control, prediction of defects, and optimization of designs [57]. Various algorithms have been used to predict the quality of in situ sensors in machine learning methods used to identify defects and predict quality, with several studies utilizing this method to study sensor data. Support vector machines, random forests, and neural networks are supervised learning techniques that have been trained to predict quality of part and the occurrence of defects based on the signature of a process. Convolutional neural networks represent deep learning methods which have demonstrated outstanding results in processing image data used in optical monitoring systems. The process optimization based on AI can adjust parameters automatically to remove initial defects and enhance process stability. Optimal control strategies can be learned by reinforcement learning algorithms when interacting with the AM process, changing conditions, and materials. These strategies will provide an opportunity to minimize the huge experimentation in trials that is being practiced in the development of process parameters of new materials and geometries [58].

ML is used to optimize design that could be topology optimization or generative design, which allows the production of structures that are optimized according to the desired performance criteria. AI algorithms have the capacity to search through large areas of design and find solutions that are hard to imagine by human designers. The combination of AM constraints into the design optimization process will guarantee that the produced designs can be produced using the existing AM technologies. The use of digital twins—virtual models of real-life AM systems that are constantly updated with real-time data—is one of the new applications of AI and IoT technologies. Digital twins facilitate AM process optimization, prediction, and simulation, and can be used in decision-making and process control. Correct construction of the digital twins demands the combination of physics-based and data-driven methods of machine learning. It is estimated that the integration of AI and ML into AM processes will enhance the rates of AM uptake as it will minimize the knowledge needed to run systems, enhance the reliability of processes, and allow the creation of new materials and applications faster. Nonetheless, there are obstacles to the collection of data, model validation, and interpretability of ML models. This requires the creation of standard datasets and benchmarks of AM ML research which will contribute to advancement in this field [59].

6.3 Industry 4.0 and Smart Manufacturing

Additive manufacturing and Industry 4.0 technologies integration is one of the most significant trends, which is changing manufacturing paradigms. Industry 4.0 is based on cyber–physical systems, the IoT, cloud computing, and big data analytics, which offer the structure of smart, connected, and adaptive manufacturing systems. AM is in a good position to both gain out of it and be involved in Industry 4.0 because of its digital nature and flexibility. The IoT can facilitate connecting AM systems, sensors, and enterprise systems and allows the collection of data, remote monitoring, and links all aspects to the supply chain management. The AM systems connected by IoT are able to report status, notify operators on issues and get updated build files and process parameters remotely. This connectivity allows the AM equipment to be used more productively and to foster the distributed manufacturing paradigm [61].

6.4 Sustainability and New Materials

The concerns related to sustainability are becoming more and more significant when it comes to the creation and implementation of the additive manufacturing technologies. AM has a number of sustainability benefits inherent to it, such as the waste of materials when compared to subtractive manufacturing, the possibility of creating lightweight structures that lead to less energy usage during operation, and the possibility of localized production that lowers transportation needs. Nevertheless, AM also introduces sustainability issues such as energy use during the construction course, environmental impact of powder production, and the necessity to provide effective recycling policies. One of the main sustainability benefits of AM is material efficiency because the technology generally just utilizes the amount of material required to create the part and support structures, and any powder that is not utilized can be recycled. On the contrary, subtractive manufacturing may waste more material, especially of complex components machined out of solid billets. Topology optimization and lightweight fabrication are also useful in material efficiency in that less material is required to attain desired performance [62].

Energy consumption in AM is a multifaceted problem because the amount of energy per unit of mass of material processed is usually greater than the traditional manufacturing processes. The overall energy impact has to, however, consider the whole product lifecycle, such as the production of materials, production itself, the use phase, and end-of-life. In the more extreme cases, where weight saved translates into large energy savings when in operation, as is the case with aerospace and automotive applications the greater manufacturing energy can be compensated with lower operational energy [63]. Sustainable materials of AM, bio-based polymers, recycled materials, as well as materials that can be recycled are actively under research. The use of renewable-based bio-based polymers presents an opportunity of decreasing the use of petroleum-based polymers. Recycled materials can be used in AM to minimize environmental effects, but there is a problem of maintaining uniform material properties [64].

Recycling and reuse of powder (recycling) will be necessary in enhancing the sustainability of powder bed AM processes. The key to successful powder management is the need to know how powder changes with repeated use and devising quality control procedures to ensure that recycled powder is within specifications. Technological trends in the development of closed-loop powder handling systems that reduce the amount of waste powder and contamination are noteworthy. The ability to create products in localized and on-demand fashion with the help of AM can lower transportation needs and stock levels, leading to sustainability. AM models of distributed manufacturing allow manufacturing of parts near their formational location, lowering the environmental footprint of global supply chains. Nevertheless, such advantages will not be achieved without proper business models, quality assurance systems, and regulatory frameworks [65]. The development of new materials further widens the opportunities of AM. Developments in research and design into improved materials are high-entropy alloys [66], metal matrix composites, and functional graded materials. Such materials are potentially the source of improved properties and new uses, but are also problematic in processing and characterizing. This is because the creation of materials specifically engineered to work with AM, instead of modifying the already existing materials, is likely to open up new possibilities and uses.

7. Conclusion

The recent developments, key uses, and future directions of additive manufacturing were briefly reviewed in this chapter, with primary focus on technological development, industrial implementation, quality control, and material issues. Additives manufacturing now is no longer a prototyping technology, but it has become a feasible production technology to manufacture end-use components in aerospace, biomedical, automotive, and other related industries. Advances in laser and electron beam systems, automation of processes and digital integration have significantly contributed to increased accuracy of fabrication processes, productivity, and material utilization. The introduction of AM-specific materials and enhanced and more advanced post-processing methods has continued to enhance part performance and expanded design choices. However, the problems of process stability, defect formation, and reproducibility are serious limitations. Large scale and safety-critical applications are still confronted with the challenge of quality control and standardization. In spite of the recent developments in in situ monitoring, data-driven control, and machine-learning-based approaches, the absence of common certification systems hinders the broad implementation in industries. Similarly, the material-related issues, such as powder nature, microstructure regulation, and defects alleviation, require a more organized research. It is expected that the further development of additive manufacturing will be oriented on high-speed and large-scale production, combined manufacturing, process control, and sustainable production. The fusion of artificial intelligence and Industry 4.0 technologies and digital manufacturing environments will be a key to increased efficiency, quality, and automation of design. Overall, the future of additive manufacturing will be based on the alignment of technological, material, process modeling, and quality assurance advancements. The three parties (academia, industry, and regulatory bodies) should work together multidisciplinary in order to deal with the current challenges and enable reliable industrial implementation. As the result of continuous research and development, additive manufacturing will become the primary part of the modern manufacturing system, contributing to flexible, efficient, and sustainable manufacturing in the industrial environment of the future.

References

[1] A. A et al., "A comprehensive review of emerging additive manufacturing (3D printing technology): Methods, materials, applications, challenges, trends and future potential," Materials Today: Proceedings, vol. 52, pp. 1309–1313, 2021. doi: 10.1016/J.MATPR.2021.11.059

[2] K. Kanishka and K. Acherjee, "Revolutionizing manufacturing: A comprehensive overview of additive manufacturing processes, materials, developments, and challenges," Journal of Manufacturing Processes, vol. 107, pp. 574–619, 2023. doi: 10.1016/j.jmapro.2023.10.024

[3] M. Grasso and B.M. Colosimo, "Process defects and in situ monitoring methods in metal powder bed fusion: A review," Measurement Science and Technology, vol. 28, no. 4, 2017. doi: 10.1088/1361-6501/AA5C4F

[4] G. Tapia and A. Elwany, "A review on process monitoring and control in metal-based additive manufacturing," Journal of Manufacturing Science and Engineering, vol. 136, no. 6, 2014. doi: 10.1115/1.4028540

[5] A.A. Tesfaye, D.B. Kitaw and E.G. Beshah, "Review on Advances in 3D Printing: Processes, Materials, Industrial Applications, and Emerging Trends in Metal Printing and Metallic Implants," Preprints, 2025. doi: 10.20944/preprints202506.1898.v1

[6] A.C. Bănică, D.G. Pătroi and I. Dumitru, "Printing the Future Layer by Layer: A Comprehensive Exploration of Additive Manufacturing in the Era of Industry 4.0," Applied Sciences, vol. 14, no. 21, 9919, 2024. doi: 10.3390/app14219919

[7] A. Molotnikov, R. Lapovok, C.F. Ng and Y. Estrin, "Current state and future trends in laser powder bed fusion technology," in Fundamentals of Laser Powder Bed Fusion of Metals, Elsevier, 2021, pp. 399–437. doi: 10.1016/B978-0-12-824090-8.00011-1

[8] S.K. Parupelli and S. Desai, "A comprehensive review of additive manufacturing (3D printing): Processes, applications and future potential," American Journal of Applied Sciences, vol. 16, no. 8, pp. 244–272, 2019. doi: 10.3844/AJASSP.2019.244.272

[9] S.R. Boopathy, K. Gupta and S. Khedkar, "Trends and Developments in Additive Manufacturing Technologies," in Additive Manufacturing Technologies, IGI Global, 2025, pp. 368–391. doi: 10.4018/979-8-3373-0533-2.ch017

[10] J. Lv, Y. Luo and Y. Zhang, "Status of 3D Printing Technology for Metal Materials," in Proc. 2017 International Conference on Management Science and Engineering (ICMSE), 2017, pp. 242–247. doi: 10.2991/ICMSE-17.2017.46

[11] M. Grasso, V. Laguzza, Q. Semeraro and B.M. Colosimo, "Metal-based additive manufacturing condition monitoring methods: From measurement to control," ISA Transactions, vol. 120, pp. 147–169, 2022. doi: 10.1016/j.isatra.2021.03.001

[12] A. Molotnikov, R. Lapovok, C.F. Ng and Y. Estrin, "Current state and future trends in laser powder bed fusion technology," in Fundamentals of Laser Powder Bed Fusion of Metals, Elsevier, 2021, pp. 399–437. doi: 10.1016/B978-0-12-824090-8.00011-1

[13] Attariani, H., Petitjean, S.R. and Niekamp, A.M. (2024). Synchronized Multi-Laser Powder Bed Fusion (M-LPBF) Additive Manufacturing: A Technique for Controlling the Microstructure of Ti–6Al–4V. *Journal of Manufacturing and Materials Processing*, *8*(6), 242.

[14] Y. Huang, M.C. Leu, J. Mazumder and A. Donmez, "Additive manufacturing: current state, future potential, gaps and needs, and recommendations," Journal of Manufacturing Science and Engineering, vol. 137, no. 1, 2015. doi: 10.1115/1.4028725

[15] Ghosal, P., Majumder, M.C. and Chattopadhyay, A. (2018). Study on direct laser metal deposition. *Materials Today: Proceedings*, *5*(5), 12509–12518.

[16] Galati, M. (2021). Electron beam melting process: A general overview. *Additive manufacturing*, 277–301.

[17] Karlsson, J., Snis, A., Engqvist, H. and Lausmaa, J. (2013). Characterization and comparison of materials produced by Electron Beam Melting (EBM) of two different Ti–6Al–4V powder fractions. *Journal of Materials Processing Technology*, *213*(12), 2109–2118.

[18] Liu, Z., Liu, J., Zhao, Z., Wang, L., Wang, Q. and Yang, R. (2019). Microstructure and tensile property of TC4 alloy produced via electron beam rapid manufacturing. *Acta Metall Sin*, *55*(6), 692–700.

[19] Oliveira, J.P., LaLonde, A.D. and Ma, J. (2020). Processing parameters in laser powder bed fusion metal additive manufacturing. *Materials & Design*, *193*, 108762.

[20] D. Herzog, M. Brandt and A. Hitzler, "Process monitoring and machine learning for defect detection in laser-based metal additive manufacturing," *Journal of Intelligent Manufacturing*, vol. 35, pp. 1407–1437, 2023. doi: 10.1007/s10845-023-02119-y

[21] Fidan, I., Huseynov, O., Ali, M.A., Alkunte, S., Rajeshirke, M., Gupta, A., ... and Sharma, A. (2023). Recent inventions in additive manufacturing: Holistic review. *Inventions*, *8*(4), 103.

[22] M. Grasso, V. Laguzza, Q. Semeraro and B.M. Colosimo, "Metal-based additive manufacturing condition monitoring methods: From measurement to control," ISA Transactions, vol. 120, pp. 147–169, 2022. doi: 10.1016/j.isatra.2021.03.001

[23] M. Khalifa, S. Ahmed and R. Hassan, "Techniques and Methods for Optimizing 3D Printing Parameters," in Additive Manufacturing Technologies, IGI Global, 2025, pp. 234–259. doi: 10.4018/979-8-3373-0533-2.ch011

[24] Najmon, J.C., Raeisi, S. and Tovar, A. (2019). Review of additive manufacturing technologies and applications in the aerospace industry. *Additive Manufacturing for the Aerospace Industry*, 7–31.

[25] Herruzo Moral, D. (2019). Manufacturing process optimization of an airplane wing rib by using additive manufacturing. https://hdl.handle.net/10016/29869 Carlos III University of Madrid.

[26] Seabra, M., Azevedo, J., Araújo, A., Reis, L., Pinto, E., Alves, N., ... and Mortágua, J.P. (2016). Selective laser melting (SLM) and topology optimization for lighter aerospace componentes. *Procedia Structural Integrity*, *1*, 289–296.

[27] Ventola, C.L. (2014). Medical applications for 3D printing: Current and projected uses. *Pharmacy and Therapeutics*, *39*(10), 704.

[28] Javaid, M. and Haleem, A. (2019). Current status and challenges of additive manufacturing in orthopaedics: An overview. *Journal of Clinical Orthopaedics and Trauma*, *10*(2), 380–386.

[29] B.I. Oladapo, S.A. Zahedi and F. Ismail, "Medical additive manufacturing then, now, and will," in Additive Manufacturing Materials and Technology, Elsevier, 2024, pp. 77–109. doi: 10.1016/b978-0-323-95383-2.00005-6

[30] Hafezi, F., Kucukgul, C., Ozler, S.B. and Koc, B. (2019). Bioprinting—Application of Additive Manufacturing in Medicine. In *Additive Manufacturing, Second Edition* (pp. 355–375). CRC Press.

[31] Vasco, J.C. (2021). Additive manufacturing for the automotive industry. In *Additive Manufacturing* (pp. 505–530). Elsevier.
[32] Blakey-Milner, B., Gradl, P., Snedden, G., Brooks, M., Pitot, J., Lopez, E., ... and Du Plessis, A. (2021). Metal additive manufacturing in aerospace: A review. *Materials & Design, 209*, 110008.
[33] Ghaffar, S.H., Corker, J. and Fan, M. (2018). Additive manufacturing technology and its implementation in construction as an eco-innovative solution. *Automation in Construction, 93*, 1–11.
[34] Lipton, J.I., Cutler, M., Nigl, F., Cohen, D. and Lipson, H. (2015). Additive manufacturing for the food industry. *Trends in Food Science & Technology, 43*(1), 114–123.
[35] Everton, S.K., Hirsch, M., Stravroulakis, P., Leach, R.K. and Clare, A.T. (2016). Review of in-situ process monitoring and in-situ metrology for metal additive manufacturing. *Materials & Design, 95*, 431–445.
[36] Du Plessis, A., Yadroitsev, I., Yadroitsava, I. and Le Roux, S.G. (2018). X-ray microcomputed tomography in additive manufacturing: a review of the current technology and applications. *3D Printing and Additive Manufacturing, 5*(3), 227–247.
[37] Vandone, A., Baraldo, S., Valente, A. and Mazzucato, F. (2019). Vision-based melt pool monitoring system setup for additive manufacturing. *Procedia Cirp, 81*, 747–752.
[38] Astm, I. (2015). ASTM52900-15 standard terminology for additive manufacturing—general principles—terminology. *ASTM International, West Conshohocken, PA, 3*(4), 5.
[39] Lewandowski, J.J. and Seifi, M. (2016). Metal additive manufacturing: A review of mechanical properties. *Annual Review of Materials Research, 46*(1), 151–186.
[40] Honarvar, F. and Varvani-Farahani, A. (2020). A review of ultrasonic testing applications in additive manufacturing: Defect evaluation, material characterization, and process control. *Ultrasonics, 108*, 106227.
[41] Muthuswamy, P. (2022). Influence of powder characteristics on properties of parts manufactured by metal additive manufacturing. *Lasers in Manufacturing and Materials Processing, 9*(3), 312–337.
[42] Islam, S.F., Hawkins, S.M., Meyer, J.L. and Sharman, A.R. (2022). Evaluation of different particle size distribution and morphology characterization techniques. *Additive Manufacturing Letters, 3*, 100077.
[43] Murtaza, H.A., Mukhangaliyeva, A., Golman, B., Perveen, A. and Talamona, D. (2024). Quantitative characterization of metal powder morphology, size distribution, and flowability for additive manufacturing. *Journal of Materials Engineering and Performance, 33*(20), 10853–10867.
[44] Nguyen, H.D., Pramanik, A., Basak, A.K., Dong, Y., Prakash, C., Debnath, S., ... and Buddhi, D. (2022). A critical review on additive manufacturing of Ti-6Al-4V alloy: Microstructure and mechanical properties. *Journal of Materials Research and Technology, 18*, 4641–4661.
[45] Hakami, A., Ojo, S. A., Abere, D.V., Uzuh, F.D. and Robert, R.A. (2025). Advancements in metal additive manufacturing: opportunities, limitations, impact on properties, and potential solutions: A review. *Progress in Additive Manufacturing, 10*(8), 4447–4495.
[46] Mostafaei, A., Zhao, C., He, Y., Ghiaasiaan, S.R., Shi, B., Shao, S., ... and Rollett, A.D. (2022). Defects and anomalies in powder bed fusion metal additive manufacturing. *Current Opinion in Solid State and Materials Science, 26*(2), 100974.
[47] Zhang, T. and Yuan, L. (2022). Understanding surface roughness on vertical surfaces of 316 L stainless steel in laser powder bed fusion additive manufacturing. *Powder Technology, 411*, 117957.
[48] Zhang, T. and Liu, C.T. (2022). Design of titanium alloys by additive manufacturing: A critical review. *Advanced Powder Materials, 1*(1), 100014.

[49] Michi, R.A., Plotkowski, A., Shyam, A., Dehoff, R.R. and Babu, S.S. (2022). Towards high-temperature applications of aluminium alloys enabled by additive manufacturing. *International Materials Reviews, 67*(3), 298–345.

[50] Khodabakhshi, F., Farshidianfar, M.H., Gerlich, A.P., Nosko, M., Trembošová, V. and Khajepour, A. (2020). Effects of laser additive manufacturing on microstructure and crystallographic texture of austenitic and martensitic stainless steels. *Additive Manufacturing, 31*, 100915.

[51] Cheng, X., Xu, J., Yang, Z., Guo, Q., Li, C., Zhou, J., ... and Ma, Z. (2024). A novel nickel-based superalloy with excellent high temperature performance designed for laser additive manufacturing. *Materials Science and Engineering: A, 911*, 146926.

[52] Singh, S.R. and Khanna, P. (2021). Wire arc additive manufacturing (WAAM): A new process to shape engineering materials. *Materials Today: Proceedings, 44*, 118–128.

[53] Sharma, K.S. (2023). Artificial intelligence assisted fabrication of 3D, 4D and 5D printed formulations or devices for drug delivery. *Current Drug Delivery, 20*(6), 752–769.

[54] Lai, J. and Wang, M. (2023). Developments of additive manufacturing and 5D printing in tissue engineering. *Journal of Materials Research, 38*(21), 4692–4725.

[55] Reddy, P.R. and Devi, P.A. (2018). Review on the advancements of additive manufacturing-4D and 5D printing. *Int J Mech Prod Eng Res Dev, 8*(4), 397–402.

[56] Murphy, S.V. and Atala, A. (2014). 3D bioprinting of tissues and organs. *Nature Biotechnology, 32*(8), 773–785.

[57] Kohak, P.G., Markad, K.M. and Lal, A. (2025, January). Advancements and Challenges in Multi-material Additive Manufacturing (MMAM) with Machine Learning Integration. In *International Conference on Recent Advances in Mechanical Infrastructure* (pp. 215–238). Singapore: Springer Nature Singapore.

[58] Qi, X., Chen, G., Li, Y., Cheng, X. and Li, C. (2019). Applying neural-network-based machine learning to additive manufacturing: Current applications, challenges, and future perspectives. *Engineering, 5*(4), 721–729.

[59] Kumar, S., Gopi, T., Harikeerthana, N., Gupta, M.K., Gaur, V., Krolczyk, G.M. and Wu, C. (2023). Machine learning techniques in additive manufacturing: A state of the art review on design, processes and production control. *Journal of Intelligent Manufacturing, 34*(1), 21–55.

[60] Arif, Z.U., Khalid, M.Y., Noroozi, R., Hossain, M., Shi, H.H., Tariq, A., ... and Umer, R. (2023). Additive manufacturing of sustainable biomaterials for biomedical applications. *Asian Journal of Pharmaceutical Sciences*, 18(3), 100812.

[61] Haghnegahdar, L., Joshi, S.S. and Dahotre, N.B. (2022). From IoT-based cloud manufacturing approach to intelligent additive manufacturing: Industrial Internet of Things—An overview. *The International Journal of Advanced Manufacturing Technology*, 119(3), 1461–1478.

[62] Singh, R., Kumar, R., Farina, I., Colangelo, F., Feo, L. and Fraternali, F. (2019). Multi-material additive manufacturing of sustainable innovative materials and structures. *Polymers*, 11(1), 62.

[63] Liu, Z.Y., Li, C., Fang, X.Y. and Guo, Y.B. (2018). Energy consumption in additive manufacturing of metal parts. *Procedia Manufacturing*, 26, 834–845.

[64] Singh, R., Kumar, R., Farina, I., Colangelo, F., Feo, L. and Fraternali, F. (2019). Multi-material additive manufacturing of sustainable innovative materials and structures. *Polymers*, 11(1), 62.

[65] Nagulin, K.Y., Terentev, A.A., Vasilev, I.S., Gilmutdinov, A.K. and Moiseev, R.E. (2025). Recycling of used powder materials for additive manufacturing by processing in an inductively coupled plasma. *Powder Technology*, 453, 120586.

[66] Zhang, X., Sun, Y., Shuai, C. and Gao, C. (2025). Progress on additive manufacturing of eutectic high entropy alloys: Coupling correlations and manipulation strategies for process-microstructure-mechanical performance. *Journal of Alloys and Compounds*, 182496.

3 Hybrid Aluminum Matrix Composites: Advancing Mechanical and Tribological Performance with Additive Manufacturing Insights

Adavayya Sangayya Chikkamath[1], Vinod Kumar V. Meti[2*] and Siddhalingeshwar I. G.[3]

This review provides a comprehensive analysis of 3D printing in the context of hybrid metal matrix composites (HMMCs), which have garnered significant attention across various engineering fields, including aerospace, marine, and defense. Their widespread adoption is attributed to their superior mechanical properties, achieved by incorporating suitable reinforcements. Aluminum alloys, commonly used as the base material, offer a combination of lightweight strength and excellent conductivity, making them particularly suitable for space applications. Further enhancements are realized through secondary processing techniques and the inclusion of ceramic reinforcements such as TiB_2 and ZrO_2, along with other reinforcing materials, enabling their integration into advanced manufacturing processes.

HMMCs are particularly valued in these industries due to their high specific strength, low density, and excellent corrosion resistance—key attributes for producing lightweight components. However, challenges such as macro and micro defects can emerge during forging due to the unique deformation behavior of HMMCs, which exhibit high flow resistance and a limited forgeable temperature range. This review explores advancements in precision secondary processing techniques to refine forging methodologies for HMMCs. Special attention is given to established techniques like stir casting and additive manufacturing (AM), which have gained prominence for their ability to fabricate complex geometries with tailored material properties.

AM has revolutionized modern manufacturing by facilitating the production of intricate three-dimensional components across diverse material systems, significantly influencing material design and application. The review also underscores the necessity for ongoing research to mitigate manufacturing challenges associated with HMMCs and to explore innovative approaches for enhancing their mechanical performance and processing efficiency. With its potential to transform product design, production, and distribution, AM remains a crucial focus for

[1-3] KLE Technological University, Hubballi 580031, Karnataka, India.
Email: chikkamathadavayya@gmial.com[1]; vinod_meti@kletech.ac.in[2]; igs@kletech.ac.in[3]
* Corresponding author: vinod_meti@kletech.ac.in

DOI: 10.1201/9781003635673-3

academic and industrial advancements, offering improved safety, versatility, and customization for end users. This review is valuable for future researchers who are selecting appropriate manufacturing methods and reinforcement strategies for hybrid composites.

1. Introduction

Recent studies have demonstrated the growing interest in aluminum matrix composites (AMCs) for various applications. These lightweight, high-performance materials have an aluminum base reinforced with fibers, whiskers, or particulates, making them highly suitable for aerospace, defense, automotive, thermal management, and sports industries. Their ability to reduce weight while enhancing efficiency has driven their adoption; however, achieving an optimal balance between superior mechanical properties and cost-effective production remains a challenge for widespread implementation. Figure 1 depicts the metal matrix composites fabrication techniques and application.

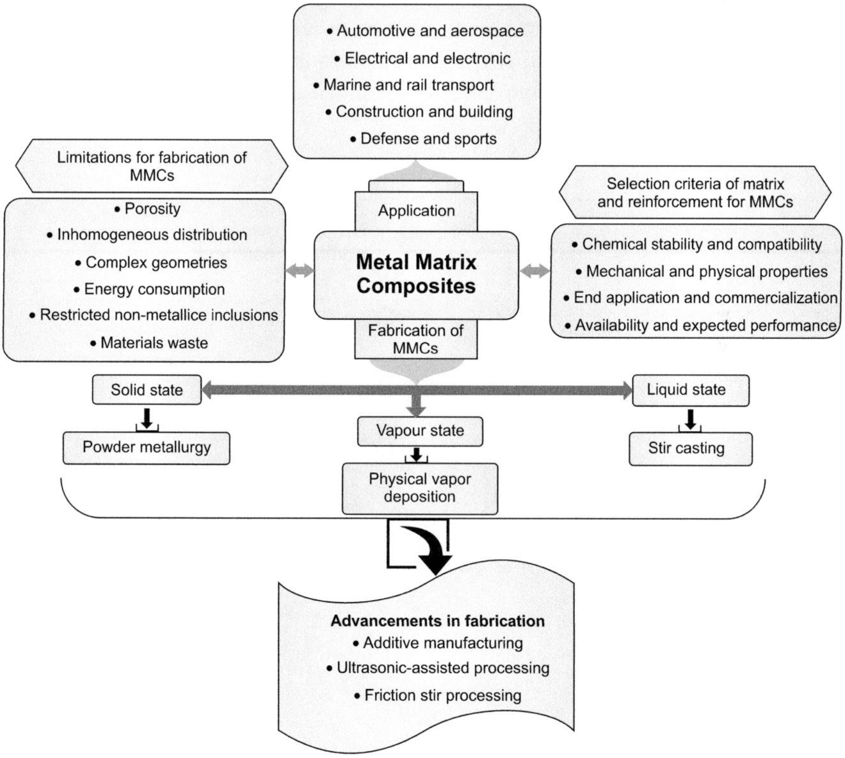

Fig. 1 A summary of metal matrix manufacturing techniques [1].

Advancements in 3D printing have facilitated the development of AMCs using additive manufacturing (AM) techniques. Over the past three decades,

AM processes have evolved from primarily serving as rapid prototyping tools to becoming viable solutions for direct manufacturing. These methods construct metallic components by layering metal alloy powder or depositing material directly onto a substrate to form functional parts. Continuous research aims to refine these technologies further and expand their capabilities [2].

Maintaining uniform reinforcement distribution is essential for producing high-quality material, including casting and property enhancement processes, across various fabrication techniques. Selective laser melting (SLM), for example, offers precise control over fiber distribution, leading to improved mechanical performance. Alloy composition, processing techniques, and heat treatment influence aluminum alloys' structural characteristics and mechanical behavior [3].

AM, commonly known as 3D printing, has gained significant traction among researchers and industry professionals in recent years. Although its widespread popularity is relatively recent, its origins can be traced back to the 19th century, with early printing and geomorphological techniques contributing to its foundation. Innovations such as the "print-icon recording" method, patented in 1951, played a crucial role in shaping modern AM by capturing object cross-sections through layer-by-layer exposure of transparent material, refer Fig. 2. These technological advancements have positioned AM as a revolutionary manufacturing approach, enabling the production of intricate, customized components with enhanced precision and efficiency [4, 5].

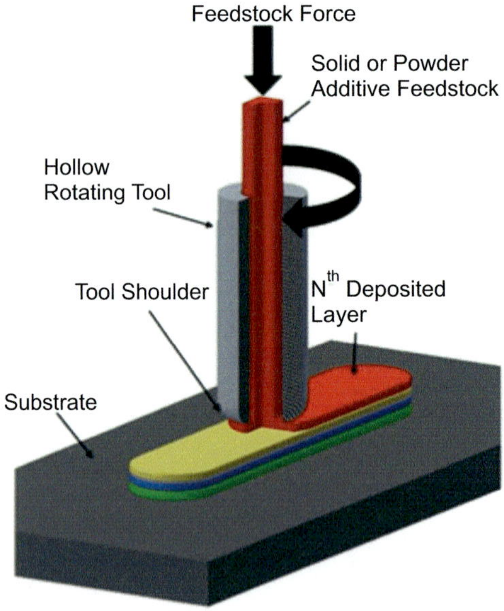

Fig. 2 The admixture process in additive manufacturing employs a concave, rotating tool to eliminate solid feedstock material [6].

The unique characteristics of aluminum, such as its low density and exceptional corrosion resistance, make it an essential material in the aerospace, automotive, and construction sectors, refer Fig. 3. This foundational knowledge paves the way for further exploration of aluminum's potential in advanced engineering applications [7, 8]. Hybrid compositions, incorporating ceramics and nanoparticles, further improve these alloys' mechanical and thermal properties. Techniques such as in situ processing, AM, and powder metallurgy integrate nanoparticles into the metal matrix, leading to advanced metal matrix nanocomposites with superior performance [9–11].

The chapter highlights the importance of intelligent control systems and real-time monitoring to enhance manufacturing surface quality, dimensional accuracy, and process stability. These efforts aim to create precise models with superior accuracy, paving the way for the future of advanced materials and manufacturing processes. Recent advancements in AMCs focus on integrating fibers, whiskers, or particulates to boost aluminum's strength while preserving its lightweight nature. AM has evolved from rapid prototyping to direct manufacturing, with advanced techniques like SLM ensuring precise reinforcement distribution. Innovations in alloy composition, processing methods, and heat treatment have greatly enhanced these composites' structural integrity and mechanical properties. Current research also explores hybrid compositions incorporating ceramics and nanoparticles to optimize material performance [12, 13].

2. Hybrid Metal Matrix Composites

This section delves into composites and hybrid materials, defining their various forms based on dimensions and classifications. These materials have become a central research area, driven by rapid progress in engineering and materials science, especially nanoscience and nanotechnology. We will examine the profound influence of innovative composite technologies on daily life. However, challenges remain, particularly in metal matrix composites (MMCs), where fusion welding creates weak zones along joints, restricting component size. While traditional methods like stir casting and powder metallurgy are prevalent in MMC fabrication, advanced manufacturing techniques, such as AM, are crucial for overcoming these limitations and enabling large-scale production [15, 16].

To fully leverage AM processes, it is essential to understand both the manufacturing capabilities and constraints that MMCs must respect. These aspects will be discussed in detail, followed by a brief literature review to examine how the design process can be adapted using AM technologies for specific compositions, refer Fig. 4. Two key constraints for direct AM processes with hybrid materials have been identified. First, the material deposition nozzle must maintain alignment with the vertical axis, and the part should be mounted on a rotating platform to avoid interference between the nozzle and the part. Additionally, the speed of material deposition and the height of the deposited material are significantly influenced by the nozzle's speed and the material's dispersion rate. Variations

Fig. 3 Examples of aluminum alloy applications in everyday life [14].

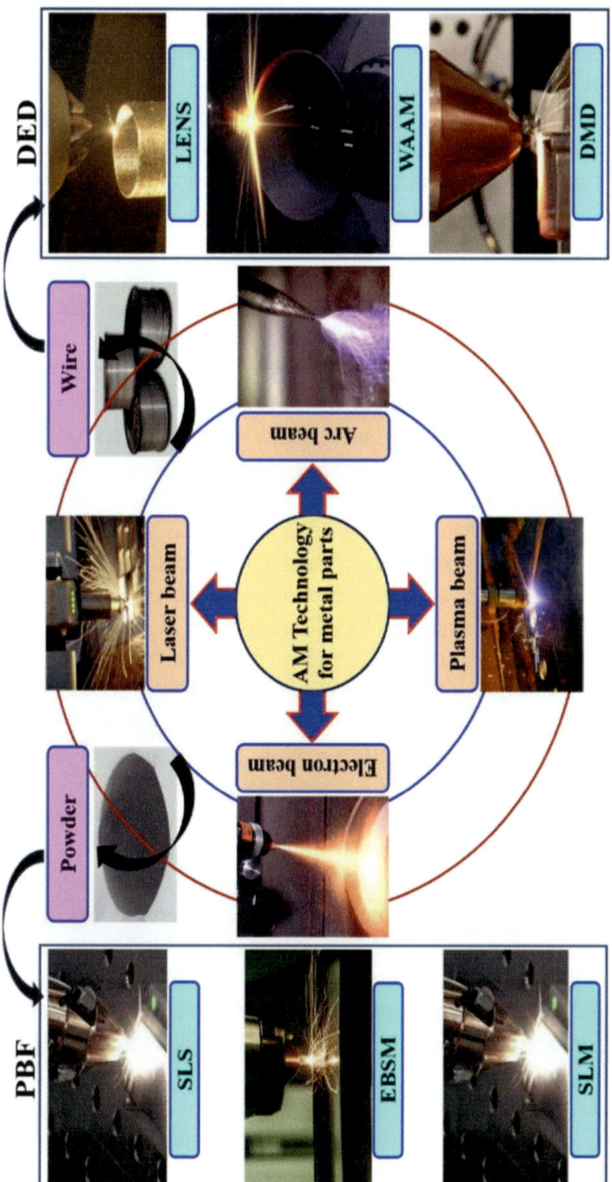

Fig. 4 Various additive manufacturing technologies are used regularly across different industries [18].

in deposition speed can lead to inconsistencies in material height, with rapid acceleration or deceleration causing issues. If the distance between the nozzle and the part becomes too large, molten droplets can solidify before reaching the surface, necessitating the minimization of such acceleration and deceleration phases. This can be achieved by avoiding sharp corners and replacing them with smoother angles [17–20].

Additionally, AM processes are utilized in essence–matrix composites, which are enhanced with hard particles, short fibers, and solid lubricants to improve the tribological properties of aluminum alloys for automotive applications. Aluminum alloys generally exhibit poor resistance to seizure and wear, limiting their use in the automotive industry. The incorporation of hybrid fibers enhances performance for specialized applications. Another area of focus is fiber-hybrid composites, examining their properties, manufacturing methods, and material selection challenges compared to traditional single-fiber composites. This analysis explores the synergistic effects guiding the selection of fiber and matrix, highlighting recent trends and prospects. It also differentiates between academic research and industrial applications, discussing current uses of these materials [21–26].

AM offers significant design flexibility and technical advantages, enabling the production of complex structures that are challenging to achieve with traditional manufacturing methods. MMCs are increasingly used for their unique benefits, particularly in the rapid production of medical devices, addressing the growing demand for lighter, stronger, safer, more time-efficient, and energy-efficient designs, refer Fig. 5. For example, MMCs reinforced with ceramic materials are commonly applied in the automotive industry for their high strength, hardness, wear resistance, and durability. Carbon-based materials, such as graphene and carbon nanotubes, reinforce MMCs in aerospace applications, refer Fig. 6. They offer high specific strength and stiffness while also being used in electronics for their low thermal expansion and high thermal conductivity. Bioactive ceramics incorporated into MMCs are utilized in biomedical devices due to their favorable biological properties [28–32].

The biomedical potential of MMCs is expanding, especially with materials like magnesium-based alloys, which are gaining attention for use in implants and sutures due to their excellent biocompatibility, biodegradability, and low cytotoxicity. Adding components such as bioactive glass (BG), mesoporous bioactive glass (MBG), and graphene oxide (GO) enhances crystallization. It strengthens the material by providing nucleation sites and improving strength and plasticity during deformation. Using AM technology to develop biodegradable MMCs with improved corrosion resistance could overcome current limitations in magnesium alloys, creating more effective materials for medical applications, including bone regeneration and wound healing [33–37].

Furthermore, MMCs are emerging as promising materials for hydrogen storage owing to their high storage capacity and favorable hydrogenation kinetics compared to other materials. The performance characteristics, including strength, hardness, and wear resistance, demonstrate that particle-reinforced MMCs

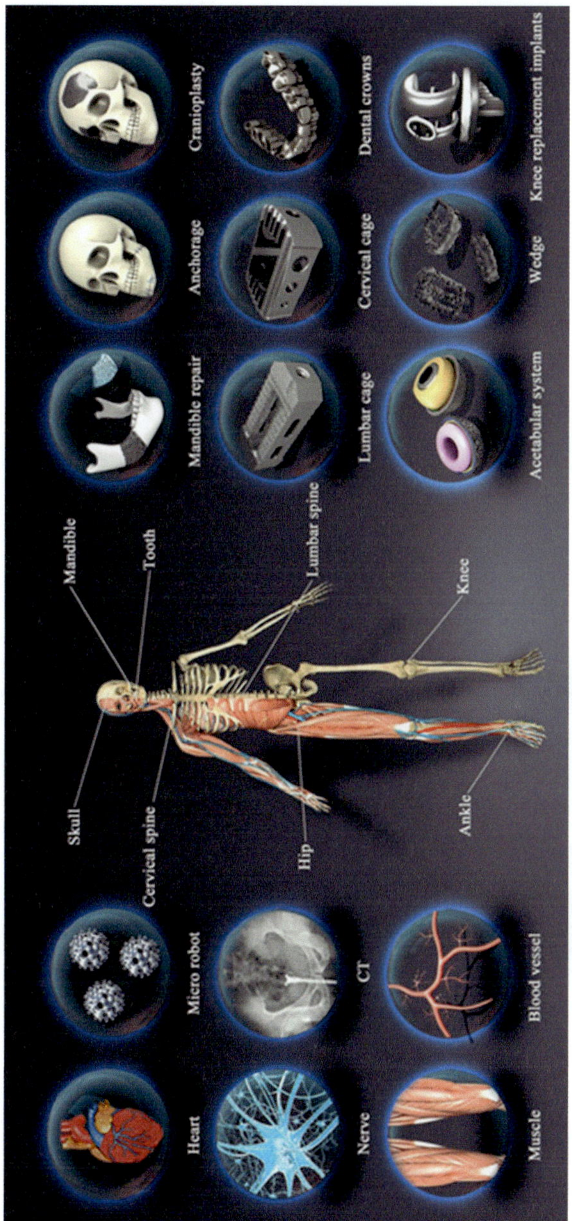

Fig. 5 Various applications of additive manufacturing technology in the biomedical field [27].

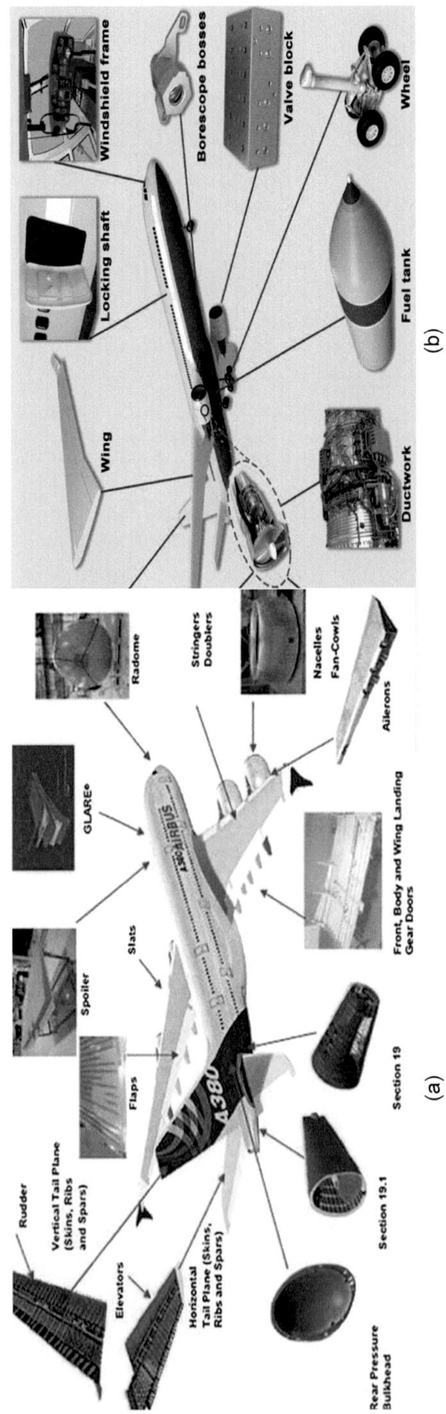

Fig. 6 (a) and (b) Application of hybrid composite materials in aerospace [26, 27].

significantly enhance the properties of aluminum alloys, making them ideal for applications in aerospace, automotive, and rail industries [38–42].

Recent advancements in composites and hybrid materials have reshaped their classification and broadened their applications in engineering and nanotechnology, refer Fig. 7. One of the primary challenges in MMCs is the development of strength-depleted zones during fusion welding, which restricts the achievable component size. AM overcomes these limitations by enabling precise deposition control and seamless integration of reinforcements, thereby improving mechanical performance. Innovations in microstructural characterization and reinforcement techniques continue to drive progress in material design, paving the way for scalable and efficient manufacturing solutions.

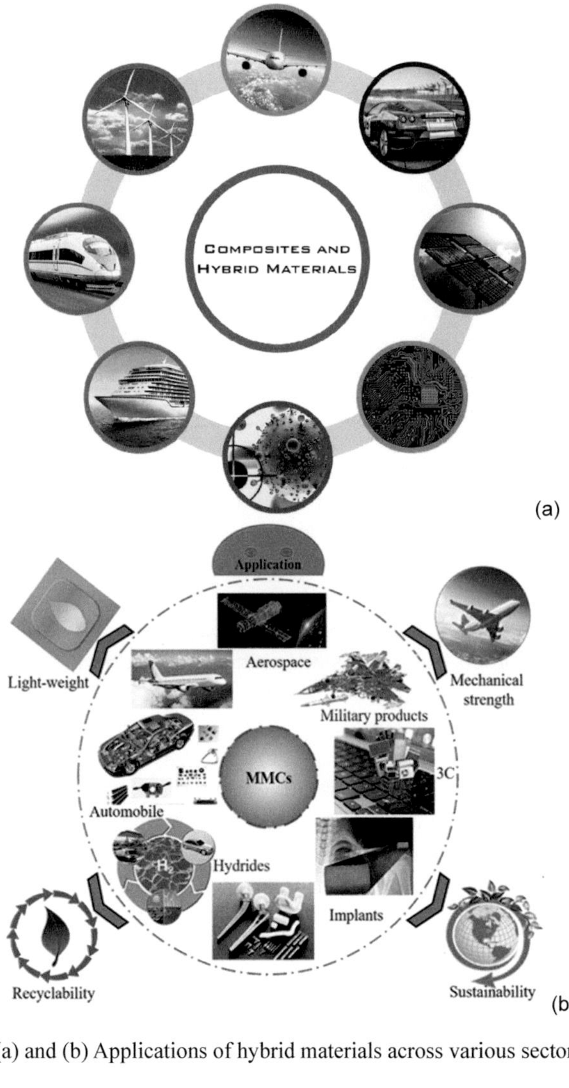

Fig. 7 (a) and (b) Applications of hybrid materials across various sectors [15–19].

2.1 Nanoparticles

Advanced hybrid metal matrix composites (HMMCs) reinforced with nano-sized particles offer remarkable mechanical and thermal properties, creating new opportunities for industries such as automotive and aerospace. To optimize the performance of these composites, it is crucial that the matrix material, typically magnesium, maintains its structural integrity, including the interface between the magnesium matrix and the reinforcement. The reinforcement design plays a pivotal role in enhancing the overall mechanical properties of the magnesium matrix while preserving its lightweight characteristics. The ideal reinforcement for HMMCs should possess three key traits: low viscosity, excellent mechanical properties, and stability that prevents damage during processing; it should also bond effectively with the magnesium matrix to form a strong interface. This interface is vital for transferring stress from the matrix to the reinforcement, achieving superior mechanical properties, minimizing energy dissipation, and avoiding issues such as agglomeration and poor densification [43–45].

Various types of reinforcements, including carbides (e.g., SiC, B_4C, TiC), borides (e.g., TiB_2), nitrides (e.g., AlN, Si_3N_4), and oxides (e.g., Al_2O_3, ZrO_2), as well as carbon-based materials such as nanotubes, have been used in HMMCs to enhance their mechanical and tribological properties. Fabrication methods like powder metallurgy and in situ processing have been explored to incorporate nanoparticles and improve strength, hardness, and wear resistance. Research has focused on nanocrystalline aluminum alloys' mechanical and microstructural characteristics, providing valuable insights into their potential applications across various industries [46–48].

Functionally graded composites, such as Al–Al_3 Ni in situ intermetallic composites, are also explored for their structural and compositional gradients, influencing their mechanical properties. This research investigates how the distribution and form of Al_3 Ni particles along the gradient affect the material's performance. Tribological properties, mainly wear resistance and friction characteristics, are emphasized in AMCs for automotive applications, showcasing the benefits of nanoparticle reinforcement [49–52].

HMMCs reinforced with nano-sized particles exhibit remarkable mechanical and thermal properties, making them highly suitable for advanced engineering applications. From an AM perspective, optimizing the nonferrous-based matrix in these composites is crucial to achieving a robust interface between the matrix and nano-reinforcements. This ensures effective stress transfer, uniform densification, and enhanced performance in functional components [53, 54].

In AM-based fabrication, reinforcements such as carbides, borides, nitrides, oxides, and carbon nanotubes can be precisely incorporated into the matrix through advanced techniques like powder bed fusion (PBF), directed energy deposition (DED), and in situ processing. These methods enable superior control over particle dispersion, reducing defects such as agglomeration or uneven reinforcement distribution that can negatively impact mechanical integrity. Unlike traditional powder metallurgy or casting methods, AM can tailor microstructures layer by layer, enhancing strength, hardness, and tribological properties [55–57].

Furthermore, state-of-the-art characterization techniques such as scanning electron microscopy (SEM) and X-ray diffraction are vital in analyzing microstructural evolution during AM processing. These analyses guide the refinement of reinforcement dispersion, interfacial bonding, and phase transformations, ensuring optimized material performance. The integration of AM in the fabrication of next-generation HMMCs opens new avenues for manufacturing lightweight, high-strength components with complex geometries, making them ideal for aerospace, defense, and high-performance engineering applications. By leveraging AM techniques, the production of HMMCs with nano-reinforcements can be further optimized to enhance durability, wear resistance, and overall mechanical performance, setting a new benchmark for advanced material development in modern manufacturing industries [58–60].

2.2 Casting Process

Aluminum alloy casting is a highly adaptable manufacturing process that shapes molten aluminum into various forms. Selecting the appropriate fabrication technique is crucial, as it influences the microstructural properties, such as the distribution, alignment, and bonding between the reinforcement and matrix. With the increasing demand for products made from MMCs, several advanced methods are being employed to fabricate them. These techniques, including wire arc additive manufacturing (WAAM), friction stir processing (FSP), SLM, and ultrasonic-assisted processes, aim to enhance the mechanical properties and processability of MMCs. For example, reducing weight by 10% in the automotive industry can lead to a 7% decrease in energy consumption. If MMCs, especially those reinforced with flyspeck, replace conventional materials, 22–70% weight reductions are possible [61–63].

In the automotive sector, MMCs reinforced with oxide and carbide particles like SiC and Al_2O_3 offer high strength, hardness, and wear resistance. These materials are used in steering shafts, piston rings, drive shafts, and gearbox housings. Carbon-based materials like carbon fiber and graphene, known for their high strength and modulus, are incorporated into magnesium matrices to create lightweight, high-strength materials ideal for aerospace applications. MMCs' excellent thermal and dimensional stability suits them for high-temperature applications, particularly in aerospace. Their potential to reduce energy consumption and CO_2 emissions positions them as a viable material for satellite components, aircraft propellers, and military equipment, contributing to mobility and battlefield survivability [64–66].

The unique solidification behavior and thermophysical phenomena in AM processes differ from traditional casting methods, influencing the microstructure and properties of AM-produced MMCs. Among standard casting techniques, sand casting is ideal for large, complex parts like automotive components, while die casting is used for precise, intricate components in consumer electronics and automotive applications. The process typically involves pattern creation, mold formation, melting, pouring, solidification, cooling, ejection, and post-casting treatments like heat treatment or machining to enhance material properties [67–69].

Despite the benefits of aluminum alloy-based MMCs, a significant challenge lies in the high cost of ceramic reinforcements and difficulties in achieving uniform distribution. Stir casting has become an effective, cost-efficient method for producing MMCs, especially those reinforced with particulate materials. This method involves mechanically stirring ceramic reinforcements into molten aluminum. While it offers advantages like ease of manufacturing and isotropic properties, challenges remain, such as the potential segregation of reinforcement particles during solidification. Controlling process parameters, including temperature, stirring speed, and duration, is essential for achieving uniform dispersion in the molten matrix [70–72].

FSP is another innovative technique for creating surface nanocomposites, refer Fig. 8. This method uniformly disperses nanoparticles within a metallic matrix through mechanical stirring and frictional heat, enhancing mechanical properties and wear resistance. Experimental validation indicates its potential for use in industries requiring durable surface materials. Similarly, vacuum-assisted block mold investment casting has been explored to infuse SiC-B_4C hybrid preforms with A6063 aluminum alloy, aiming for uniform infiltration and reinforcement particle distribution. This process impacts the composite's microstructure and mechanical properties, as studied in detail [73, 74].

Furthermore, ultrasonic and stir-casting methods have strengthened nanocomposite foams by adding varying percentages of silicon dioxide (SiO_2) nanoparticles. The addition of SiO_2 significantly increased the hardness of the nanocomposites, with the highest hardness observed in the Al-0.75 wt% SiO_2 nanocomposite. SEM analysis confirmed the uniform distribution of SiO_2 nanoparticles throughout the matrix, and the compressive behavior of the foams

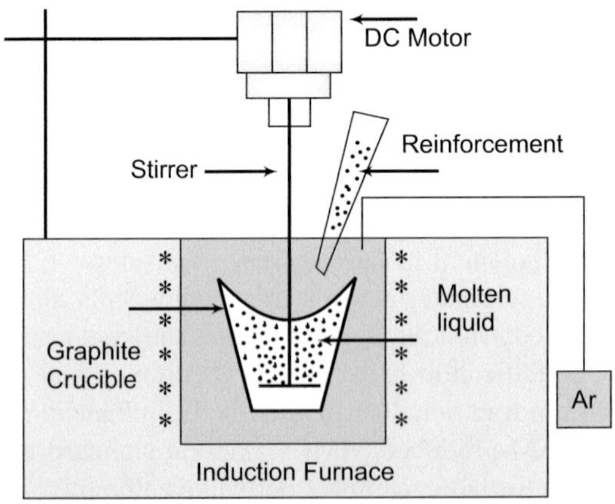

Fig. 8 Automated motorized stir casting process [76].

was found to depend on relative density and structural characteristics. These findings highlight the effectiveness of these techniques in improving material properties and performance [75].

In one study, nanosized SiC particles were introduced into commercially pure aluminum using stir casting assisted by ultrasonic vibration at varying volume fractions (0%, 1%, 3%, 5%, 7%, and 9%). Results showed that up to a volume fraction of 7%, a uniform dispersion of nanosized SiC particles was achieved throughout the matrix. The solidified microstructure of the nanocomposites exhibited α-Al dendritic diameters averaging 90 μm, a significant reduction from the original 270 μm in the matrix. Thermal analysis during solidification revealed that, while decalescence disappeared, nanosized SiC particles increased the nucleation temperature of α-Al. Furthermore, at ambient temperatures of 453°K, the yield and ultimate tensile strength of the SiC/Al composites were significantly improved, while fracture strain remained satisfactory [77, 78]. Additionally, AM enhances the hybrid composition of the material, with wear resistance improving by 88% as the graphite (Gr) content increases. Microhardness and compressive strength peak at 1% Gr content, after which they decline [79, 80]. This work presents an innovative approach to fabricating aluminum alloy-based MMCs by integrating multiple advanced processing techniques, including WAAM, FSP, SLM, and ultrasonic-assisted stir casting. From an AM perspective, these techniques collectively enhance microstructural properties, enabling precise control over reinforcement dispersion and interfacial bonding—key factors in optimizing mechanical performance.

A two-step mixing strategy ensures uniform distribution of ceramic reinforcements, effectively reducing porosity and enhancing matrix–reinforcement interactions. Integrating nanosized SiC and graphite (Gr) nanoparticles is crucial in refining the α-Al dendritic structure, significantly improving tensile strength, microhardness, and wear resistance. Unlike conventional casting methods, AM-based approaches such as SLM and WAAM allow for the layer-by-layer deposition of MMCs, providing enhanced control over phase distribution, microstructure refinement, and mechanical property tailoring. Furthermore, AM enables the in situ synthesis of reinforced composites, where controlled energy input from laser-based or arc-based systems promotes homogeneous reinforcement dispersion. The incorporation of ultrasonic-assisted processing further aids in breaking particle agglomerations, ensuring a more consistent composite structure with superior load-bearing capacity.

2.3 Properties

Hybrid aluminum alloys, composed of various elements and reinforcements, undergo improvements in their properties through a synergistic approach, leading to enhanced overall performance. These alloys typically display superior strength, increased stiffness, and improved mechanical properties compared to their monolithic counterparts. Including reinforcing materials, such as ceramics or fibers, boosts tensile and yield strengths, making them ideal for applications requiring robust structural integrity. Additionally, hybrid aluminum alloys exhibit better high-temperature performance, allowing them to endure challenging environmental conditions. Their controlled thermal expansion coefficient and efficient thermal management also enhance their versatility [81, 82].

The temperature-dependent tensile strength of boride ceramics like ZrB_2, HfB_2, and TiB_2 is analyzed to understand how their mechanical properties change with temperature variations. Through experimental testing and theoretical modeling, the research investigates the behavior of these materials under different thermal conditions, offering insights into their suitability for high-temperature applications across multiple industries. The study also explores the development of these materials, which involves combining various components to achieve improved thermal performance. These hybrid materials' thermal conductivity and coefficient of thermal expansion (CTE) properties are examined through experimental analysis and numerical simulations, providing valuable information for designing advanced heat sink materials for efficient heat dissipation [83].

Furthermore, advancements in hybrid alloys often result in lighter materials, addressing concerns about fuel consumption and efficiency in industries such as aerospace and automotive. The benefits of composites, such as reduced weight, enhanced strength, and increased corrosion resistance, are discussed, particularly in power transmission infrastructure. The study emphasizes the role of advanced composites in solving challenges related to aging infrastructure and environmental issues. Researchers continue to explore new processing techniques to improve microstructure, reduce porosity, and ensure even distribution of reinforcements, optimizing both mechanical and tribological properties. As hybrid aluminum alloys evolve, their distinctive characteristics continue to drive innovations in various engineering fields [84].

The impact of coatings on the interface bonding between silicon nitride ($Si_3 N_4$) and AMCs is also examined. Experimental analysis highlights the strength, toughness, and wear resistance enhancements provided by coated reinforcement particles. Additionally, neural networks are being explored as effective tools for optimizing MMCs' design and fabrication processes, facilitating more accurate material property prediction and optimization. The study also addresses the unique fracture behavior of AM MMCs, which differs from traditional materials. As these alloys evolve, understanding their fracture mechanisms is essential for further optimizing their mechanical properties, such as fracture toughness, fatigue resistance, tensile strength, and ductility. Specific reinforcements can tailor these properties, improving characteristics such as high-temperature performance, wear resistance, electrical conductivity, erosion resistance, and biocompatibility, thereby broadening the application range of MMCs [85, 86].

Beyond material selection, the performance of AM MMCs is primarily determined by the processing parameters. Properly optimized parameters ensure a uniform distribution of the reinforcements and a strong matrix–reinforcement interface. In the automotive industry, the evolving development of aluminum hybrid

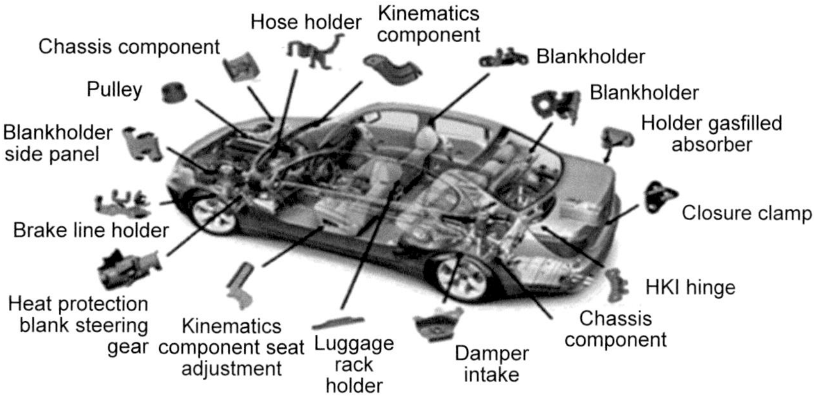

Fig. 9 Various applications within the automotive industry [87].

composites is gaining attention due to their potential to enhance performance. These composites offer improvements in mechanical properties such as strength, stiffness, and reduced weight, making them suitable for lightweight applications. Incorporating reinforcements like ceramics and fibers significantly enhances these properties [88–90].

Secondary processes, including heat treatment, surface finishing, and machining, also play a critical role in optimizing the performance of these materials. Heat treatment refines the microstructure and improves mechanical characteristics, while surface finishing ensures desired aesthetics and corrosion resistance, contributing to durability. Machining is essential for the precision shaping of components used in automotive applications, refer Fig. 9. These processes are integral to fine-tuning the properties of aluminum-based hybrid composites for optimal automotive performance. Additionally, understanding the microstructural behaviors, such as solid-state phase transformation and interdiffusion between the matrix and reinforcements, under multiple thermal cycles during AM, is essential for further optimizing these materials for specific applications [91, 92].

Integrating hybrid aluminum alloys and MMCs with advanced reinforcement strategies is crucial in enhancing matrix–reinforcement interfacial bonding. Significant mechanical and thermal properties improvements are achieved by employing reinforcements such as boride ceramics (ZrB_2, HfB_2, TiB_2) and silicon nitride ($Si_3 N_4$) coatings. AM techniques provide a unique advantage in this context, allowing for precise control over the distribution of these advanced reinforcements, which can be challenging to achieve with traditional methods. AM processes such as SLM and DED allow for the layer-by-layer deposition of reinforced aluminum alloys, providing fine control over reinforcement integration. This capability ensures uniform dispersion of ceramic and nitride particles within the matrix, enhancing the interfacial bonding and improving mechanical properties. The study optimizes powder flowability and reinforcement distribution by carefully managing the thermodynamics and kinetics of reactive magnesium alloys during the AM process, further enhancing the material's performance [93, 94].

Comprehensive experimental and theoretical analyses of temperature-dependent tensile strength and thermal properties provide valuable insights into the behavior of these hybrid composites at elevated temperatures. These insights are particularly important for applications where materials are subjected to high thermal stresses, such as in the aerospace and automotive industries. The ability to manipulate the thermal history of the material during AM fabrication allows for precise control of phase formation and stress distribution, improving the composite's performance under extreme conditions [95, 96].

In addition to the primary AM processes, secondary processes such as heat treatment, surface finishing, and machining are employed to refine the microstructure and optimize mechanical and tribological properties. For instance, heat treatment processes in AM allow for solid-state phase transformations and interdiffusion mechanisms, contributing to improved strength, hardness, and wear resistance. Surface finishing techniques further enhance the surface integrity of printed components, ensuring they meet the rigorous demands of high-performance applications, refer Fig. 10.

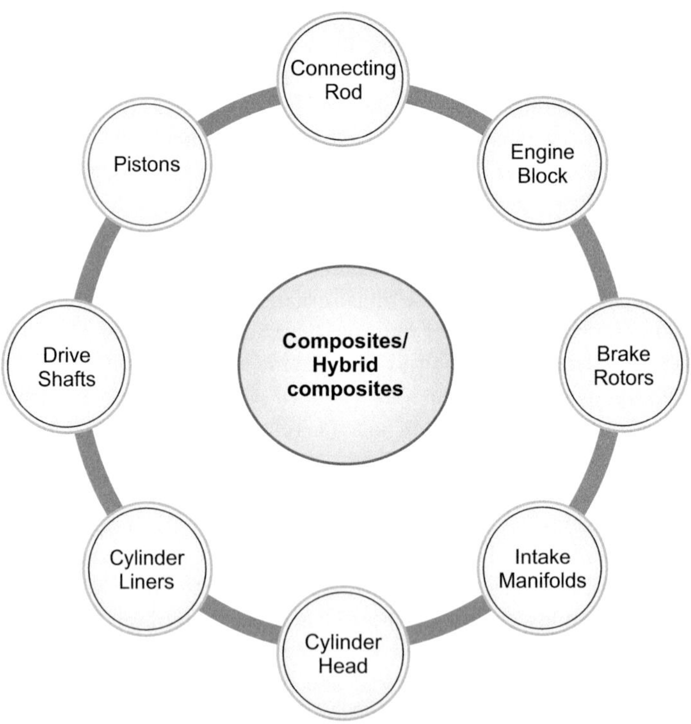

Fig. 10 Hybrid alloy applications in the automobile sector following secondary processing [97].

This integration of AM with advanced reinforcement strategies, thermodynamic control, and post-processing techniques paves the way for developing hybrid aluminum alloys and MMCs with superior mechanical and thermal properties, offering significant advantages in applications requiring high performance under extreme conditions. By leveraging AM's flexibility and precision, these materials can be tailored to meet the evolving needs of the aerospace, automotive, and defense industries, refer Fig. 11.

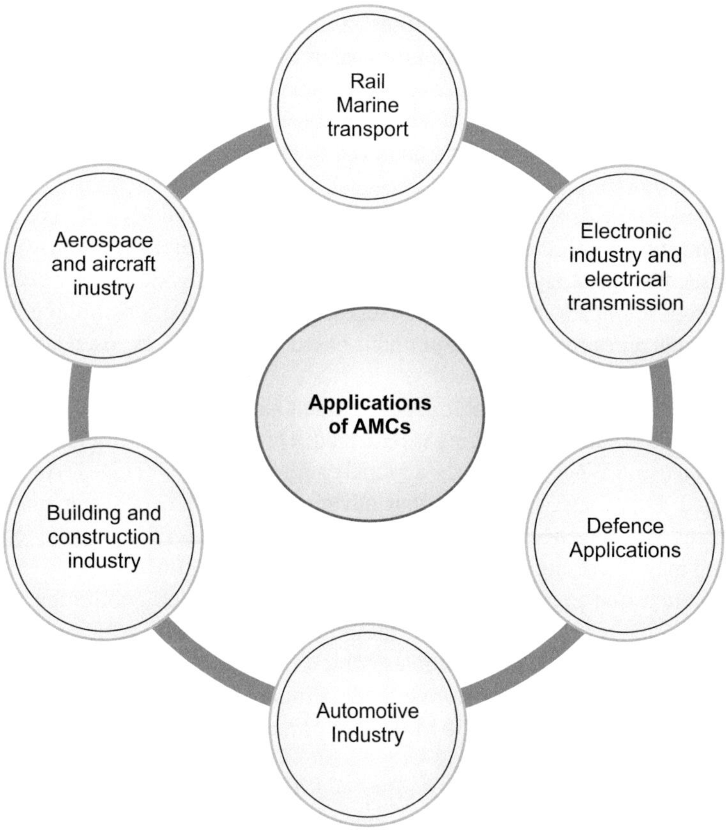

Fig. 11 Application of forged hybrid alloys across various industries [97].

3. Methodology

3.1 Secondary Process

The research investigates the mechanical and microstructural properties of both as-cast and T6 heat-treated samples. AM processes require follow-up heat treatments to enhance performance. However, the existing heat treatment protocols for AM parts are primarily based on traditional casting or forging standards, which are not fully applicable to the unique requirements of the AM process. This limitation highlights the need for specific post-processing treatments such as hot isostatic pressing (HIP), open-die forging, and other densification methods to regulate microstructure and properties. These approaches aim to overcome the challenges where the mechanical properties of AM parts align with castings but fail to match those of forged materials and where high-temperature treatments meet forging requirements but not casting standards [98–100].

The study demonstrates that the mechanical properties of $Al_2 O_3$ nanoparticle-dispersed alloys are significantly improved. Optical microscopy reveals a uniform distribution of nanoparticles within the material. Additionally, the study explores the mechanical and physical changes that occur during the forging of aluminum alloys, specifically Al 6061 and Al 7075. Variables such as temperature, deformation rate, and die shape are thoroughly analyzed to understand their effects on forging behavior [100–102].

The research also examines the wear resistance of Al 6061–TiO_2 composites during hot forging, focusing on how different amounts of TiO_2 affect wear rate and surface morphology under abrasive conditions. These findings are crucial for engineering applications where wear resistance is essential, positioning TiO_2 as a promising reinforcement in aluminum-based composites exposed to abrasive environments [103, 104].

Furthermore, the potential for adding functional components via AM to semi-finished parts produced by forging is redefined. This study, which includes material deposition using WAAM and laser directed energy deposition (L-DED) techniques on forged T-sections, concludes that advanced WAAM deposition rates offer significant advantages for hybrid manufacturing of Ti-6Al-4V aerospace forgings, reducing manufacturing costs and processing time [105, 106].

This research explores an innovative hybrid approach combining stir and squeeze casting techniques to fabricate A356 aluminum alloy samples reinforced with $Al_2 O_3$ nanoparticle nanoparticles, which substantially improves mechanical properties. From an AM perspective, integrating these casting methods with AM offers new avenues for achieving high-performance hybrid composites that can meet the demanding requirements of industries such as aerospace.

AM allows for the precise layering and deposition of reinforced materials, providing enhanced control over the incorporation of $Al_2 O_3$ nanoparticles into the A356 alloy matrix. This can help achieve uniform distribution and optimal particle–matrix bonding, which can be challenging in traditional casting processes. AM facilitates precise microstructure control, enhancing tensile strength, wear resistance, and overall mechanical performance when combined with the stir and squeeze casting methods.

Post-processing treatments, such as HIP and open-die forging, are critical for optimizing the properties of AM parts. These methods are typically used in conventional casting and forging but are equally important in the AM context to bridge the gap between the two approaches. For instance, applying HIP improves the densification of AM components, eliminating voids and enhancing structural integrity, while open-die forging allows further refinement of the material's microstructure. By controlling temperature, deformation rates, and die geometry, these post-processing steps optimize the mechanical properties of the hybrid aluminum composites fabricated through AM and casting techniques [107, 108].

Furthermore, integrating functional components via AM onto forged T-sections demonstrates the potential for cost-effective hybrid manufacturing. By combining AM's capability to create intricate geometries and functional features with traditional forging's strength and toughness, manufacturers can produce highly efficient, complex components with tailored material properties. This hybrid manufacturing approach offers significant advantages in aerospace applications, where material performance and cost-efficiency are critical.

In summary, combining stir and squeeze casting with AM and strategic post-processing techniques represents a powerful approach to fabricating hybrid aluminum composites with enhanced mechanical properties. This method holds great promise for aerospace and other high-performance industries, enabling the production of components with complex geometries, superior material properties, and cost-effective manufacturing solutions.

This study systematically examines the significant effects of AMCs produced through hot extrusion and powder metallurgy on their mechanical and physical properties. The results revealed a refined grain size in the matrix and a uniform dispersion of alumina nanoparticles. SEM analysis showed that these nanoparticles enhance the nanocomposites by impeding dislocation movement. The study also explored the processing of Al/SiC composites, initially fabricated through powder metallurgy, with FSP. It evaluated how various processing conditions influence the composites' mechanical properties and microstructure. Experimental results indicated that FSP improves the dispersion of SiC particles within the aluminum matrix, thereby enhancing mechanical properties such as tensile strength and hardness [110, 111].

To address challenges in AMC forging, the research delves into advanced precision forging technologies and hybrid forming methods as potential solutions. It stresses the importance of precise control over forging parameters to optimize the production of high-precision aluminum alloy components. Real-time in situ techniques, such as high-speed X-ray imaging, can be employed to observe defects formation and collective effects during AM. By integrating numerical simulation techniques, researchers can achieve precise control over defects, microstructure, and properties of AM MMCs [112, 113].

A study on Ti-6Al-4V deposited via laser PBF analyzed the impact of heat treatment and compression on spherical test samples, observing the microstructural evolution. Results suggested that the deposited material's microstructure could be enhanced during hot working, enabling the future manufacturing of complex titanium forgings in a single step using shape-optimized pre-forms produced by AM. Recent advances in preparation strategies focus on in situ reaction mechanisms and microstructural modifications. These strategies reveal the mechanical properties of in situ oxide ceramic-reinforced AMCs, facilitating a deeper understanding and providing guidance for developing high-performance AMCs [114, 115].

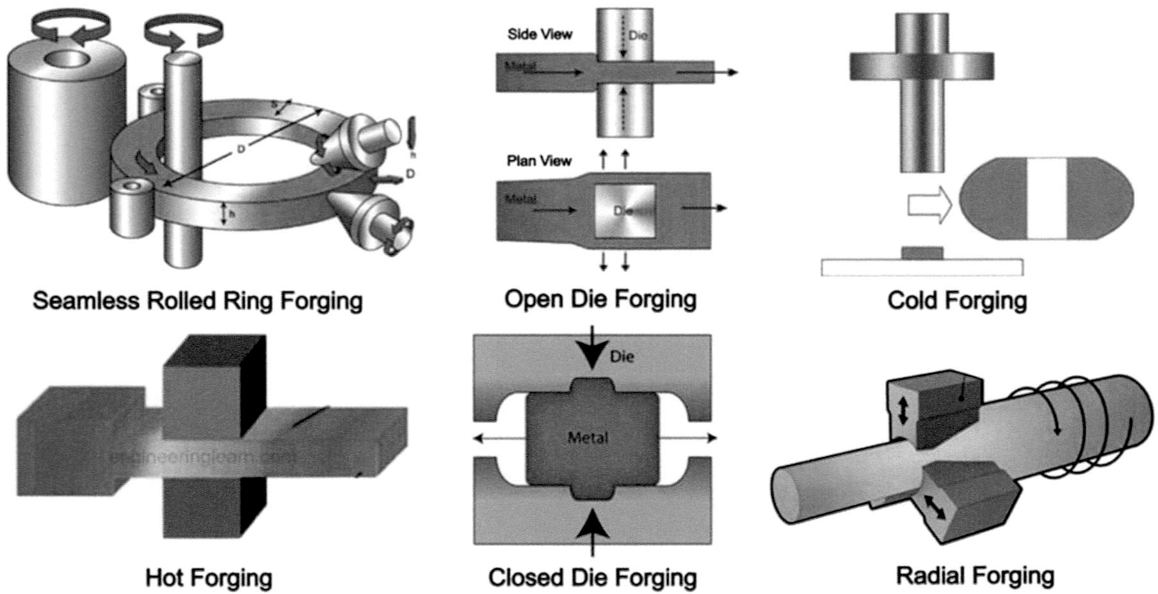

Fig. 12 Various types of forging techniques [109].

The study also examined the relationship between various forging conditions and the resulting mechanical properties of aluminum alloys. Researchers utilized a WAAM (GMA-DED) system installed in a robotic arm to explore the potential of incorporating functional elements into semi-finished parts produced by forging. The experiment involved depositing aluminum AlSi12 on a T-section forged from AW6082 to reduce the number of forging stages, minimize material loss, and decrease tool wear and forging defects. The results confirmed the feasibility of the hybrid manufacturing route but indicated the need for further research to improve the material adhesion of the initial additive layer [116–120].

The research also investigated incorporating TiB_2 and TiC particles into a steel matrix to enhance wear resistance, analyzing the evolution of the composite's microstructure under various synthesis conditions. The findings provided valuable insights into the relationship between microstructural changes and wear performance, aiding the optimization of these composites for high-wear applications. Additionally, the paper discusses the formation of an ultrafine-grained microstructure in steel produced through multiple forging passes at 500°C, which significantly improved its mechanical properties, including hardness and wear resistance [121–124].

Another study examined the effect of nickel (Ni) content on the microstructure and mechanical properties of $TiC–TiB_2–Ni$ composites. The experimental analysis highlighted how varying Ni concentrations influence the composite's structural integrity and performance, offering guidance for optimizing Ni content to achieve the desired properties for industrial applications. The impact of self-propagating high-temperature synthesis (SHS) in the $Ni–Ti–B_4C$ system during casting was also explored, revealing the formation of TiB_2 and TiC particles that enhance the steel's mechanical properties [125–127].

The study systematically analyzed the effects of temperature, deformation rate, and die geometry on forging processes, offering insights into optimal forging parameters. It emphasized the importance of designing and controlling the forging process based on material characteristics, tool/workpiece interface conditions, and equipment. Furthermore, the paper discussed the mechanical and tribological properties of MMCs fabricated through powder metallurgy, with compositions ranging from 2.5% to 10% weight percentage of nanoparticles coated with silver. SEM analysis confirmed a homogeneous distribution within the copper matrix, and although density slightly decreased, microhardness significantly increased, reaching 175 HV with 10% nanoparticles [128, 129].

A structured examination of the latest developments in AMC forging, including recent advances, forging parameters, microstructural changes induced by forging techniques, and dynamic effects during cold forging, is presented in Fig. 12. The content is organized logically, with clear sectioning for ease of understanding. SEM micrographs reveal microstructural changes occurring during dynamic recrystallization, showing significant grain refinement and the formation of new grains. Cold deformation results in extensive dislocation structures, while severe plastic deformation promotes the development of ultrafine grains. The

SEM analysis captures the evolution of recrystallized grains post-deformation, illustrating their distribution and size variations. This study underscores the critical impact of deformation conditions on recrystallization mechanisms and microstructures, offering valuable insights for optimizing material processing and strengthening techniques [130, 131].

Table 1 depicts the researchers worked with the materials along with the methods, composition and findings.

Table 1 Various compositions, their significance, and the enhancements they bring.

Scholars	*Method*	*Material*	*Findings*
[132]	Centrifugal and spraying and hot forging	Al–23Si–8Fe–5Mn Alloy	The rapid compaction method produces a dense, pore-free material demonstrating exceptional compressive mechanical properties. This process notably enhances the material's hardness and yield strength, contributing significantly to its structural integrity.
[133]	Using and not using lubricant in cold forging	6082 Aluminum alloy	The microstructure of aluminum alloy 6082 undergoes alterations during the cold forging process. Applying lubrication minimizes surface damage and lowers the friction coefficient during cold forging.
[134]	Hot forging	Al6061-Al_2O_3 Nano Composites	When analyzing the microstructure of the Al6061/nano Al_2O_3 composite, it was observed that the reinforcement and matrix exhibited excellent bonding, with alumina particles distributed evenly in both cast and hot forged states.
[135]	Warm Forging	Al 6061 Alloy during	Homogenizing the raw material improves the formability of Al 6061, thereby enhancing its forging performance.
[136]	Hot upset forging	2XXX series aluminum alloy	Hot upset forging led to notable enhancements in mechanical and physical properties, including densification, which increased from 98.7% to 100.0%.
[137]	Multidirectional forging (MF)	2024 aluminum alloy	In a multidirectional forged alloy, the distribution of flow stress was found to be more uniform than the plastic strain distribution.
[138]	Wire and arc additive manufacturing (WAAM)	Ti/WC	The economical WAAM process and paraxial WC powder injection resulted in strong metallurgical bonding between the particles and the matrix.

A hybrid manufacturing approach integrating stir and squeeze casting with advanced post-processing techniques such as T6 heat treatment, WAAM, and L-DED offers an effective solution for fabricating A356 and Ti-6Al-4V components reinforced with Al_2O_3 nanoparticles and TiB_2/TiC particles. AM enhances MMC production by optimizing reinforcement dispersion, interfacial bonding, and densification. The precise control over material deposition in WAAM and L-DED facilitates uniform dispersion of reinforcements, allowing for layer-by-layer construction of complex geometries while maintaining tailored microstructures. AM's localized energy input control refines reinforcement distribution, improving mechanical strength and wear resistance.

Post-processing techniques like T6 heat treatment contribute to solid solution strengthening and precipitation hardening, enhancing tensile strength and resistance to thermal and mechanical stresses. Integrating AM with conventional casting methods enables real-time monitoring and optimization of deformation parameters using advanced characterization tools such as SEM, electron backscatter diffraction (EBSD), high-speed X-ray imaging, and numerical simulations. These tools ensure precise microstructural control, facilitating adjustments in temperature, deposition rate, and particle dispersion to achieve superior material properties.

The combination of AM, casting, and heat treatment significantly improves the mechanical properties, energy efficiency, and wear resistance of MMCs, making them highly suitable for aerospace and automotive applications. By leveraging AM's precision with traditional fabrication methods, hybrid manufacturing paves the way for developing next-generation composite materials with enhanced durability, performance, and reliability in demanding environments.

4. Results and Discussion

4.1 Analysis of SEM

This research explores the metallurgical and mechanical properties of forged aluminum alloy hybrid composites through SEM analysis, showing improvements in strength, hardness, and wear resistance due to a more uniform distribution of reinforcing phases. A thorough investigation into the impact of relevant processing parameters on the microstructure and properties of AM MMCs is crucial. This research aims to clarify how microstructure characteristics, such as phase type, distribution, and size of defects, influence the overall mechanical properties of AM parts. Additionally, the solidification processes in the non-equilibrium state during AM are examined. The study covers the fabrication of boron nitride (B_4N) and fly ash-based LM6 MMCs, demonstrating enhanced mechanical properties, including increased hardness and tensile strength. Dynamic and post-dynamic recrystallization phenomena under different deformation conditions are analyzed, with SEM results indicating significant grain refinement and the formation of new grains. The study on the refinement of AZ61 magnesium alloy through multiaxial forging highlights the impact of dynamic recrystallization, revealing improved grain structure via SEM analysis. Similarly, ultrafine-grained steel produced by up to nine forging passes at 500°C shows a microstructure of fragmented cementite particles with enhanced mechanical properties compared to coarse-grained steel [139, 140].

Mechanical properties of aluminum-based HMMCs, particularly A356 aluminum alloy hybrid composites reinforced with Gr–Fe_3O_4–B_4C nanoparticles synthesized by SLM, are explored through SEM analysis, which confirms successful nanoparticle dispersion and effective interfacial bonding. The paper also examines how strain from past deformation affects grain refinement in coarse-grained 7475 Al alloy during hot multidirectional forging, showing that pass strain significantly influences grain refinement kinetics. The formation of micro shear bands and the subsequent evolution of new fine grains are discussed, along with the role of nanoparticles such as SiC in strengthening nanocomposites by obstructing dislocation movement [141, 142].

The study further analyzes Al 5083 composites processed via cryomilling and consolidation, revealing grain sizes of 115 nm and B_4C reinforcement of 38 nm. The nanoparticles' interaction with dislocations enhances strength and flexibility compared to composites reinforced with micrometric particles. The study of AA7175 aluminum alloy with SiC, B_4C, and graphite reinforcements reveals enhanced mechanical properties with uniformly dispersed nanoparticles, as confirmed by SEM. The effect of pass strain on grain refinement in 7475 Al alloy during hot multidirectional forging is further explored, with SEM analysis highlighting microstructural changes [143–146].

The fabrication of AA1070 and AA6063 composites reinforced with B_4C_p particles, analyzing how different compositions influence the microstructure. SEM images of medium carbon low-alloy steel subjected to multidirectional forging and annealing reveal enhanced wear resistance. The microstructure of A356 aluminum alloy-based hybrid composites processed by SLM demonstrates improved mechanical properties, with SEM confirming a refined microstructure. Experimental findings on TiC reinforcement in SLM-processed Al matrix composites indicate a change from aggregation to uniform dispersion as laser energy per unit length increases [147–150].

The study investigates hybrid composites' mechanical and thermal properties with nano-sized reinforcements, showing yield strength, hardness, and interfacial bonding improvements. Boron nitride nanotubes (BNNTs) reinforced aluminum composites exhibit refined grain structures and enhanced strength, while spark plasma sintering (SPS) significantly increases yield and fracture strength. The impact of SPS on nanocrystalline TiC for tool applications is also explored, showing changes in grain structure and densification. The study also examines TiB_2–TiC–WC composite ceramic tool materials, highlighting the influence of composition on hardness, toughness, and wear resistance [151–154].

Ultrasonic vibration-assisted fabrication of nanosized SiC_p/Al composites results in enhanced mechanical properties due to uniform dispersion, microstructure refinement, and improved interfacial bonding. The study also provides insights into optimizing the fabrication process of SiC_p/Al composites for applications such as heat exchangers and electronic packaging. The effect of diamond addition on thermal conductivity and CTE is discussed, with findings suggesting improvements for applications requiring enhanced thermal properties [155–158].

The integration of advanced forging and AM techniques, such as stir and squeeze casting, SLM, WAAM, and ultrasonic-assisted processing, facilitates the fabrication of aluminum alloy hybrid composites and AZ61 magnesium alloys with significantly refined microstructures. From an AM perspective, these methods provide enhanced control over material deposition, reinforcement integration, and microstructural evolution, which are essential for optimizing the properties of MMCs [159, 160].

AM techniques like SLM and WAAM play a crucial role in the process. SLM allows for the layer-by-layer construction of components, enabling precise control over the distribution of reinforcements such as Al_2O_3, TiC, B_4C, and BNNTs. The control over laser power and scanning speed during SLM ensures the careful melting and solidification of the matrix material, improving the interfacial bonding between the matrix and reinforcements. WAAM, in contrast, offers a higher deposition rate, making it suitable for large-scale production of hybrid composites with complex geometries, which is one of the key advantages of AM.

Ultrasonic-assisted processing enhances the uniformity of reinforcement dispersion by preventing particle agglomeration during the deposition process. This further improves the overall quality of the composites, enhancing their tensile strength, hardness, and wear resistance [161, 162].

Advanced characterization techniques, such as SEM, confirm that the reinforcements are uniformly dispersed, enhancing interfacial bonding between the matrix and reinforcement materials. These enhancements contribute to a significant increase in mechanical properties compared to conventional casting methods. In addition, the dynamic recrystallization and nonequilibrium solidification processes under varied thermal and deformation conditions are investigated to understand how microstructures evolve during fabrication. These insights are crucial for mitigating defects like porosity or weak bonding, ensuring the final product is defect-free and performs optimally [163, 164].

Combining AM with traditional forging and casting methods can optimize the hybrid composites' microstructural characteristics and material properties. This approach enables the scalable production of high-performance parts well-suited for demanding applications in aerospace, automotive, and advanced tooling industries. Creating complex, high-strength components with tailored material properties offers a significant advantage in these industries, providing durable, lightweight, and high-performance solutions for modern engineering challenges.

4.2 Analysis of Vickers Hardness Test

The study explores the impact of secondary processes on material hardness, aiming for optimization. It discusses material selection and processing in aircraft advancements, focusing on how secondary treatments influence properties and hardness values [165]. The role of temper and surface condition in the exfoliation behavior of Al–Zn–Mg–Cu alloys is highlighted, emphasizing hardness testing as a critical assessment tool. The research also examines secondary processes and their effects on resistance and hardness [166], emphasizing the importance of these processes in aerospace materials. The study delves into quenching and heating rates' influence on aging in Al–Zn–Mg–(Zr) alloys, with hardness tests playing a key role in evaluating performance [167]. It analyzes aluminum's response to intermediate strain rates and temperatures, ensuring structural integrity by avoiding fractures, with hardness values serving as an indicator of process effectiveness. The work investigates the effects of secondary treatments, such as heat treatment and surface finishing, on hardness and mechanical properties, which are crucial for material enhancement [168–170]. The influence of forging parameters on the mechanical behavior of aluminum alloys is also assessed, demonstrating how traditional and advanced forging methods affect hardness [171]. The study underlines the importance of temper and surface conditions for aerospace applications, revealing how machining or polishing impacts corrosion resistance and hardness properties. Heat treatment conditions are shown to significantly alter material performance. Investigations into quenching rates in Zr-containing Al–Zn–Mg alloys highlight the microstructural influence on aging and hardness [172–175]. The research incorporates advanced techniques like planetary ball milling to refine aluminum powder and integrate nano-reinforcements, yielding significant hardness and tensile strength improvements at optimal reinforcement levels [176]. Studies also address the fabrication of high-performance thermoelectric materials using advanced sintering techniques, emphasizing electrical conductivity and thermal performance [177, 178]. Rapid heating methods like infrared radiation are evaluated for their effects on mechanical properties and grain refinement in forged AA2618 alloys, demonstrating significant benefits for high-temperature applications [179, 180]. The integration of CNTs in AMCs is explored, showing enhancements in hardness through processes like spark plasma sintering and hot rolling, which refine microstructures and ensure uniform reinforcement distribution [181–183]. Silicon carbide-based and aluminum nitride-reinforced composites undergo secondary processing to improve hardness, wear resistance, and mechanical properties, which are crucial for engineering applications. Vacuum heat treatment is identified as a critical method for enhancing the microhardness of aluminum alloys reinforced with TiN particles, improving their application potential in lightweight and wear-resistant components [184–189]. Using hybrid reinforcements in LM6 composites via stir casting demonstrates significant hardness and tensile strength improvements, showcasing the method's efficiency. Additionally, insights into dynamic material modeling and advanced processing techniques underline the importance of secondary processes in tailoring mechanical properties for demanding applications [190, 191]. Secondary processes, including heat treatments, surface modifications, and advanced sintering techniques, are critical in optimizing materials' hardness and mechanical performance. These processes become even more significant when viewed from an AM perspective, offering precise control over material properties after the initial fabrication stage. In aerospace applications, where high-performance materials are essential, the influence of secondary processes such as quenching rates, aging treatments, and temper conditions on materials like Al–Zn–Mg–Cu alloys is particularly important. These treatments significantly affect mechanical properties like tensile strength and hardness, making them vital for ensuring that AM-produced parts meet stringent performance requirements.

AM techniques such as SLM or DED allow for the fabrication of complex aluminum alloy parts with the potential for optimized material properties. After the AM process, secondary techniques such as quenching and aging can further refine the microstructure, ensuring uniform hardness and improved mechanical performance. For example, controlling quenching rates and aging treatments on AM-produced parts can help achieve the desired strength and hardness, which are critical for aerospace components.

Advanced techniques such as SPS and infrared heating are fundamental in AM. These methods, combined with AM techniques like PBF or WAAM, promote the densification of the material while preserving or enhancing the microstructural properties. Spark plasma sintering, for instance, allows for the uniform sintering of the material while preventing grain growth, leading to improved wear resistance and mechanical properties. AM allows for fine control over the powder flowability and deposition rate, which enhances the effectiveness of these post-processing techniques in achieving high-quality components.

Planetary ball milling, another key process, can refine the distribution of reinforcement particles such as carbon nanotubes (CNTs), SiC, and TiN within AMCs. When integrated with AM, this technique enhances the uniformity of reinforcement dispersion, improving mechanical properties such as hardness, wear resistance, and overall durability. In particular, CNTs and SiC, known for their superior mechanical properties, contribute to the enhancement of composite materials, making them suitable for demanding engineering applications like aerospace.

The combination of AM techniques and these advanced secondary processes makes manufacturing complex, high-performance materials with tailored properties possible. These processes enhance the material's mechanical performance and enable the production of components that meet the stringent demands of aerospace, automotive, and other high-performance sectors. By leveraging AM's capabilities with secondary treatments, engineers can push the boundaries of material performance, creating lightweight, durable, and highly functional components for advanced applications.

4.3 Tensile Behavior

The unique properties of matrix composites are achieved through direct squeeze casting of aluminum–copper alloys enhanced with silicon carbide particles. This study primarily evaluates the composite's wear performance under different operating conditions. The results highlight its exceptional wear resistance, suggesting its suitability for applications in demanding environments. The research underscores the importance of composite materials in improving wear resistance while offering valuable insights for future industrial applications and optimization strategies. Furthermore, hybrid composites produced via squeeze casting, incorporating WO_3 particles and $Al_{18}B_4O_3$ whiskers, demonstrate how combining reinforcements can significantly enhance the mechanical and microstructural properties of the composites [192, 193].

The investigation also utilizes AA7175 aluminum alloy reinforced with varying weight percentages (3%, 6%, 9%, and 12%) of SiC, B_4C, and graphite particles to create HMMCs. Mechanical properties such as tensile strength, hardness, and elongation were analyzed, revealing a decline in elongation with increased reinforcement content, while tensile strength and hardness improved notably, especially at 12% reinforcement. The study validates stir casting as a cost-effective and efficient method for fabricating high-strength components. Maximum tensile strength was recorded at a feed rate of 0.6 rpm, increasing the AMC-deposited layer's tensile strength from 140 MPa to 169 MPa. Additionally, wear resistance improved with a feed rate of 1 rpm, reducing adhesion and frictional wear [194–196].

Nanocrystalline aluminum 5083 alloys with enhanced strength and ductility were fabricated using cryomilling and SPS. Cryomilling induced plastic deformation, improving hardness, while SPS facilitated densification, yielding superior mechanical properties for structural applications. Research into Al_2O_3–ZrO_2–SiC ceramic composites produced via SPS explored how compositional and sintering variations affect microstructure, flexural strength, fracture toughness, and hardness. Experimentally, the interactions between reinforcing phases and the matrix were analyzed, revealing their influence on mechanical performance [197, 198].

The microstructure and tensile characteristics of AZ91D/AlN nanocomposites were studied, showing that introducing 1% AlN nanoparticles into the AZ91D matrix at 200°C significantly enhanced tensile properties, including yield strength and ductility. Ultrasonic cavitation-based dispersion of nanoparticles proved effective, outperforming traditional mechanical stirring methods. Magnesium-based nanocomposites reinforced with nano-Al_2O_3, fabricated through melt deposition and hot extrusion, exhibited improved stiffness, strength, and ductility without compromising flexibility. Similarly, incorporating nano-Y_2O_3 particles into magnesium matrices resulted in grain refinement, increased stiffness, and enhanced fracture resistance [199].

AMCs reinforced with nanometric Al_2O_3 particles showcased improved tensile strength, stiffness, and ductility, with secondary processes like heat treatment enhancing particle–matrix bonding. Research on AMCs reinforced with SiC and carbon nanotubes (CNTs) demonstrated significant improvements in tensile properties attributed to the mismatch in thermal expansion, grain refinement, and dispersive strengthening. The study also explored how tribological properties, such as wear resistance and hardness, are influenced by ceramic reinforcements in automotive applications [200, 201].

SPS produced Al–CNT composites with enhanced thermal conductivity, electrical performance, and tribological behavior. These materials address the stringent demands of the automotive sector, including lower emissions and improved fuel efficiency. Further research into reducing CNT agglomeration through dispersion techniques revealed the role of grain refinement and microstructural adjustments in enhancing material properties. Finally, dynamic recrystallization and severe plastic deformation techniques were identified as promising methods for producing ultrafine-grained materials with superior mechanical and wear properties. This review also highlights advancements in nanostructured materials, offering insights into aerospace and automotive engineering innovations [202, 203].

Advanced fabrication techniques combined with AM significantly enhance the tensile behavior of matrix composites by optimizing reinforcement dispersion, microstructural integrity, and mechanical strength. Processes like direct squeeze casting, stir casting, cryomilling, and SPS, when integrated with AM techniques such as SLM and DED, enable precise control over reinforcement distribution, leading to improved tensile strength and wear resistance. Reinforcements like SiC, B_4C, carbon nanotubes (CNTs), and Al_2O_3 further enhance mechanical properties, allowing for the development of lightweight yet highly durable components for aerospace and automotive applications.

Hybrid composites incorporating tungsten oxide (WO_3) and boron oxide ($Al_{1\ 8}B_4O_3$) whiskers demonstrate superior structural integrity, benefiting from AM's capability to tailor reinforcement patterns and refine grain structures. When integrated with AM, ultrasonic cavitation and severe plastic deformation promote fine-grained microstructures, enhancing mechanical performance. Additionally, AM methods like PBF precisely regulate heating and cooling rates, further optimizing grain refinement and tensile properties.

The combination of AM and advanced processing methods enables the fabrication of high-strength, lightweight composites with tailored reinforcement distribution and scalable, cost-effective production. The ability to precisely control microstructure and mechanical properties ensures that these materials meet the rigorous demands of aerospace, automotive, and other high-performance industries.

4.4 Fractography

Nanoparticles play a crucial role in influencing fracture morphology by facilitating particle pull-out, crack deflection, and grain refinement, leading to enhanced mechanical properties in aluminum matrix hybrid composites. Fractography analysis provides insights into these mechanisms, demonstrating the improved tensile strength, hardness, and stiffness achieved through boron carbide reinforcement. The effective distribution of boron carbide within the aluminum matrix contributes to crack bridging and load transfer efficiency, reinforcing the composite structure.

Mathematical modeling further aids in predicting mechanical behavior, offering a quantitative understanding of material performance [204, 205].

Secondary processes, including heat treatment and surface modifications, significantly enhance the tensile properties of these composites by optimizing microstructure and interfacial bonding. A combination of fractography and mathematical modeling provides a comprehensive approach to tailoring mechanical properties for specific applications. The tensile fracture behavior of AMC layers deposited at different powder feed rates reveals challenges such as ceramic particle agglomeration, leading to pore formation and stress concentration. Finite element modeling (FEM) is employed to assess the impact of nanoparticle clustering on mechanical performance, offering valuable insights into stress distribution and deformation mechanisms in in situ ZrB_2-reinforced AA6016 composites [205, 206].

Additionally, the fabrication and characterization of TiCx–TiB_2/Al composites for heat sink applications highlight the importance of thermal conductivity and heat dissipation efficiency. Experimental evaluations explore the effects of processing parameters on microstructural evolution and thermal performance, aiding in developing advanced heat sink materials for thermal management in electronic and mechanical systems. The synthesis of TiB_2/TiC composites is also analyzed, focusing on composition ratios and processing conditions that influence final properties. These investigations contribute to optimizing fabrication techniques and performance enhancement strategies for next-generation composite materials [207–210].

Nanoparticles play a significant role in enhancing the properties of aluminum matrix hybrid composites, particularly when processed through AM techniques. Incorporating nano-sized reinforcements, such as ZrB_2, TiCx, and TiB_2, into aluminum alloys substantially improves mechanical and thermal properties. The mechanisms through which these nanoparticles enhance composite performance include particle pull-out, crack deflection, and grain refinement. These mechanisms contribute to the material's overall strength, durability, and wear resistance, which are critical for aerospace, automotive, and electronics applications.

AM techniques such as SLM and WAAM enable the precise control of particle distribution within the matrix. The ability to deposit material layer by layer allows for uniform dispersion of nanoparticles, which optimizes the mechanisms mentioned above and enhances the overall composite performance. AM can precisely tailor the reinforcement distribution by controlling the heat input, deposition rates, and scanning paths to ensure that the composite material exhibits superior mechanical properties.

Secondary processes such as heat treatment and surface modification are crucial for optimizing the mechanical performance of these nano-reinforced composites. In AM, post-processing techniques, including heat treatments (e.g., solutionizing and aging) and surface modifications (e.g., laser hardening or coating), further improve interfacial bonding between the matrix and nano-reinforcements. These secondary processes help refine the microstructure, reduce residual stresses, and enhance the overall mechanical behavior of AM-produced parts.

FEM is a powerful tool used to analyze the effects of nanoparticle agglomeration in AMCs. In situ studies of ZrB_2 nanoparticle-reinforced AA6016 composites reveal the impact of particulate distribution on stress distribution and deformation behavior. Through AM, the precise control over reinforcement dispersion achieved by techniques like SLM allows for the minimization of agglomeration, leading to better stress management and improved structural integrity of the composite material. FEM simulations guide the optimization of fabrication parameters, ensuring that the composites maintain optimal performance under real-world loading conditions.

In addition to enhancing mechanical properties, nanoparticle-reinforced aluminum composites also show promise in thermal management applications. For instance, TiCx–TiB_2/Al composites are being explored for heat sink applications. Integrating TiB_2 and TiC nanoparticles into aluminum matrices via AM improves the composites' thermal conductivity and overall thermal stability, making them ideal for applications where heat dissipation is critical. Similarly, synthesizing TiB_2/TiC composites through AM offers valuable insights into optimizing fabrication parameters such as powder flowability, laser power, and scanning speeds to improve the final component's thermal and mechanical properties. By leveraging the precision of AM and combining it with secondary processes, it is possible to produce aluminum matrix hybrid composites with nanoparticles that exhibit enhanced mechanical and thermal performance. The ability to tailor the material properties by controlling reinforcement dispersion, heat treatment, and surface modifications makes AM an ideal method for fabricating advanced composites for high-performance engineering applications. These advancements open the door for more efficient, cost-effective manufacturing solutions in the aerospace, automotive, and electronics industries, where strength and thermal management are critical.

4.5 *Wear a Test*

The abrasive wear resistance of steel demonstrates that the microstructure is significantly refined through the multidirectional forging (MF) process. This refinement leads to a notable improvement in ductility without compromising tensile strength or hardness after controlled annealing. Wear tests were conducted to analyze the relationship between grain refinement and wear behavior, offering valuable insights into optimizing aluminum alloys for superior wear resistance. In this study, researchers investigated the effect of multidirectional forging on high-purity aluminum, aiming to refine grains through compressive deformation. The findings were evaluated using Vickers hardness tests and SEM. Furthermore, tensile and wear properties of an aluminum nanostructured matrix fabricated via hot extrusion and mechanical milling with B_4C nanoparticles (2–4 wt%) revealed significant improvements in wear rate, yield strength, tensile strength, and hardness when compared to commercial-grade aluminum. Composites like Al–Cu–2%B_4C and 4%B_4C exhibited superior wear resistance compared to unreinforced aluminum, with wear influenced by load and sliding speed, as evidenced through optical microscopy of worn surfaces [211–215].

Mechanical milling has shown significant potential in developing high-performance aluminum/ceramic nanocomposites tailored for industrial applications. The wear and tribological behavior of Al–B_4C and Al–SiC composites under varying contact pressures were examined, highlighting the impact of contact conditions on wear mechanisms and surface morphology. Comparative analysis confirmed the suitability of these composites for specific industrial applications based on their wear resistance and mechanical properties.

Furthermore, the mechanical and wear characteristics of Al–Cu matrix composites produced through direct squeeze casting were evaluated, demonstrating uniform dispersion of SiC particles within the matrix. Among the tested variants, the composite containing 15% SiC exhibited superior mechanical properties and wear resistance compared to those reinforced with 5% and 10% SiC. Dry sliding wear tests revealed an oxidative wear mechanism, with wear loss following a linear relationship with applied load, emphasizing the importance of reinforcement content in enhancing material durability and performance [216–218].

Thermal conductivity in Al–SiC composites was analyzed concerning particle size distribution, emphasizing the impact of monomodal and bimodal distributions. A separate study on Al6063 alloy incorporated 25 nm aluminum oxide nanoparticles through mechanical alloying (MA) and hot powder extrusion (HPE), producing ultra-fine-grained (UFG) materials with unique microstructures. These materials exhibited enhanced hardness, yield strength, and fracture toughness due to bimodal grain structures and oxide dispersion. The UFG alloy also showed a ductile–brittle fracture mode influenced by Al_2O_3 nanoparticle dispersion. SPD processes, coupled with precipitation hardening, enhanced the strength of Al 7000 series alloys, while phase precipitation kinetics during annealing highlighted the relationship between grain structure and mechanical properties [219, 220].

Investigations into aluminum-based composites, such as Al_2O_3-reinforced 6061 alloys, assessed wear behavior through pin-on-disk tests under varying sliding speeds and loads. Findings indicated that wear rates increased with higher Al_2O_3 concentrations and applied loads, while the influence of sliding speed varied depending on reinforcement content. Morphological analysis revealed that adhesion and delamination were dominant wear mechanisms at higher speeds, whereas abrasion was more prevalent at lower speeds.

Similarly, studies on Al–Si alloys reinforced with graphite demonstrated improved wear resistance, particularly under high-temperature conditions, making them suitable for applications such as engine components. Additionally, an analysis of the delamination wear mechanism underscored the significance of minimizing large second-phase particles or inclusions to enhance wear resistance, as illustrated by comparisons between annealed zone-refined iron and AISI 1020 steel. These insights contribute to developing advanced wear-resistant materials for high-performance applications [221–225].

The abrasive wear characteristics of Al–SiC particle-reinforced composites were analyzed using block-on-disc abrasion tests, identifying that increased particle volume fraction and size reduced specific wear rates. The dominant wear mechanisms included adhesion and delamination at higher sliding speeds and abrasion at lower speeds. Under deformation-heat treatments, grain size refinement's impact on wear resistance was further examined in copper-based alloys, including CuAlFe bronze and $CuZnP_b$ brass. Equal-channel angular pressing (ECAP) emerged as a key technique for grain refinement, enhancing wear resistance and mechanical properties across aerospace, automotive, and energy sectors, providing valuable pathways for material optimization in demanding applications [226–228].

Integrating MF and advanced reinforcement strategies has significantly refined the microstructure of materials, resulting in enhanced mechanical properties such as improved ductility, wear resistance, and strength. When applied to AM, these processes take on new potential for producing high-performance MMCs, including aluminum and steel alloys. Through AM, MF can be further optimized, allowing for the precise control of material deformation and microstructural refinement, improving the end parts' performance without compromising strength.

In the context of AMCs, the addition of reinforcing particles such as boron carbide (B_4C) and silicon carbide (SiC) significantly boosts the mechanical and tribological properties of the material. When using AM techniques such as SLM and DED, precisely controlling the distribution and dispersion of these reinforcements results in superior wear resistance, tensile strength, and overall performance. The thermal conductivity and wear behavior of these aluminum composites, particularly those reinforced with SiC and alumina (Al_2O_3), can be optimized using AM to tailor the material properties for specific applications, such as in automotive or aerospace industries where both wear resistance and thermal performance are crucial.

One of the main advantages of AM is its ability to directly influence the reinforcement distribution and bonding in MMCs. For instance, reinforcement particles like SiC and B_4C can be accurately placed within the matrix through layer-by-layer deposition, ensuring uniform distribution and enhancing the mechanical and thermal properties of the composite. These materials, combined with advanced processing techniques such as heat treatment or laser sintering, exhibit notable improvements in wear resistance, thermal conductivity, and structural integrity, making them ideal for high-performance applications.

Moreover, AM can be employed to incorporate additional refinement techniques such as equal channel angular pressing (ECAP) and severe plastic deformation (SPD) into the manufacturing process. These methods are traditionally used to refine the grain structure of materials, improving properties like wear resistance and strength. Combined with AM, these grain refinement techniques allow for the production of parts with ultra-fine microstructures and optimized mechanical properties. The flexibility of AM enables real-time adjustments during the process, offering greater control over the refinement of the grain structure at each deposition stage.

In addition to the benefits of reinforcement and grain refinement, the hybrid approach of combining MF with AM technologies provides an effective solution for enhancing the wear resistance of aluminum and copper-based alloys. The resulting parts achieve a more uniform microstructure by employing multidirectional loading during the AM process, directly impacting their wear behavior and mechanical performance. This is particularly advantageous in industries requiring superior durability, such as aerospace, automotive, and manufacturing, where components must withstand harsh operating conditions while maintaining their structural integrity.

In summary, by leveraging the power of AM and combining it with multidirectional forging, reinforcement strategies, and grain refinement techniques like ECAP and SPD, it is possible to significantly enhance materials' mechanical, thermal, and tribological properties. These advancements allow the development of next-generation MMCs with superior performance characteristics tailored to demanding industrial applications, opening new possibilities for high-performance, cost-effective manufacturing solutions.

5. Conclusion

- Producing MMCs involves diverse fabrication techniques, each with cost, efficiency, and final product quality trade-offs. The optimal method is chosen based on the intended application's specific needs. AM of MMCs is a nascent field, lagging behind AM advancements in other metal matrix systems like aluminum, nickel, and titanium. Key challenges include sourcing high-quality raw materials, particularly powders, and wires, essential for performance enhancement. For example, magnesium alloys, while attractive for their low melting points, present AM challenges due to compositional instability and high vapor pressure. Controlling volatilization and improving powder flowability is critical. Establishing material standards will accelerate AM MMC adoption in the automotive, aerospace, and biomedical sectors, where their superior properties are valuable.
- Understanding the reinforcement phase and matrix interface is crucial. Advanced characterization techniques such as TEM and 3D atom probe tomography and computational modeling can reveal interfacial behavior and atomic interactions. Powder metallurgy (P/M) offers precise control and enhanced properties but faces cost and machinery limitations. Stir casting is cost-effective for micro-sized reinforcements but struggles with nanoparticles and porosity. Diffusion bonding creates complex shapes but requires clean surfaces. Infiltration provides near-net shaping but has high tooling costs. In situ reinforcement formation offers uniform, stable phases, though reaction control is vital. Deposition processes yield high-strength composites but are

expensive and limited in shape. Spray deposition offers a more affordable alternative for specific applications.

- Forging enhances AMCs by refining grain structures and increasing dislocation density, improving strength and hardness, though potentially reducing ductility. HMMCs also significantly enhance properties, but issues like agglomeration and porosity must be addressed.
- Transitioning MMCs from lab to industry necessitates continuous innovation. While stir casting remains cost-effective for large-scale production, advanced methods like P/M and in situ reinforcement enhance mechanical properties for demanding applications. Secondary processes like forging further refine component performance. By combining established practices with modern techniques, the next generation of MMCs can achieve improved efficiency and performance.

References

[1] Sarmah, P. and K. Gupta. "Recent Advancements in the Fabrication of Metal Matrix Composites: A Systematic Review." *Materials* 17, no. 18 (2024): 4635.

[2] Surappa, M.K. "Aluminium Matrix Composites: Challenges and Opportunities." *Sadana* 28 (2003).

[3] Vayre, B., F. Vignat and F. Villeneuve. "Metallic Additive Manufacturing: State-of-the-Art Review and Prospects." *Mechanics & Industry* 13, no. 2 (2012): 89–96.

[4] Kim, D.S., S.W. Bae and K.H. Choi. "Development of Industrial SFF System Using Dual Laser and Optimal Process." *Robotics and Computer-Integrated Manufacturing* 23, no. 6 (2007): 659–66.

[5] John, M. "Photo-Glyph Recording." US Patent 2,775,758, December 1956.

[6] Escobar Jr, R., Fraser, K., Jordon, J.B. and Alison, P.G. *Meshfree Simulation of Oxide Dispersion in Meld of Aluminum Alloys.* Proceedings of the 12th International Symposium on Friction Stir Welding, Saguenay, QC, Canada, June 2018, pp. 26–28.

[7] Li, N., et al. "Laser Additive Manufacturing on Metal Matrix Composites: A Review." *Chinese Journal of Mechanical Engineering* 34 (2021): 1–16.

[8] You, X., et al. "A Review of Research on Aluminum Alloy Materials in Structural Engineering." *Developments in the Built Environment* (2024): 100319.

[9] Kumar, G.V., et al. "Studies on Al6061-SiC and Al7075-Al_2O_3 Metal Matrix Composites." *Journal of Minerals & Materials Characterization & Engineering* 9, no. 1 (2010): 43–55.

[10] Dao, M., et al. "Toward a Quantitative Understanding of the Mechanical Behavior of Nanocrystalline Metals." *Acta Materialia* 55, no. 12 (2007): 4041–65.

[11] Mortensen, A. and I. Jin. "Solidification Processing of Metal Matrix Composites." *International Materials Reviews* 37, no. 1 (1992): 101–28.

[12] Qian, W., et al. "In Situ X-Ray Imaging of Fatigue Crack Growth from Multiple Defects in Additively Manufactured AlSi10Mg Alloy." *International Journal of Fatigue* 155 (2022): 106616.

[13] Gotoh, R., et al. "Experimental and Analytical Investigations of AlSi10Mg, Stainless Steel, Inconel 625, and Ti-6Al-4V Porous Materials Printed via Powder Bed Fusion." *Progress in Additive Manufacturing* 7, no. 5 (2022): 943–55.

[14] Zhu, J., Li, Z., Su, M.-N. and Young, B. (2019). "Numerical study and design of aluminium alloy channel section columns with welds." *Thin-Walled Structures* 139, 139–150.

[15] Gu, H., et al. "Introducing Advanced Composites and Hybrid Materials." *Advanced Composites and Hybrid Materials* 1 (2018): 1–5.

[16] Ezatpour, H.R., et al. "Microstructure, Mechanical Analysis, and Optimal Selection of 7075 Aluminum Alloy-Based Composite Reinforced with Alumina Nanoparticles." *Materials Chemistry and Physics* 178 (2016): 119–27.

[17] Prakash, K.S., T. Nancharaih and V.S. Rao. "Additive Manufacturing Techniques in Manufacturing: An Overview." *Materials Today: Proceedings* 5, no. 2 (2018): 3873–82.

[18] Zhang, C., et al. "Additive Manufacturing of Magnesium Matrix Composites: Comprehensive Review of Recent Progress and Research Perspectives." *Journal of Magnesium and Alloys* 11, no. 2 (2023): 425–61.

[19] Sakai, T., et al. "Dynamic and Post-Dynamic Recrystallization under Hot, Cold, and Severe Plastic Deformation Conditions." *Progress in Materials Science* 60 (2014): 130–207.

[20] Swolfs, Y., I. Verpoest and L. Gorbatikh. "Recent Advances in Fiber-Hybrid Composites: Materials Selection, Opportunities, and Applications." *International Materials Reviews* 64, no. 4 (2019): 181–215.

[21] Cheng, E.J., H. Katsui, R. Tu and T. Goto. "Long-Range Ordered Structure of Ti–B–C–N in a TiB_2–TiCxN1−x Eutectic Composite." *Journal of the American Ceramic Society* 97, no. 8 (2014): 2423–26.

[22] Sinha, A.K., Narang, H.K. and Bhattacharya, S. (2020). Mechanical properties of hybrid polymer composites: A review. Journal of the Brazilian Society of Mechanical Sciences and Engineering, 42(4), Article 170, 1–21.

[23] Khelge, S., V. Kumar, V. Shetty and J. Kumaraswamy. "Effect of Reinforcement Particles on Aluminum Alloy Composites' Mechanical and Wear Properties." *Materials Today: Proceedings* 52 (2022): 571–76.

[24] Fanani, E.W.A., E. Surojo, A.R. Prabowo and H.I. Akbar. "Recent Progress in Hybrid Aluminum Composite: Manufacturing and Application." *Metals* 11, no. 12 (2021): 1919.

[25] Pedneault, J., V. Desjardins, M. Margni, D. Conciatori, M. Fafard and L. Sorelli. "Economic and Environmental Life Cycle Assessment of a Short-Span Aluminum Composite Bridge Deck in Canada." *Journal of Cleaner Production* 310 (2021): 127405.

[26] Raheem, A. and Subbaya, D.K. (2020). A review on hybrid composites used for marine propellers. *Material Science Research India*, ISSN 0973-3469.

[27] Liu, G., Zhang, X., Chen, X., He, Y., Cheng, L., Huo, M., Yin, J., Hao, F., Chen, S. and Wang, P. (2021). Progress in understanding and advancing magnesium-based materials for biomedical and structural applications. *Materials Science and Engineering: R: Reports*, 145, p.100596.

[28] Haghshenas, M. (2017). Mechanical characteristics of biodegradable magnesium matrix composites: A review. *Journal of Magnesium and Alloys*, 5(2), pp. 189–201.

[29] Bommala, V.K., Krishna, M.G. and Rao, C.T. (2019). Magnesium matrix composites for biomedical applications: A review. *Journal of Magnesium and Alloys*, 7(1), pp. 72–79.

[30] Su, J.L., Teng, J., Xu, Z.L. and Li, Y. (2020). Biodegradable magnesium-matrix composites: A review. *International Journal of Minerals, Metallurgy and Materials*, 27, pp. 724–744.

[31] Yang, Y., Lu, C., Peng, S., Shen, L., Wang, D., Qi, F. and Shuai, C. (2020). Laser additive manufacturing of Mg-based composite with improved degradation behaviour. *Virtual and Physical Prototyping*, 15(3), pp. 278–293.

[32] Yang, Y., Lu, C., Shen, L., Zhao, Z., Peng, S. and Shuai, C. (2023). *In-situ* deposition of apatite layer to protect Mg-based composite fabricated via laser additive manufacturing. *Journal of Magnesium and Alloys*, 11(2), pp. 629–640.

[33] Shuai, C., Wang, B., Bin, S., Peng, S. and Gao, C. (2020). TiO_2-induced in situ reaction in graphene oxide-reinforced AZ61 biocomposites to enhance the interfacial bonding. *ACS Applied Materials & Interfaces*, 12(20), pp. 23464–23473.

[34] Tao, J.X., Zhao, M.C., Zhao, Y.C., Yin, D.F., Liu, L., Gao, C. and Atrens, A. (2020). Influence of graphene oxide (GO) on microstructure and biodegradation of ZK30-xGO composites prepared by selective laser melting. *Journal of Magnesium and Alloys*, 8(3), pp. 952–962.

[35] Telang, V.S., Pemmada, R., Thomas, V., Ramakrishna, S., Tandon, P. and Nanda, H.S. (2021). Harnessing additive manufacturing for magnesium-based metallic bioimplants: Recent advances and future perspectives. *Current Opinion in Biomedical Engineering*, 17, p. 100264.

[36] Zamani, Y., Ghazanfari, H., Erabi, G., Moghanian, A., Fakić, B., Hosseini, S.M. and Mahammod, B.P. (2021). A review of additive manufacturing of Mg-based alloys and composite implants. *Journal of Composites and Compounds*, 3(6), pp. 71–83.

[37] Sezer, N., Evis, Z. and Koc, M. (2021). Additive manufacturing of biodegradable magnesium implants and scaffolds: Review recent advances and research trends. *Journal of Magnesium and Alloys*, 9(2), pp. 392–415.

[38] Li, Q., Lu, Y., Luo, Q., Yang, X., Yang, Y., Tan, J. and Pan, F. (2021). Thermodynamics and kinetics of hydriding and dehydriding reactions in Mg-based hydrogen storage materials. *Journal of Magnesium and Alloys*, 9(6), pp. 1922–1941.

[39] Feng, D., Zhou, D., Zhao, Z., Zhai, T., Yuan, Z., Sun, H. and Zhang, Y. (2021). Progress of graphene and loaded transition metals on Mg-based hydrogen storage alloys. *International Journal of Hydrogen Energy*, 46(67), pp. 33468–33485.

[40] Ok, M. (2005). Production and mechanical properties of Al_2O_3 particle-reinforced 2024 aluminum alloy composites. *Journal of Materials Processing Technology*, 161(3), pp. 381–387.

[41] Skillingberg, M. and Green, J. (2007). Aluminum applications in the rail industry. Light Metal Age, 65(5), 8–11.

[42] Ithers, G. and De Waas Tilakaratna, P. (2005). Performance evaluation of ULTALITE® low-cost aluminum metal matrix composite-based brake drums. SAE Transactions, 114(6), 902–907.

[43] Gupta, V.N.B., Ramana Rao, P.S.V. and Raju, B.N. (2020). A review on advanced hybrid metal matrix composites reinforced with nanoparticles. *International Journal of Advanced Research in Engineering and Technology*, 11(9), 228–244.

[44] Zhang, C., Li, Z., Zhang, J., Tang, H. and Wang, H. (2023). Additive manufacturing of magnesium matrix composites: Comprehensive review of recent progress and research perspectives. *Journal of Magnesium and Alloys*, 11(2), 425–461.

[45] Guan, H., Xiao, H., Ouyang, S., Tang, A., Chen, X., Tan, J. and Pan, F. (2022). A review of the design, processes, and properties of Mg-based composites. *Nanotechnology Reviews*, 11(1), pp. 712–730.

[46] Jdelsztajn, L., Schoenung, J.M., Jodoin, B. and Kim, G.E. (2005). Cold spray deposition of nanocrystalline aluminum alloys. *Metallurgical and Materials Transactions A*, 36, pp. 657–666.

[47] Rajan, T.P.D., Pillai, R.M. and Pai, B.C. (2008). Functionally graded Al-Al3Ni in situ intermetallic composites: Fabrication and microstructural characterization. *Journal of Alloys and Compounds*, 453(1-2), pp. L4–L7.

[48] Prasad, S.V. and Asthana, R. (2004). Aluminum metal-matrix composites for automotive applications: Tribological considerations. *Tribology Letters*, 17(3), pp. 445–453.

[49] Yadav, D. and Bauri, R. (2011). Processing, microstructure, and mechanical properties of nickel particles embedded aluminum matrix composite. *Materials Science and Engineering: A*, 528(3), 1326–1333.

[50] Zhou, X., Su, D., Wu, C. and Liu, L. (2012). Tensile mechanical properties and strengthening mechanism of hybrid carbon nanotube and silicon carbide nanoparticle-reinforced magnesium alloy composites. Journal of Nanomaterials, Article ID 851862, 7 pages. https://doi.org/10.1155/2012/851862.

[51] Cao, G., Choi, H., Oportus, J., Konishi, H. and Li, X. (2008). Study on tensile properties and microstructure of cast AZ91D/AlN nanocomposites. *Materials Science and Engineering: A, 494*(1-2), 127–131.

[52] Hassan, S.F. and Gupta, M. (2005). Development of high-performance magnesium nano-composites using nano-Al_2O_3 as reinforcement. *Materials Science and Engineering: A, 392*(1-2), 163–168.
[53] Wong, E.W.L. and Gupta, M. (2006). Simultaneously improving strength and ductility of magnesium using nano-size SiC particulates and microwaves. *Advanced Engineering Materials, 8*(8), 735–740.
[54] Hassan, S.F. and Gupta, M. (2007). Development and characterization of ductile Mg/Y_2O_3 nanocomposites. *Materials Science and Engineering: A, 392*(1-2), 163–168.
[55] Tun, K.S. and Gupta, M. (2007). Improving mechanical properties of magnesium using nano-yttria reinforcement and microwave-assisted powder metallurgy method. *Composites Science and Technology, 67*(13), 2657–2664.
[56] Alizadeh, A. and Taheri-Nassaj, E. (2012). Mechanical properties and wear behavior of Al–2 wt.% Cu alloy composites reinforced by B_4C nanoparticles and fabricated by mechanical milling and hot extrusion. *Materials Characterization, 67*, 119–128.
[57] Kerti, I. and Toptan, F. (2008). Microstructural variations in cast B_4C-reinforced aluminum matrix composites (AMCs). *Materials Letters, 62*(8-9), 1215–1218.
[58] Díaz, L.A., Valdés, A.F., Díaz, C., Espino, A.M. and Torrecillas, R. (2003). Alumina/molybdenum nanocomposites obtained in organic media. *Journal of the European Ceramic Society, 23*(15), 2829–2834.
[59] Lee, K.B., Sim, H.S., Kwon, H., Kwon, H. and Cho, S.Y. (2001). Tensile properties of 5052 Al matrix composites reinforced with B4C particles. *Metallurgical and Materials Transactions A, 32*(8), 2142–2147.
[60] Glazoff, M.V., Zolotorevsky, V.S. and Belov, N.A. (2010). *Casting aluminum alloys.* Elsevier.
[61] Sarmah, P. and Gupta, K. (2024). Recent advancements in fabrication of metal matrix composites: A systematic review. *Materials, 17*(18), 4635.
[62] Kumar, D., Phanden, R.K. and Thakur, L. (2021). A review on environment-friendly and lightweight magnesium-based metal matrix composites and alloys. *Materials Today: Proceedings, 38*, 359–364.
[63] Simsir, H. and Akgul, Y. (2022). Using a new sustainable carbon reinforcement in magnesium matrix composites. *Materials Chemistry and Physics, 281*, 125886.
[64] Elias, L. and Hegde, A.C. (2023). Electrodeposited Metallic/Nanocomposite Coatings. In *Advances in Corrosion Control of Magnesium and Its Alloys: Metal Matrix Composites and Protective Coatings* (p. 243).
[65] Singh, H., Brar, G.S., Kumar, H. and Aggarwal, V. (2021). A review on metal matrix composites for automobile applications. *Materials Today: Proceedings, 43*, 320–325.
[66] Kumar, S., Suman, K.N.S., Ravindra, K., Poddar, P. and SB, V.S. (2017). Microstructure, mechanical response, and fractography of $AZ91E/Al_2O_3$(p) nanocomposite fabricated by semi-solid stir casting method. *Journal of Magnesium and Alloys, 5*(1), 48–55.
[67] Srinivasan, V., Kunjiappan, S. and Palanisamy, P. (2021). A brief review of carbon nanotube-reinforced metal matrix composites for aerospace and defense applications. *International Nano Letters, 11*(4), 321–345.
[68] Au, M. (2005). Hydrogen storage properties of magnesium-based nanostructured composite materials. *Materials Science and Engineering: B, 117*(1), 37–44.
[69] Singh, L., Singh, B. and Saxena, K.K. (2020). Manufacturing techniques for metal matrix composites (MMC): An overview. *Advances in Materials and Processing Technologies, 6*(2), 441–457.
[70] Pawar, P.B., Wabale, R.M. and Utpat, A.A. (2018). A comprehensive study of aluminum-based metal matrix composites: Challenges and opportunities. *Materials Today: Proceedings, 5*(11), 23937–23944.

[71] Hashim, J., Looney, L. and Hashmi, M.S.J. (1999). Metal matrix composites: Production is done using the stir casting method. *Journal of Materials Processing Technology, 92*, 1–7.
[72] Zahmatkesh, B. and Enayati, M.H. (2010). A novel approach for the development of surface nanocomposite by friction stir processing. *Materials Science and Engineering: A, 527*(24-25), 6734–6740.
[73] Prakrathi, S., Ravikumar, M., Udupa, K.R. and Udaya Bhat, K. (2013). Fabrication of hybrid surface composite through friction stir processing and its impression creep behavior. ISRN Materials Science, 2013, Article ID 615460, 8 pages. https://doi.org/10.1155/2013/615460.
[74] Kisasoz, A., Guler, K.A. and Karaaslan, A. (2012). Infiltration of A6063 aluminum alloy into SiC–B_4C hybrid preforms using vacuum-assisted block mold investment casting technique. *Transactions of Nonferrous Metals Society of China, 22*(7), 1563–1567.
[75] Salehi, A., Babakhani, A. and Zebarjad, S.M. (2015). Microstructural and mechanical properties of Al–SiO_2 nanocomposite foams produced by an ultrasonic technique. *Materials Science and Engineering: A, 638*, 54–59.
[76] Bharat, N. and Bose, P.S.C. (2022). A study on conventional and non-conventional machining behavior of metal matrix composites: a review. *International Journal of Ambient Energy, 43*(1), 7600–7616.
[77] Vaezi, M.R. (2008). Two-step solochemical synthesis of ZnO/TiO_2 nano-composite materials. *Journal of Materials Processing Technology, 205*(1-3), 332–337.
[78] Hashim, J., Looney, L. and Hashmi, M.S.J. (1999). Metal matrix composites: Production is done using the stir casting method. *Journal of Materials Processing Technology, 92*, 1–7.
[79] Li, Q., Qiu, F., Dong, B.X., Geng, R., Lv, M.M., Zhao, Q.L. and Jiang, Q.C. (2018). Fabrication, microstructure refinement, and strengthening mechanisms of nanosized SiC_P/Al composites assisted ultrasonic vibration. *Materials Science and Engineering: A, 735*, 310–317.
[80] Zhilyaev, A.P. and Langdon, T.G. (2008). Using high-pressure torsion for metal processing: Fundamentals and applications. *Progress in Materials Science, 53*(6), 893–979.
[81] Caceres, C.H. (2007). Economic and environmental factors in light alloys automotive applications. *Metallurgical and Materials Transactions A, 38*, 1649–1662.
[82] Cheng, T. and Li, W. (2015). The temperature-dependent ideal tensile strength of ZrB_2, HfB_2, and TiB_2. *Journal of the American Ceramic Society, 98*(1), 190–196.
[83] Kitzmantel, M. and Neubauer, E. (2015). Innovative hybrid heat sink materials with high thermal conductivities and tailored CTE. In *Components and Packaging for Laser Systems* (Vol. 9346, pp. 51–61). SPIE.
[84] Ujah, C.O., Kallon, D.V.V., Aikhuele, D.O. and Aigbodion, V.S. (2022). Advanced composite materials: A panacea for improved electricity transmission. *Applied Sciences, 12*(16), 8291.
[85] Ramesh, C.S., Keshavamurthy, R., Channabasappa, B.H. and Ahmed, A. (2009). Microstructure and mechanical properties of Ni–P coated Si3N4 reinforced Al6061 composites. *Materials Science and Engineering: A, 502*(1-2), 99–106.
[86] Altinkok, N. and Koker, R. (2006). Modeling tensile and density properties prediction in particle-reinforced metal matrix composites using neural networks. *Materials & Design, 27*(8), 625–631.
[87] Buchmayr, B. and Stampfl, J. (2017). Austrian conference highlights innovations in metal Additive Manufacturing. Metal AM, 3(1), 52–54.
[88] Perry, M.E.J. (2021). Material flow and microstructure evolution during additive friction stir deposition of aluminum alloys (Doctoral dissertation). Virginia Polytechnic Institute and State University. https://vtechworks.lib.vt.edu/handle/10919/104253.
[89] Luo, S., Wang, H., Gao, Z., Wu, Y. and Wang, H. (2021). Interaction between high-velocity gas and liquid in gas atomization revealed by a new coupled simulation model. *Materials & Design, 212*, 110264.

[90] Shi, Y., Lu, W., Sun, W., Zhang, S., Yang, B. and Wang, J. (2022). Impact of gas pressure on particle feature in Fe-based amorphous alloy powders via gas atomization: Simulation and experiment. *Journal of Materials Science & Technology, 105*, 203–213.
[91] Zhang, J., Yuan, W., Song, B., Yin, S., Wang, X., Wei, Q. and Shi, Y. (2022). Towards understanding metallurgical defect formation of selective laser melted wrought aluminum alloys. *Advanced Powder Materials, 1*(4), 100035.
[92] Gök, D.A., Bayraktar, C. and Hoşkun, M. (2024). A review on processing, mechanical, and wear properties of Al matrix composites reinforced with Al_2O_3, SiC, B_4C, and MgO by powder metallurgy method. *Journal of Materials Research and Technology,* 31, 1132–1150. https://doi.org/10.1016/j.jmrt.2024.06.110.
[93] Chandel, R., Sharma, N. and Bansal, S.A. (2021). A review of recent developments of aluminum-based hybrid composites for automotive applications. *Emergent Materials, 4*(5), 1243–1257.
[94] Zhang, H., Loukus, J. and Loukus, A. (2009). Improvement of the bonding interface in hybrid fiber/particle preform reinforced Al matrix composite. *Materials Letters, 63*(2), 310–312.
[95] Du Plessis, A., Razavi, N., Benedetti, M., Murchio, S., Leary, M., Watson, M., et al. (2022). Properties and applications of additively manufactured metallic cellular materials: A review. *Progress in Materials Science, 125*, 100918.
[96] Sames, W.J., List, F.A., Pannala, S., Dehoff, R.R. and Babu, S.S. (2016). The metallurgy and processing science of metal additive manufacturing. *International Materials Reviews, 61*(5), 315–360.
[97] Chandel, R., Sharma, N. and Bansal, S.A. (2021). A review on recent developments of aluminum-based hybrid composites for automotive applications. *Emergent Materials, 4*(5), 1243–1257.
[98] Sudmanns, M., Birnbaum, A.J., Gu, Y., Iliopoulos, A.P., Callahan, P.G., Michopoulos, J.G. and El-Awady, J.A. (2022). The interplay of local chemistry and plasticity in controlling microstructure formation during laser powder bed fusion of metals. *Additive Manufacturing, 55*, 102791.
[99] Fang, Z.C., Wu, Z.L., Huang, C.G. and Wu, C.W. (2020). Review on residual stress in selective laser melting additive manufacturing of alloy parts. *Optics & Laser Technology, 129*, 106283.
[100] Krakhmalev, P., Fredriksson, G., Thuvander, M., Åsberg, M., Vilardell, A.M., Oikonomou, C. and Kazantseva, N. (2020). Influence of heat treatment under hot isostatic pressing (HIP) on the microstructure of intermetallic-reinforced tool steel manufactured by laser powder bed fusion. *Materials Science and Engineering: A, 772*, 138699.
[101] Liverani, E., Lutey, A.H., Ascari, A. and Fortunato, A. (2020). The effects of hot isostatic pressing (HIP) and solubilization heat treatment on the density, mechanical properties, and microstructure of austenitic stainless steel parts produced by selective laser melting (SLM). *The International Journal of Advanced Manufacturing Technology, 107*, 109–122.
[102] Logesh, K., Hariharasakthisudhan, P., Moshi, A.A.M., Rajan, B.S. and Sathickbasha, K. (2019). Mechanical properties and microstructure of A356 alloy reinforced AlN/MWCNT/graphite/Al composites fabricated by stir casting. *Materials Research Express, 7*(1), 015004.
[103] Shibata, K. and Ushio, H. (1994). Tribological application of MMC for reducing engine weight. *Tribology International, 27*(1), 39–44.
[104] Balamugundan, B., Karthikeyan, L., Karthik, K. and Keerthi, C. (2017). Enhancement of mechanical properties on aluminum alloys–A review. *IOSR Journal of Mechanical and Civil Engineering (IOSR-JMCE), e-ISSN: 2278–1684*, 2014.
[105] Ramesh, C.S., Bharathesh, T.P., Verma, S.M. and Keshavamurthy, R. (2012). Sand abrasive wear behavior of hot forged Al 6061-TiO_2 composites. *Journal of Materials Engineering and Performance, 21*, 74–82.

[106] Bambach, M., Sizova, I., Sydow, B., Hemes, S. and Meiners, F. (2020). Hybrid manufacturing of components from Ti-6Al-4V by metal forming and wire-arc additive manufacturing. *Journal of Materials Processing Technology, 282*, 116689.

[107] Bobba, S., Rao, M.S. and Raj, P. (2021). Experimental analysis on the forging process of Al 6061 and Al 7075 aluminum alloys. *SunText Review of Material Science, 2*(1), 1–14.

[108] Shin, J.H., Choi, H.J. and Bae, D.H. (2014). The structure and properties of 2024 aluminum composites reinforced with TiO_2 nanoparticles. *Materials Science and Engineering: A, 607*, 605–610.

[109] Di Chen, Q., Pan, Z., Liu, Z., Zeng, S., Shi, Q., Peng, J. and Li, Y. (2024). Microstructural evolution, mechanical properties, and tribological behavior of SiC reinforced 316L matrix composites fabricated by additive manufacturing. *Journal of Materials Research and Technology, 30*, 889–899.

[110] Mohapatra, S.K. and Maity, K. (2017). Synthesis and characterization of hot extruded aluminum-based MMC developed by powder metallurgy route. *International Journal of Mechanical and Materials Engineering, 12*, 1–9.

[111] Izadi, H., Nolting, A., Munro, C., Bishop, D.P., Plucknett, K.P. and Gerlich, A.P. (2013). Friction stir processing of Al/SiC composites fabricated by powder metallurgy. *Journal of Materials Processing Technology, 213*(11), 1900–1907.

[112] Fanani, E.W.A., Surojo, E., Prabowo, A.R., Ariawan, D. and Akbar, H.I. (2021). Recent development in aluminum matrix composite forging affects mechanical and physical properties. *Procedia Structural Integrity, 33*, 3–10.

[113] Mostafaei, A., Zhao, C., He, Y., Ghiaasiaan, S.R., Shi, B., Shao, S. and Rollett, A.D. (2022). Defects and anomalies in powder bed fusion metal additive manufacturing. *Current Opinion in Solid State and Materials Science, 26*(2), 100974.

[114] Pragana, J.P.M., Sampaio, R.F., Bragança, I.M.F., Silva, C.M.A. and Martins, P.A.F. (2021). Hybrid metal additive manufacturing: A state–of-the-art review. *Advances in Industrial and Manufacturing Engineering, 2*, 100032.

[115] Babu, P.A., Saraf, M.R., Vora, K.C., Chaurasiya, S.M. and Kuppan, P. (2015). Influence of forging parameters on aluminum alloy's mechanical behavior and hot forgeability. *Materials Today: Proceedings, 2*(4-5), 3238–3244.

[116] Akbar, H.I., Surojo, E., Ariawan, D., Putra, G.A. and Wibowo, R.T. (2020). Effect of reinforcement material on manufactured aluminum matrix composite properties using stir casting route. *Procedia Structural Integrity, 27*, 62–68.

[117] Hirtler, M., Jedynak, A., Sydow, B., Sviridov, A. and Bambach, M. (2018). Investigation of microstructure and hardness of a rib geometry produced by metal forming and wire arc additive manufacturing. *MATEC Web of Conferences, 190*, 2005.

[118] Rajan, T.P.D., Pillai, R.M. and Pai, B.C. (2008). Functionally graded Al-Al_3Ni in situ intermetallic composites: fabrication and microstructural characterization. *Journal of Alloys and Compounds, 453*(1-2), L4–L7.

[119] Padap, A.K., Chaudhari, G.P. and Nath, S.K. (2010). Mechanical and dry sliding wear behavior of ultrafine-grained AISI 1024 steel processed using multiaxial forging. *Journal of Materials Science, 45*, 4837–4845.

[120] Yue, X., Cai, Z., Lü, X., Wang, J. and Ru, H. (2016). Effect of Ni content on microstructures and mechanical properties of hot-pressed TiC–TiB_2–Ni composite. *Materials Science and Engineering: A, 668*, 208–214.

[121] Wang, H.Y., Huang, L. and Jiang, Q.C. (2005). In situ synthesis of TiB_2–TiC particulates locally reinforced medium carbon steel–matrix composites via the SHS reaction of Ni–Ti–B_4C system during casting. *Materials Science and Engineering: A, 407*(1–2), 98–104.

[122] Ferkel, H. (1999). Properties of copper reinforced by laser-generated Al_2O_3-nanoparticles. *Nanostructured Materials, 11*(5), 595–602.

[123] Ravi, K.R., Sreekumar, V.M., Pillai, R.M., Mahato, C., Amaranathan, K.R. and Pai, B.C. (2007). Optimization of mixing parameters through a water model for metal matrix composites synthesis. *Materials & Design, 28*(3), 871–881.

[124] Lashgari, H.R., Zangeneh, S., Shahmir, H., Saghafi, M. and Emamy, M. (2010). Heat treatment affects the microstructure, tensile properties, and dry sliding wear behavior of A356–10% B_4C cast composites. *Materials & Design, 31*(9), 4414–4422.

[125] Sun, L.G., Wu, G., Wang, Q. and Lu, J. (2020). Nanostructural metallic materials: Structures and mechanical properties. Materials Today, 38, 114–135. https://doi.org/10.1016/j.mattod.2020.04.005.

[126] Almeida, N.G.S., Pereira, P.H.R., De Faria, C.G., Aguilar, M.T.P. and Cetlin, P.R. (2020). Mechanical behavior and microstructures of aluminum processed by low strain amplitude multi-directional confined forging. *Journal of Materials Research and Technology, 9*(3), 3190–3197.

[127] Rezayat, M., Karamimoghadam, M., Ashkani, O. and Bodaghi, M. (2023). Characterization and optimization of Cu–Al_2O_3 nanocomposites synthesized via high energy planetary milling: A morphological and structural study. *Journal of Composites Science*, *7*(7), 300. https://doi.org/10.3390/jcs7070300.

[128] Chandrasekhar, P., Chand, S. and Singh, S. (2018). Investigation of Dynamic Effects during Cold Upset Forging of Silicon Carbide Particulate-Reinforced Aluminium Metal Matrix Composite Preforms. *Materials Today: Proceedings, 5*(9), 20201–20209.

[129] Dám, K., Průša, F. and Vojtěch, D. (2014). Structural and mechanical characteristics of the Al–23Si–8Fe–5Mn alloy prepared by a combination of centrifugal spraying and hot die forging. *Materials Science and Engineering: A, 610*, 197–202.

[130] Meti, V.K.V., Konaraddi, R. and Siddhalingeshwar, I.G. (2018). Mechanical and tribological properties of AA7075 based MMC processed through ultrasound-assisted casting technique. *Materials Today: Proceedings, 5*, 25677–25687.

[131] Asgharzadeh, H., Simchi, A. and Kim, H.S. (2011). Microstructure and mechanical properties of oxide-dispersion strengthened Al6063 alloy with ultra-fine grain structure. *Metallurgical and Materials Transactions A, 42*, 816–824.

[132] Deng, L., Wang, X., Jin, J. and Xia, J. (2018). Precision forging technology for aluminum alloy. *Frontiers of Mechanical Engineering, 13*, 25–36.

[133] Bouquerel, J., Diawara, B., Dubois, A., Dubar, M., Vogt, J.B. and Najjar, D. (2015). Investigations of the microstructural response to a cold forging process of the 6082-T6 alloy. *Materials & Design, 68*, 245–258.

[134] Purohit, R., Qureshi, M.M.U. and Kumar, B. (2017). Effect of forging on aluminum matrix nanocomposites: A review. *Materials Today: Proceedings, 4*, 5357–5360.

[135] Babu, P.K.A., Saraf, M.R., Vora, K.C. and Kuppan, P. (2018). Forgeability study of Al 6061 alloy during warm forging of cylindrical billets. *Materials Today: Proceedings, 5*, 11950–11961.

[136] Sweet, G.A.W., Shalchi, B., Williams, B.W., Taylor, A., Hexemer, R.L., Donaldson, I.W. and Paul, D. (2019). Microstructural evolution of a forged 2XXX series aluminum powder metallurgy alloy. *Materials Characterization, 151*, 342–350.

[137] Nouri, S., Kazeminezhad, M. and Shadkam, A. (2020). Flow stress of 2024 aluminum alloy during multi-directional forging process and natural aging after plastic deformation. *Materials Chemistry and Physics, 123446*.

[138] Zhang, H., Sun, M., Liu, Y., Ma, D., Xu, B., Huang, M. and Li, Y. (2021). Ultrafine-grained dual-phase maraging steel with high strength and excellent cryogenic toughness. *Acta Materialia, 211*, 116878.

[139] Peng, X., Wu, S., Qian, W., Bao, J., Hu, Y., Zhan, Z. and Withers, P.J. (2022). The potency of defects on fatigue of additively manufactured metals. *International Journal of Mechanical Sciences, 221*, 107185.

[140] Zhu, Y.T. and Liao, X. (2004). Retaining ductility. *Nature Materials, 3*(6), 351–352.

[141] Clement, C.D. and Kabir, A.S. (2024). Effects of SLM process parameters and TiC nanoparticles on the microstructure and mechanical properties of AlSi10Mg. *Journal of Materials Engineering and Performance*, 1–15.

[142] Anoop, J., Suyamburajan, V.A., Babu, P.S. and Filketu, S.A. (2022). Analysis and experimental investigation of A356 aluminium alloy hybrid composites reinforced with Gr-Fe_3O_4-B_4C nanoparticles synthesised by selective laser melting (SLM). *Journal of Nanomaterials*, 2022, Article ID 7510499, 8 pages. https://doi.org/10.1155/2022/7510499.

[143] Sitdikov, O., Sakai, T., Goloborodko, A., Miura, H. and Kaibyshev, R. (2004). Effect of pass strain on grain refinement in 7475 Al alloy during hot multidirectional forging. *Materials Transactions, 45*(7), 2232–2238.

[144] Shu, S., Yang, H., Tong, C. and Qiu, F. (2016). Fabrication of TiCx-TiB_2/Al composites for application as a heat sink. *Materials, 9*(8), 642.

[145] Zhang, Z., Topping, T., Li, Y., Vogt, R., Zhou, Y., Haines, C. and Lavernia, E.J. (2011). Mechanical behavior of ultrafine-grained Al composites reinforced with B_4C nanoparticles. *Scripta Materialia, 65*(8), 652–655.

[146] Kumar, N., Khanduja, D. and Singh, R.P. (2023). A study of mechanical properties on aluminum-based hybrid metal matrix composite (AA7175/B_4C/SiC/Gr). *Journal of The Institution of Engineers (India): Series D*, 1–8.

[147] Toptan, F., Kilicarslan, A., Karaaslan, A., Cigdem, M. and Kerti, I. (2010). Processing and microstructural characterization of AA 1070 and AA 6063 matrix B_4C_p reinforced composites. *Materials & Design, 31*, S87–S91.

[148] Zhu, Q.F., Lei, L.I., Ban, C.Y., Zhao, Z.H., Zuo, Y.B. and Cui, J.Z. (2014). Structure uniformity and limits of grain refinement of high purity aluminum during multi-directional forging process at room temperature. *Transactions of Nonferrous Metals Society of China, 24*(5), 1301–1306.

[149] Subbiah, R., Lokesh, K., Singh, S.K., Chatterjee, S. and Eswaraiah, D. (2019). Investigation on microstructure and mechanical properties of treated AISI 440 steel by tempering process - A review. *Materials Today: Proceedings, 18*, 2802–2805.

[150] Yuan, P., Gu, D. and Dai, D. (2015). Particulate migration behavior and mechanism during selective laser melting of TiC reinforced Al matrix nanocomposites. *Materials & Design, 82*, 46–55.

[151] Rajan, T.P.D., Pillai, R.M. and Pai, B.C. (1998). Reinforcement coatings and interfaces in aluminum metal matrix composites. *Journal of Materials Science, 33*, 3491–3503.

[152] Teber, A., Schoenstein, F., Têtard, F., Abdellaoui, M. and Jouini, N. (2012). Effect of SPS process sintering on the microstructure and mechanical properties of nanocrystalline TiC for tools application. *International Journal of Refractory Metals and Hard Materials, 30*(1), 64–70.

[153] Vallauri, D. and Deorsola, F.A. (2009). Synthesis of TiC–TiB_2–Ni cermets by thermal explosion under pressure. *Materials Research Bulletin, 44*(7), 1528–1533.

[154] Soundararajan, R., Ramprakash, M.P., Vel, T.M., Ratteesh, M. and Kumar, S.P. (2021). Investigation of metallurgical and mechanical behavior of forged aluminum alloy hybrid composites. *Materials Today: Proceedings, 45*, 2179–2185.

[155] Chu, K., Jia, C., Tian, W., Liang, X., Chen, H. and Guo, H. (2010). Thermal conductivity of spark plasma sintering consolidated SiC_p/Al composites containing pores: Numerical study and experimental validation. *Composites Part A: Applied Science and Manufacturing, 41*(1), 161–167.

[156] Hong, G., Han, Y., Zhang, X., Jia, C. and Xu, J. (2015). Microstructure and thermophysical properties of SiC/Al composites mixed with diamond. *Transactions of Nonferrous Metals Society of China, 25*(1), 170–174.

[157] Lashgari, H.R., Sufizadeh, A.R. and Emamy, M. (2010). The effect of strontium on the microstructure and wear properties of A356–10% B_4C cast composites. *Materials & Design, 31*(4), 2187–2195.

[158] Mendoza-Duarte, J.M., Robles-Hernandez, F.C., Estrada-Guel, I. and Martínez-Sánchez, R. (2017). Aluminum sintering in air atmosphere using high-frequency induction heating. *Microscopy and Microanalysis, 23*(S1), 1950–1951.

[159] Yao, X., Zhang, Z., Zheng, Y.F., Kong, C., Quadir, M.Z., Liang, J.M. and Zhang, D.L. (2017). Effects of SiC nanoparticle content on the microstructure and tensile mechanical properties of ultrafine-grained AA6063-SiCnp nanocomposites fabricated by powder metallurgy. *Journal of Materials Science & Technology, 33*(9), 1023–1030.

[160] Bouquerel, J., Diawara, B., Dubois, A., Dubar, M., Vogt, J.B. and Najjar, D. (2015). Investigations of the microstructural response to a cold forging process of the 6082-T6 alloy. *Materials & Design, 68*, 245–258.

[161] Salehi, A., Babakhani, A. and Zebarjad, S.M. (2015). Microstructural and mechanical properties of Al–SiO_2 nanocomposite foams produced by an ultrasonic technique. *Materials Science and Engineering: A, 638*, 54–59.

[162] El-Sabbagh, A., Soliman, M., Taha, M. and Palkowski, H. (2012). Hot rolling behaviour of stir-cast Al 6061 and Al 6082 alloys–SiC fine particulates reinforced composites. *Journal of Materials Processing Technology, 212*(2), 497–508.

[163] Ravindran, P., Manisekar, K., Kumar, S.V. and Rathika, P. (2013). Investigating microstructure and mechanical properties of aluminum hybrid nano-composites with solid lubricant additions. *Materials & Design, 51*, 448–456.

[164] Williams, J.C. and Starke Jr, E.A. (2003). Progress in structural materials for aerospace systems. *Acta Materialia, 51*(19), 5775–5799.

[165] Georgantzia, E., Gkantou, M. and Kamaris, G.S. (2021). Aluminum alloys as structural material: A review of research. *Engineering Structures, 227*, 111372.

[166] Wloka, J., Hack, T. and Virtanen, S. (2007). Influence of temper and surface condition on the exfoliation behavior of high strength Al–Zn–Mg–Cu alloys. *Corrosion Science, 49*(3), 1437–1449.

[167] Deschamps, A. and Bréchet, Y. (1998). Influence of quench and heating rates on the aging response of an Al–Zn–Mg–(Zr) alloy. *Materials Science and Engineering: A, 251*(1-2), 200–207.

[168] Reddy, A.C. (2018). Effects of Rapid Preheating and Solution Heat Treatment on Mechanical Properties of AA2618 Forged Samples. *International Journal of Applied Research in Mechanical Engineering (IJARME)*, ISSN 2231–5950.

[169] Raj, R. (1981). Development of a processing map for use in warm-forming and hot-forming processes. *Metallurgical Transactions A, 12*, 1089–1097.

[170] Meti, V. K., Shirur, S., Nampoothiri, J., Ravi, K.R. and Siddhalingeshwar, I.G. (2018). Synthesis, Characterization, and Mechanical Properties of AA7075 Based MMCs Reinforced with TiB2 Particles Processed Through Ultrasound-Assisted In-Situ Casting Technique. *Transactions of the Indian Institute of Metals, 71*(4), 841–848.

[171] Vani, V.V. and Chak, S.K. (2018). The effect of process parameters in aluminum metal matrix composites with powder metallurgy. *Manufacturing Review, 5*, 7.

[172] Sun, G., Du, W., Chen, C. and Zhao, X. (2024). Hybrid additive manufacturing of ceramic/aluminum-titanium gradient composites via arc and laser cladding eutectic pools. *Virtual and Physical Prototyping, 19*(1), e2419950.

[173] Zhao, Y., Topping, T., Bingert, J.F., Thornton, J.J., Dangelewicz, A.M., Li, Y. and Lavernia, E.J. (2008). High tensile ductility and strength in bulk nanostructured nickel. *Advanced Materials, 20*(16), 3028–3033.

[174] Manu, K.S., Kumar, S.A., Rajan, T.P.D., Mohammed, M.R. and Pai, B.C. (2017). Effect of alumina nanoparticle on strengthening of Al-Si alloy through dendrite refinement, interfacial bonding, and dislocation bowing. *Journal of Alloys and Compounds, 712*, 394–405.

[175] Wloka, J., Hack, T. and Virtanen, S. (2007). Influence of temper and surface condition on the exfoliation behavior of high strength Al–Zn–Mg–Cu alloys. *Corrosion Science, 49*(3), 1437–1449.

[176] Kang, Y.C. and Chan, S.L.I. (2004). Tensile properties of nanometric Al_2O_3 particulate-reinforced aluminum matrix composites. *Materials Chemistry and Physics, 85*(2–3), 438–443.

[177] Wang, H., Li, J.F., Nan, C.W., Zhou, M., Liu, W., Zhang, B.P. and Kita, T. (2006). High-performance $Ag_{0.8}Pb_{18}$+$xSbTe_{20}$ thermoelectric bulk materials fabricated by mechanical alloying and spark plasma sintering. *Applied Physics Letters, 88*(9).

[178] Li, J.F., Wang, K., Zhang, B.P. and Zhang, L.M. (2006). Ferroelectric and piezoelectric properties of fine-grained $Na_{0.5}K_{0.5}NbO_3$ lead-free piezoelectric ceramics prepared by spark plasma sintering. *Journal of the American Ceramic Society, 89*(2), 706–709.

[179] Majta, J. and Bator, A. (2002). Mechanical behavior of hot and warm formed micro-alloyed steels. *Journal of Materials Processing Technology, 125*, 77–83.

[180] Park, N.K., Yeom, J.T. and Na, Y.S. (2002). Characterization of deformation stability in hot forging of conventional Ti–6Al–4V using processing maps. *Journal of Materials Processing Technology, 130*, 540–545.

[181] Ye, J., Ajdelsztajn, L. and Schoenung, J.M. (2006). Bulk nanocrystalline aluminum 5083 alloys fabricated by a novel technique: Cryomilling and spark plasma sintering. *Metallurgical and Materials Transactions A, 37*, 2569–2579.

[182] Guo, B., Ni, S., Yi, J., Shen, R., Tang, Z., Du, Y. and Song, M. (2017). Microstructures and mechanical properties of carbon nanotubes reinforced pure aluminum composites synthesized by spark plasma sintering and hot rolling. *Materials Science and Engineering: A, 698*, 282–288.

[183] Equbal, M.I., Talukdar, P., Kumar, V. and Ohdar, R.K. (2014). Deformation behavior of micro-alloyed steel using thermomechanical simulator and finite element method. *Procedia Materials Science, 6*, 674–681.

[184] Vijay, S.J. and Murugan, N. (2010). Influence of tool pin profile on the metallurgical and mechanical properties of friction stir welded Al–10 wt.% TiB_2 metal matrix composite. *Materials & Design, 31*(7), 3585–3589.

[185] Prasad, Y.V.R.K., Gegel, H.L., Doraivelu, S.M., Malas, J.C., Morgan, J.T., Lark, K.A. and Barker, D.R. (1984). Modelling of Dynamic Material Behaviour in Hot Deformation: Forging of Ti-6242 *Metallurgical Transactions A, 15*, 1883–1892.

[186] Malas, J.C. and Seetharaman, V. (1992). Using material behavior models to develop process control strategies. *JOM, 44*, 8–13.

[187] Liu, Y.Q., Cong, H.T., Wang, W., Sun, C.H. and Cheng, H.M. (2009). AlN nanoparticle-reinforced nanocrystalline Al matrix composites: Fabrication and mechanical properties. *Materials Science and Engineering: A, 505*(1–2), 151–156.

[188] Tapasztó, O. and Balázsi, C. (2010). The effect of milling time on the sintering kinetics of Si_3N_4-based nanocomposites. *Ceramics International, 36*(7), 2247–2251.

[189] Li, W.Y., Yang, C. and Liao, H. (2011). Effect of vacuum heat treatment on microstructure and microhardness of cold-sprayed TiN particle-reinforced Al alloy-based composites. *Materials & Design, 32*(1), 388–394.

[190] Nampoothiri, J., Harini, R.S., Nayak, S.K., Raj, B. and Ravi, K.R. (2016). Post-in-situ reaction ultrasonic treatment for generation of Al–4.4 Cu/TiB_2 nanocomposite: A route to enhance the strength of metal matrix nanocomposites. *Journal of Alloys and Compounds, 683*, 370–378.

[191] Long, J., Li, X., Fang, D., Peng, P. and He, Q. (2013). Fabrication of diamond particles reinforced Al-matrix composites by hot-press sintering. *International Journal of Refractory Metals and Hard Materials, 41*, 85–89.

[192] Shanthi, M., Lim, C.Y.H. and Lu, L. (2007). Effects of Grain Size on the Wear of Recycled AZ91 Mg *Tribology International, 40*(2), 335–338.

[193] Lengauer, W. and Dreyer, K. (2006). Tailoring hardness and toughness gradients in functional gradient hard metals (FGHMs). *International Journal of Refractory Metals and Hard Materials, 24*(1–2), 155–161.

[194] Feng, Y.C., Geng, L., Fan, G.H., Li, A.B. and Zheng, Z.Z. (2009). The properties and microstructure of hybrid composites reinforced with WO3 particles and $Al_{18}B_4O_{33}$ whiskers by squeeze casting. *Materials & Design, 30*(9), 3632–3635.

[195] Li, Y., Zhao, Y.H., Ortalan, V., Liu, W., Zhang, Z.H., Vogt, R.G. and Schoenung, J.M. (2009). Investigation of aluminum-based nanocomposites with ultra-high strength. *Materials Science and Engineering: A, 527*(1–2), 305–316.

[196] Chai, J., Zhu, Y., Wang, Z., Shen, T., Liu, Y., Niu, L. and Liu, C. (2020). Microstructure and mechanical properties of SPS sintered Al_2O_3–ZrO_2 (3Y)–SiC ceramic composites. *Materials Science and Engineering: A, 781*, 139197.

[197] Chung, K.H., He, J., Shin, D.H. and Schoenung, J.M. (2003). Mechanisms of microstructure evolution during cryomilling in the presence of hard particles. *Materials Science and Engineering: A, 356*(1–2), 23–31.

[198] Jayaprakash, D., Niranjan, K. and Vinod, B. (2023). Studies on mechanical and microstructural properties of aluminum hybrid composites: Influence of SiC/Gr particles by double stir-casting approach. *Silicon, 15*(3), 1247–1261.

[199] Nemati, N., Khosroshahi, R., Emamy, M. and Zolriasatein, A. (2011). Investigation of microstructure, hardness, and wear properties of Al–4.5 wt.% Cu–TiC nanocomposites produced by mechanical milling. *Materials & Design, 32*(7), 3718–3729.

[200] Thomson, K.E., Jiang, D., Yao, W., Ritchie, R.O. and Mukherjee, A.K. (2012). Characterization and mechanical testing of alumina-based nanocomposites reinforced with niobium and carbon nanotubes fabricated by spark plasma sintering. *Acta Materialia, 60*(2), 622–632.

[201] Rohatgi, P. (1991). Cast aluminum–matrix composites for automotive applications. *JOM, 43*, 10–15.

[202] Rubel, R.I., Ali, M.H., Jafor, M.A. and Alam, M.M. (2019). Carbon nanotubes agglomeration in reinforced composites: A review. *AIMS Materials Science, 6*(5), 756–780.

[203] Hashim, J. (2001). The production of cast metal matrix composite by a modified stir casting method. *Jurnal Teknologi, 35*(1), 9–20.

[204] Straumal, B., Martynenko, N., Temralieva, D., Serebryany, V., Tabachkova, N., Shchetinin, I. and Estrin, Y. (2020). The effect of equal-channel angular pressing on microstructure, mechanical properties, and biodegradation behavior of magnesium alloyed with silver and gadolinium. *Crystals, 10*(10), 918.

[205] Wang, D., Wang, H., Sun, S., Zhu, X. and Tu, G. (2014). Fabrication and characterization of TiB_2/TiC composites. *International Journal of Refractory Metals and Hard Materials, 45*, 95–101.

[206] Wang, H., Sun, S., Wang, D. and Tu, G. (2012). Characterization of the structure of TiB_2/TiC composites prepared via mechanical alloying and subsequent pressureless sintering. *Powder Technology, 217*, 340–346.

[207] Khalil, K.A., Sherif, E.S.M., Nabawy, A.M., Abdo, H.S., Marzouk, W.W. and Alharbi, H.F. (2016). Titanium carbide nanofibers-reinforced aluminum compacts, a new strategy to enhance mechanical properties. *Materials, 9*(5), 399.

[208] Mao, J., Yi, Y., Huang, S., He, H., Fu, Y. and Tang, J. (2024). Effect of multidirectional forging on the microstructures and mechanical properties of the Al–Mg–Si alloy. *Metals and Materials International*, 1–18.

[209] Poirier, D., Drew, R.A., Trudeau, M.L. and Gauvin, R. (2010). Fabrication and properties of mechanically milled alumina/aluminum nanocomposites. *Materials Science and Engineering: A, 527*(29–30), 7605–7614.
[210] Onat, A. (2010). Mechanical and dry sliding wear properties of silicon carbide particulate reinforced aluminum–copper alloy matrix composites produced by direct squeeze casting. *Journal of Alloys and Compounds, 489*(1), 119–124.
[211] Molina, J.M., Narciso, J., Weber, L., Mortensen, A. and Louis, E. (2008). Thermal conductivity of Al–SiC composites with monomodal and bimodal particle size distribution. *Materials Science and Engineering: A, 480*(1–2), 483–488.
[212] Al-Qutub, A.M., Allam, I.M. and Abdul Samad, M.A. (2008). Wear and friction of Al–Al_2O_3 composites at various sliding speeds. *Journal of Materials Science, 43*, 5797–5803.
[213] Clarke, J. and Sarkar, A.D. (1979). Wear characteristics of as-cast binary aluminium–silicon alloys. *Wear, 54*(1), 7–16.
[214] Suh, N.P. (1973). The delamination theory of wear. *Wear, 25*(1), 111–124.
[215] Alpas, A.T. and Embury, J.D. (1990). Sliding and abrasive wear behavior of aluminum (2014)-SiC particle reinforced composite. *Scripta Metallurgica et Materialia, 24*(5), 931–934.
[216] Godet, M. (1984). The Third-Body Approach: A Mechanical View of Wear. *Wear, 100*(1–3), 437–452.
[217] Valiev, R.Z. and Langdon, T.G. (2006). Principles of equal-channel angular pressing as a processing tool for grain refinement. *Progress in Materials Science, 51*(7), 881–981.
[218] Sezer, N., Evis, Z. and Koc, M. (2021). Additive manufacturing of biodegradable magnesium implants and scaffolds: Review recent advances and research trends. *Journal of Magnesium and Alloys, 9*(2), 392–415.
[219] Li, N., Huang, S., Zhang, G., Qin, R., Liu, W., Xiong, H. and Blackburn, J. (2019). Progress in additive manufacturing on new materials: A review. *Journal of Materials Science & Technology, 35*(2), 242–269.
[220] Shi, Y., Lu, W., Sun, W., Zhang, S., Yang, B. and Wang, J. (2022). Impact of gas pressure on particle feature in Fe-based amorphous alloy powders via gas atomization: Simulation and experiment. *Journal of Materials Science & Technology, 105*, 203–213.
[221] Zhang, J., Yuan, W., Song, B., Yin, S., Wang, X., Wei, Q. and Shi, Y. (2022). Towards understanding metallurgical defect formation of selective laser melted wrought aluminum alloys. *Advanced Powder Materials, 1*(4), 100035.
[222] Peng, X., Wu, S., Qian, W., Bao, J., Hu, Y., Zhan, Z. and Withers, P.J. (2022). The potency of defects on fatigue of additively manufactured metals. *International Journal of Mechanical Sciences, 221*, 107185.
[223] Ma, G., Xiao, H., Ye, J. and He, Y. (2020). Research status and development of magnesium matrix composites. *Materials Science and Technology, 36*(6), 645–653.
[224] Mussatto, A., Ahad, I.U., Mousavian, R.T., Delaure, Y. and Brabazon, D. (2021). Advanced production routes for metal matrix composites. *Engineering Reports, 3*(5), e12330.
[225] Hegab, H.A. (2016). Design for additive manufacturing of composite materials and potential alloys: A review. *Manufacturing Review, 3*, 11.
[226] Kirka, M.M., Nandwana, P., Lee, Y. and Dehoff, R.R. (2017). Solidification and solid-state transformation sciences in metals additive manufacturing. *Scripta Materialia, 135*, 130–134.
[227] Ye, H.Z. and Liu, X.Y. (2004). Review of recent studies in magnesium matrix composites. *Journal of Materials Science, 39*, 6153–6171.
[228] Chikkamath, A.S., Meti, V.K.V. and Siddhalingeshwar, I. G. (2024). Investigating the Synergistic Impact of TiB_2 and ZrO_2 Reinforcements on Forged AA7075-Based Hybrid Metal Matrix Composites. *Materials Science Forum, 1135*, 97–111. Trans Tech Publications Ltd.

4 Mechanical Properties Prediction of Metal Additive Manufactured Parts Using Artificial Intelligence

V.S. Kathavate[1*], Nilesh Kolhe[2], M.G. Kulthe[3], P.P. Deshpande[4] and Murthy S.S.S. Chavali Yadav[5]

Additive manufacturing (AM) opens up ample avenues of fabricating a variety of materials, from polymer and composite to metals, using layer-by-layer technique. The AM process can produce the parts with complex structures that exhibit excellent mechanical, thermal, physical, and chemical properties. However, porosity and defects induced during layer-by-layer deposition of these AM parts significantly impacts their mechanical properties and overall performance. As the mechanical reliability of AM parts in structural applications is of vital importance, investigating the influence of intrinsic defects and microstructural changes on the mechanical properties of AM parts become increasingly apparent. Till date, several studies have extensively investigated the correlation between microstructural changes and mechanical properties of AM parts fabricated via different AM techniques, highlighting the potential of artificial intelligence (AI) in predicting the mechanical properties based on the microstructural changes has become a timely topic. In view of this, this chapter reports comprehensive outlooks on the machine learning and data-driven approaches for predicting the effect of microstructural changes on the mechanical properties of metal AM parts, thereby accurately forecasting mechanical and overall performance. By leveraging the advantage of AI and machine learning (ML) algorithms, different datasets derived from various AM processes can be used to identify the patterns and predict the accurate models, thereby facilitating the optimization in design parameters for improved reliability and functionality. The chapter begins with an introduction of AM and AI–ML approach in metallurgy and materials science followed by an overview of various AM techniques and parameters involved in them. A comprehensive outlook on the existing experimental studies in particular and optimization (or algorithm) studies

[1] Department of Defence System Engineering, SRM University AP, Neerukonda-Tadikonda Road, Mangalagiri Mandal, Guntur District, Neerukonda 522 502, Andhra Pradesh, India.
[2] Department of Research and Development, Dr. Vishwanath Karad MIT World Peace University, Paud Road, Kothrud, Pune 411038, Maharashtra, India.
[3,4] Department of Metallurgy and Materials Engineering, COEP Technological University, Shivajinagar, Pune 411005, Maharashtra, India.
[5] Department of Research and Development, Dr. Vishwanath Karad MIT World Peace University, Pune 411038, Maharashtra, India.
* Corresponding author: vskathavate@gmail.com

DOI: 10.1201/9781003635673-4

in general highlighting the influence of microstructure on the mechanical properties of metal AM parts is provided. While the predictive capabilities of AI–ML models and data-driven algorithms to accurately predict the mechanical properties of metal AM parts are provided in the midst of the chapter, the recent progresses in the AI–ML approach for predicting the mechanical properties of metal AM parts along with the recent challenges and futuristic outlooks are presented in the last part of the chapter. Unlike other reports, we mainly emphasize on the mechanical properties prediction via AI–ML of wide spectrum metals.

1. Additive Manufacturing (AM): A Paradigm Shift in Manufacturing

Additive manufacturing (AM) is a transformative material fabrication process/technology that builds the material layer by layer (vertical or horizontal) using digital models and 3D printing. The inherent unique features such as reduction in machining process and cost, high product customization, and flexibility in the design made AM to attract significant attention of manufacturing community in recent time [1–3]. Since its discovery in 1980s, AM has opened new avenues in manufacturing covering a large spectrum of applications including aerospace, energy sector, polymers and composites, oil and gas, and tool-making industries, hence became a crucial part of Industry 4.0. With the advent of new AM technologies, AM has also extended its applications in healthcare and biomedical devices/parts with excellent engineering and functional properties [4–6], and mass customization where patient-specific implants and prosthetics is a primary requirement. Owing to its product customization such as lightweight structures, rapid prototyping, and ease in designing process (i.e., complex shape and size), AM has become increasingly important in aerospace and defense industries. For instance, its ability to produce functional protypes of parts within minimal time significantly reduces product development cycles. In view of the above attractive features, the market for AM materials is expected to grow at a compound annual growth rate (CAGR) of ~ 27% till the end of 2025 [2]. Therefore, undoubtedly since the last decade, AM has brought major paradigm shift in the field of manufacturing and materials industries.

With the rapid evolution of new technologies, the world is continuously changing. Therefore, AM industries are trying to innovate by integrating advanced materials, production processes, and sustainable practices. This will help to meet the growing demands for customization, scalability, and efficiency across aerospace, healthcare, and automative sectors. However, from the performance point of view, the research is still in progress to fabricate the damage-tolerant AM parts with excellent mechanical properties. However, intrinsic defects such as porosities, voids, and micro cracks induced during the AM process deteriorate the functional performance, so it is difficult to maintain the synergy between damage tolerance and mechanical properties. Furthermore, mechanical properties of the same material fabricated via different AM processes are different due to the specific set of parameters associated with each process. This further demands the more effective way to tailor the mechanical properties of AM parts using pre- and post-fabrication treatments. As all the above processes are more focused toward experiments and their outcomes are mainly dependent on the experimental observations, the process parameters optimization and enhancing mechanical properties are bit tedious, require efforts, and are time-consuming, and therefore AM community is continuously looking for alternatives, such as artificial intelligence (AI) and machine learning (ML) algorithms.

The AI tools and ML algorithms are becoming increasingly important in the field of Mechanical, Metallurgy, and Materials Science and Engineering in the sense that these tools assist the materials discovery, process optimization, structure, phase, defect identification, and properties prediction. The intriguing feature of AI–ML is that it reduces time-consuming and trial-and-error experimental methods based on training and testing of previously available experimental data, thereby predicting (near) accurate models for structure, engineering properties, and performance of the material.

In view of this, the present chapter shed light on mechanical performance prediction of metal additive manufactured (MAM) parts using AI–ML techniques, thereby answering the following open questions in the AM field: (a) Are the present-day strategies well optimized for obtaining the set of optimized process parameters and mechanical properties? (b) How AI–ML overcomes the above challenge? (c) How much the AI–ML techniques in predicting mechanical properties of AM parts are efficient (or accurate)?

1.1 Additive Manufacturing vs. Conventional Manufacturing

As explained earlier, AM fabricates the materials/parts adding or depositing the layers of powders (i.e., layer by layer adding up layers, additive formation), while conventional fabrication techniques such as casting, forging, and machining fabricate the material/part in formative and subtractive manner. This further leads to high efficiency, optimal raw material usage, excellent scalability, and cost savings of AM over other conventional manufacturing processes. Following is our take on that.

One of the advantages of AM technique is that it can produce prototypes without any special tooling that further caters to the rapid product development which is unlikely in conventional manufacturing processes. Furthermore, AM is highly resource effective so that raw material waste is minimum, as only desired/pre-determined amount of powder is deposited over layer by layer, while on the other hand, traditional subtractive manufacturing process could lead to higher material waste, as the material is being removed from the raw material/workpiece. Although the conventional manufacturing processes are more compatible with mass production, AM is overcoming the gap quickly with higher production rate for low-volume and customized materials/components. One of the downsides of AM is its capability in printing the large part/devices due to the limitations in printer size; however, it is possible to fabricate comparatively larger parts/devices using conventional manufacturing techniques. On materials compatibility front, AM is flexible with many materials ranging from conventional metals to ceramics, polymers, and composites, while conventional manufacturing processes are dependent on the raw materials (powders or molten metal). AM surpasses conventional manufacturing process for the customization in fabrication of devices/parts due to ease in programming where user-specific requirements can be considered in printing or layer-by-layer deposition program. Economically, AM dominates over conventional manufacturing with the advent of technology, while high initial setup cost raises the production cost in conventional manufacturing process. Another reason why AM is preferred nowadays over conventional manufacturing process is its energy-efficient nature with less wastage of materials, thus offering reasonably higher overall efficiency.

To sum up, AM offers many advantages over conventional manufacturing process with ease in manufacturing, flexibility, materials compatibility, rapid-prototyping, and energy efficiency, therefore becoming increasingly popular. Table 1 summarizes such advantages. As the present chapter focuses on mechanical properties of MAM parts, we highlight MAM process and important parameters involved in it.

Table 1 Summary of the comparison between AM and conventional manufacturing process [4–6].

Additive Manufacturing	*Conventional Manufacturing*
Material fabrication is layer by layer.	Material fabrication is formative and subtractive.
Can rapidly produce prototypes without any special tooling.	Requires special tools and molds which slow down prototyping.
Raw material waste is minimum.	Leads to higher material wastage.
Better suited for production of low-volume and customized components.	Better suited for mass production.
Size of parts is limited by printer size hence production of large parts is difficult.	Can fabricate large parts using traditional manufacturing.
Requirements for customization can be programmed.	Customization of components is limited.
More cost-efficient and energy-efficient.	Lower energy efficiency and setup cost is higher.

1.2 Metal Additive Manufacturing (MAM) and Materials in MAM

MAM in AM fabricates the components via infinite 3D iteration of high-performance structures through the precise layering and bonding of powders. Unlike traditional manufacturing processes, MAM uses a CAD model to define the geometry of each layer along the build and scan directions, thereby offering the precise control over layer thickness. The standard that represents the MAM process is ISO/

ASTM 52900. Owing to its key benefits such as complex geometry fabrication, high customization, multi-material integration, assembly-free construction, and materials waste reduction, MAM has emerged as a viable candidate to replace conventional manufacturing process in industrial production. MAM is mainly classified on two primary criteria: (i) type of heat source and (ii) raw material supplied [7–10]. The selection of heat source electron beam, laser, and electric arc depends on the end product and its applications, while power or wire can be fed as a raw material. The MAM process can also be classified according to (i) direct energy deposition and (ii) powder bed fusion. Figure 1 represents the detailed classification of MAM process. The direct energy deposition (DED), powder bed fusion (PBF), and binder jetting (BJ) have been used in many instances. PBF typically uses laser and electron beam as heat sources, while the main heat sources for DED are laser, electric arc, and plasma.

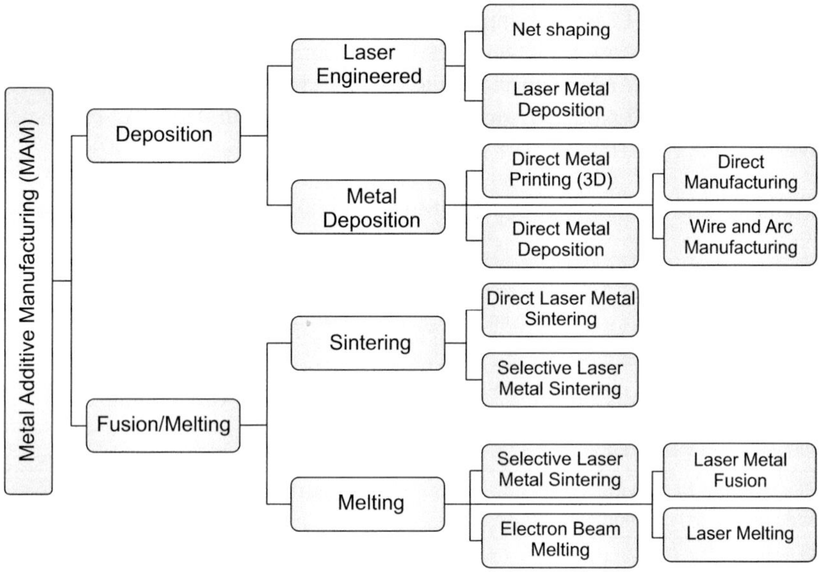

Fig. 1 A detailed classification of MAM based on metal deposition and fusion melting.

In DED the material is fed either via powder or wire, while in PBF fed material is sintered directly. Note that the size of the device/parts and its complexity associated with the design are also decisive factors in the selection of appropriate MAM process. For instance, DED can be used to fabricate the parts up to a size of ~ 200 cm, while size of parts fabricated via PBF is limited to a few centimeters (~ 30 cm). However, PBF is preferred over DED for manufacturing complex parts/components with a high resolution. It is possible to fabricate the parts/devices as small as up to or less than 10 cm with BJ that further have high resolutions and more precise control and complexities. The relative density, ρ of the finished products follows the trend; ρ_{DED} (100%) ≥ ρ_{PBF} (99%) ≥ ρ_{BJ} (70–90%), indicating the negligible porosities in parts manufactured via DED. All the above attractive features indicate the suitability of MAM for the several applications. We have highlighted the importance of heat source and size of the finished parts while identifying the features of MAM, nevertheless, material selection also plays an important role and in the following paragraph we shed light on materials in MAM.

The most intriguing feature of MAM is it can fabricate the materials in the form of sheets, plates, and tubes. There are a few instances where it is also possible to fabricate the materials in shape of wire and electrode using MAM, which is quite not possible with conventional manufacturing process. It is worth mentioning that the MAM can offer a range of material fabrication from conventional steel and aluminum (Al) to modern day titanium (Ti) and nickel (Ni) based super alloys. The Al alloys [11], particularly *7xxx* series produced via PBF (laser) [12] and DED [13, 14], possess yield strength, $\sigma_{YS} \sim 400$ MPa and employed to aerospace and automobile applications. A few studies have demonstrated the applicability of PBF (laser) [15] and DED (laser and WAAM) [16, 17] process for the fabrication of stainless steel (*SS*316) with excellent mechanical properties and corrosion resistance, while Ni-based superalloys fabricated via PBF (EBM and Laser) [18–20] and DED (laser) [21] have been found suitable for aero engines, turbines, and combustion chambers, as their creep resistance and tensile strength have been stable at elevated temperatures. The studies [18–20] also mentioned that the Ni alloys fabricated via PBF incurred the residual stresses in the material during manufacturing that are detrimental to the mechanical and thermal properties. In recent time, Ti alloys mainly Ti – 6Al – 4V and intermetallics of Ti – Al have found a unique niche in aero and bio medical implants applications owing to their high strength to weight ratio (i.e., specific strength), corrosion resistance, and biocompatibility. The AM processes PBF [22–24] and DED [25, 26] have been found suitable for the fabrication of Ti – 6Al – 4V alloys, Ti – Al intermetallics can be manufactured by PBF (EBM and laser) [27, 28] alone. The feasibility of SLM technique has also been demonstrated for fabricating Ti – 6Al – 4V alloys that are further subjected to high strain rate applications, an important aspect in dynamic/shock loading environments in aero turbines. The high entropy alloys (HEA), an alloy with more than 4 principal elements in equiatomic or near equiatomic stochiometric ratio, have attracted a significant attention in recent time due to their exception mechanical, electrical, magnetic, and oxidation properties. One of the versatile features of HEA is that their properties are stable at high temperatures, and hence they are the potential candidates in aerospace, turbine, and nuclear operations. Fabrication of AM HEAs has attracted significant attention in recent times and finds a unique niche in high temperature operations such as furnaces, reactors, and high temperature devices that include refractory elements such as niobium, tantalum, molybdenum, and tungsten. Another versatility of MAM lies in fabrication of miniaturized parts such as anti-scattering grids and beam collimators for medical devices that further replace the traditional lead (*Pb*) materials with precision. Figure 2 represents the overview of the materials fabricated via MAM on a commercial and mass (production) basis along with their processing strategies.

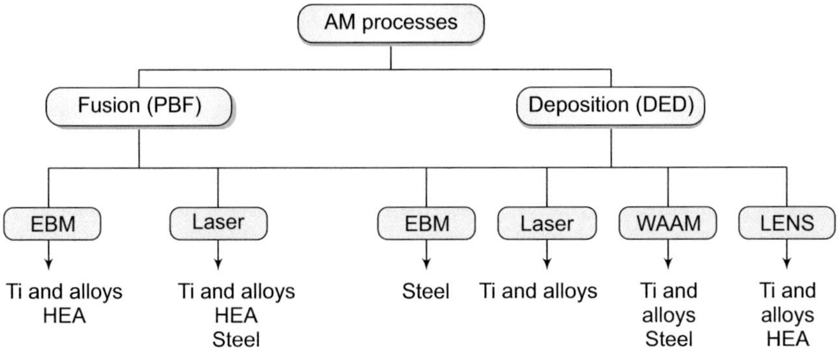

Fig. 2 Summary of the MAM processes and materials/parts of different materials fabricated via different MAM processes.

1.3 Mechanical Properties of MAM-Fabricated Parts

In this section, we focus on the mechanical properties of MAM-fabricated parts under quasistatic and dynamic loading environment for the materials steels, Ti – 6Al – 4V, and HEA for automobile, aerospace, and nuclear application point of view.

The mechanical properties of AM-fabricated steel markedly depend on the grain refinement during the process. The Carlton and co-workers [29] in case of SS316 has demonstrated to the change in failure mode from ductile to brittle with an increase in porosity from 0.1% to 2.4%. However, post-fabrication treatment such as thermal annealing has proven beneficial for obtaining the defect-free microstructure that further enhanced the elongation of SS316 up to ~ 600 MPa that was further comparable to wrought SS316 (~ 500 MPa). The higher scan speeds adopted (≥400 mm/s) during the fabrication of SS316 appears to be an important reason for the higher porosities in the material. The hardness, *H*, ultimate strength, σ_{ut}, and elongation, δ of precipitation-hardened AM steels was significantly improved by ~ 20%, ~ 13 – 15%, and ~ 18%, respectively, when these steels were subjected to the post-fabrication treatment such as solution aging and thermal annealing [30]. The excellent work hardening characteristics of AM-fabricated 18 – Ni300 steel (18% Ni, 9% Co, 4% Mo, and Bal. Fe) have been demonstrated by Kempen et al. [31], where the ageing process at 480°C for 5 h resulted in σ_{ut} (by ~ 72%) and Rockwell *H*(*HRC*) (by ~ 45%). Our understanding on the above work indicates that the presence of retained austenite and its reversion during the ageing process could be the potential reasons for the enhanced σ_{ut} and *H*. The high cycle fatigue (HCF) performance with R ~ 0.1 (in tension–tension) of LBM-processed SS316L stainless steel was investigated by Gorsse and co-workers [32] indicating that higher fatigue strength (~ 510 MPa) compared to wrought SS 316L (~ 290 MPa) is due to high surface roughness of AM-fabricated SS316L stainless steel.

The Ti alloys produced via LBM had demonstrated significant improvement in the σ_y and σ_{ut} values (approximately 2 times) than those fabricated via sheet metal processing [33, 34]. The formation of a very fine α' martensitic phase appears to be an important reason which further promotes the grain refinement possibly due to higher cooling rates associated with LBM. Owing to the distorted HCP structure of α' martensitic phase, it is possible to obtain the strength-ductility synergy in AM Ti alloys, which could be attributed to the formation of pre-dominant α' martensitic phase that makes the stronger grains as compared to α (lamellar) phase [33–36]. The MAM-processed Ti – 6Al – 4V has shown exceptional σ_y and σ_{ut} values than those wrought (ASTM-F1472-14) and cast (ASTM-F-1108-14) alloys [37, 38]. The main reason for the improved mechanical properties of MAM *Ti* alloys are the resistance offered by twins that further impedes the plastic deformation [33–38].

Besides σ_y and σ_{ut}, the ductility of LBM processes Ti – 6Al – 4V alloys do not show any significant improvement, however, with an advent of EBM, it is possible to achieve the % elongation ~ 20 – 25% in their as-build state [39]. One of the important reasons for high ductility of EBM-processed Ti – 6Al – 4V alloys is presence of acicular α lamella representing the Widmanstätten like morphology along with the segregation of retained austenite at the α-phase GB. The scan parameters in LBM are also responsible for change in microstructure of Ti – 6Al – 4V alloys that further tailors the mechanical properties [40]. The controlled process parameters in LBM such as layer thickness (30, 60, 90 μm), laser power (175 and 375 W), scan speed (686, 710, 1029 mm/s), and volume energy (33.74, 50.62, 68.47 J/mm^3) have been significant in maintaining the strength–ductility synergy [40].

All the above studies corroborate that processing parameters influence microstructure evolution that further have marked influence on the mechanical properties of MAM parts. This eventually underscores a critical challenge of multivariate, nonlinear dependencies governing material behavior under quasistatic and dynamic loads. The scalability and predictive accuracies of traditional experiments and simulations are still under scrutiny due to inherent limitations such as computational complexities, reliance on simplified models, hurdles in deep uncertainties, and inability to capture complex real-world dynamics, though they have provided foundational insights into mechanical behavior of MAM parts. Furthermore, one of the downsides associated with experimental methods is the cost, time, efforts, and materials-specific validation cycles, while computational simulations struggle to integrate the materials processing that spans across rapid solidification, dislocation dynamics, and phase transformation [41]. This gap is more intense in aerospace and automobile applications, where component reliability is utmost important and demands precise property prediction. The artificial intelligence is a transformative tool that has set a paradigm shift to alleviate such complexities and problems leveraging machine learning (ML) to map processing–structure–property relationship with unprecedented resolution.

2. Role of Artificial Intelligence in Metallurgy Engineering and Materials Science

In recent years, AI and ML have emerged as transformative techniques that cater the paradigm shifts in the field of metallurgy engineering and materials science. The role of AI and ML techniques in metallurgy engineering and materials science is gaining significant attention due to their ability of new materials discovery, materials process optimization, and materials performance predication. This computational revolution has addressed the critical challenges in handling complex multivariate relationships between parameters, microstructural evolution, and mechanical properties that often defy the traditional analytical modeling approaches.

The implementation of AI in Metallurgy and Materials Science typically includes ML architecture ranging from conventional neural networks (CNN) that further facilitates the microstructural image analysis to recurrent neural networks (RNN) for materials process optimization. The various multidimensional datasets include: (a) composition space coordinates such as alloying element and trace impurities, (b) process parameters (history of thermal cycles/treatments, energy densities, and scan strategies in case of AM), (c) characterization datasets (microstructural images acquired from optical microscope (OM), scanning electron and transmission electron microscope (SEM/TEM), electron backscattered diffraction (EBSD) images, X-ray diffraction (XRD) spectra, and nanoindentation arrays), and (d) information about mechanical properties such as tensile/compression strength, fatigue crack growth, and curves/data obtained from various mechanical tests. It is also possible to predict the properties through raw microstructural images that can be combined with physics informed regularization. The role of AI from AM point of view is vital in facilitating the real-time quality prediction through in situ monitoring of data from melt-pool sensors and layer-wise thermography.

In current scenario, many AI tools have been also proven beneficial for process–structure–property optimization that are further helpful in solving the inverse problems related to MAM parts. For instance, Gaussian process regression models find a unique niche in high-dimensional composition spaces to identify novel alloys, thereby meeting the extraordinary synergy between σ_y and toughness, K [41], a crucial step in alloy design. The parameter optimization in MAM process can be performed using Bayesian optimization loops that further minimize defect density by tuning the laser power, scan speed, and hatch speed via multi-objective loss function. Following this, the defect mitigation strategies can be implemented by enabling ML models combining with XGBoost and random forest classifiers that further have ~ 90% accuracy in porosity prediction [41]. The role of AI, particularly in MAM, is also extended in microstructure control using Graph Neural Networks (GNN) to predict solidification morphology evolution under Marangoni convection effects from thermal gradient and solidification rates. Furthermore, from finite element method/simulation viewpoint, ML surrogate model combines with FEM results to produce hybrid-physics AI framework, thereby enabling digital twin integration that further helps in virtual quantification of MAM parts through Monte-Carlo simulations of structure–property correlationship across process parameter distributions. Figure 3 depicts in wide spectrum of applications of AI–ML in metallurgy engineering and materials science with special focus on AM.

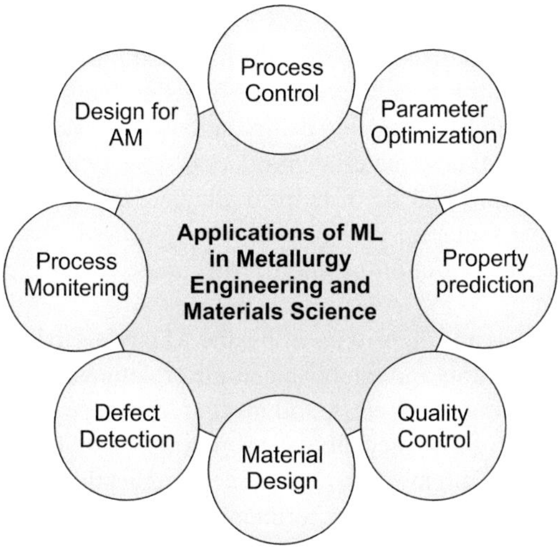

Fig. 3 Applications of AI–ML in metallurgy engineering and materials science with special focus on AM.

The recent advances in ML demonstrate its predictive capability to assimilate heterogeneous dataset that typically includes laser parameters, thermal histories, defect distributions, and mechanical properties for the predictive models and frameworks that further transcend the traditional models [41–43]. For instance, Mehrabi and co-workers [43] for AM steels have demonstrated that retained austenite fractions influence the σ_y and fatigue resistance, while SHAP analysis reveals the predominant process parameters such as energy density and layer orientations that can further affect the mechanical properties of MAM parts. All the above information reveals that the AI–ML-driven insights in case of MAM have not only opened ample avenues of alloy development but also enabled AM community for the closed loop optimization of AM process for the nuclear and automobile application. The mechanical reliability of these structures presents several critical challenges though. Therefore, in the following section we shed light on mechanical property prediction using AI–ML techniques for the engineering material in general and MAM parts in particular.

3. AI Techniques for Mechanical Property Predictions

3.1 AI Techniques and Algorithms

The material scientists and engineers are aiming at improved mechanical performance of engineering and functional materials in extreme environment that further plays a crucial role in many engineering applications. As discussed earlier, the conventional experimental techniques along with trial-and-error methods may not be sufficient to meet the current high throughput industrial demands due to labor-intensive, time-consuming, and expensive nature. In many instances, experimental methods may incur some error in the estimation of these mechanical properties, while simulations prove to be beneficial for virtual experiments; however, it requires high-performing computational workstations. Therefore, AI–ML-driven approach in new materials/product discovery and process–structure–property correlationship will be of help to alleviate this problem. Therefore, in this section we will focus on various AI techniques/tools that are used for the mechanical property prediction of engineering materials in general and MAM parts in particular.

ML-based models and algorithms allow the AI models to learn (train) the data from previous experiments and establish complex relationships between materials composition, processing parameters, and mechanical properties. Figure 4 reflects in process of prediction of mechanical properties by material using AI-based methods [41]. First and foremost, the ML models predict the mechanical properties using structured datasets (previous experimental readings/observations/dat). These structured datasets are further divided into three parts: (a) training datasets, (b) validation datasets, and (c) testing datasets. The first one allows the model to train the data patterns within the known data, while the latter ones fine-tune the parameters and evaluate the model accuracy, respectively. For mechanical property predictions, ML algorithms use four different learnings: (a) supervised learning, (b) unsupervised learning, (c) semi-supervised learning, and (d) reinforcement learning (Figure 5). The supervised learning algorithms typically use labeled data with artificial neural network (ANN), while unsupervised leaning identifies patterns in unlabeled data. The extent to these two models is semi-supervised learning that further utilizes both labeled and unlabeled data for training. The reinforcement learning model underscores the trial-and-error learning. In Figure 4, we discuss the various ML models that are important from the mechanical property prediction viewpoint, therefore, to introduce them to the readers serves the purpose of the chapter.

3.1.1 Random Forest (RF)

The random forest (RF) model is the robust technique for regression that further employs bagging and bootstrap processes [41, 44]. The RF creates multiple decision trees in parallel during bagging process that are trained on randomly sampled subsets of datasets with replacement, thereby enabling diversity among

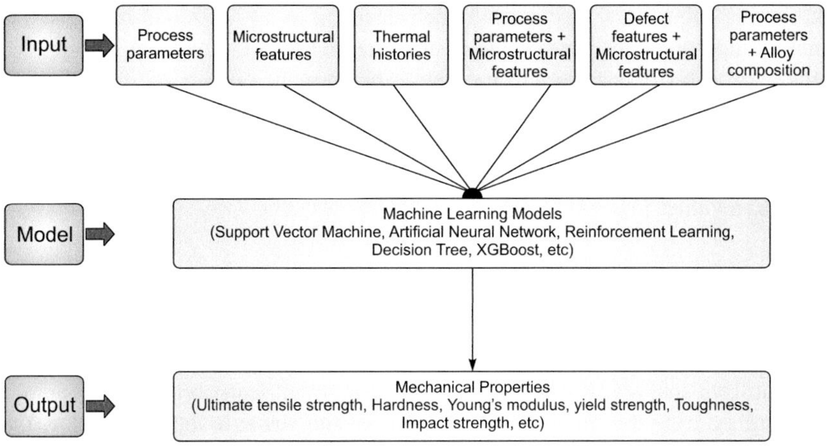

Fig. 4 Representative flow chart of the process of mechanical property prediction of engineering materials using AI–ML tools.

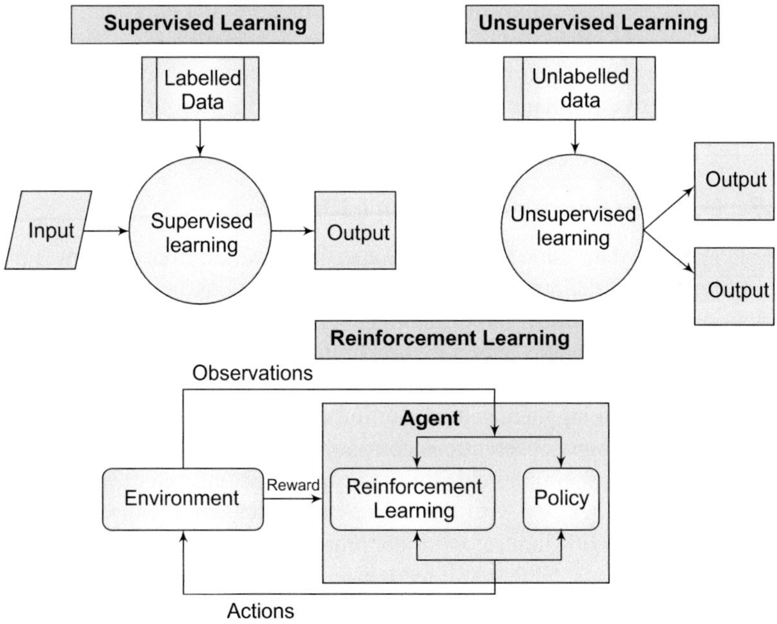

Fig. 5 Summary of the ML techniques and algorithms used in mechanical property prediction of materials.

the trees that further reduces a risk of overfitting the data. The bootstrapping involves sampling of dataset with replacement from the original dataset that further generates multiple subsets for training. This method ensures the reduced variance and strong robustness by exposing each decision tree to different datasets. The final predictive model is produced after averaging the outputs from all the trees. This approach offers excellent predictive performance compared to individual decision trees. One of the potential advantages of RF is its ability to generate high dimensional datasets with a large number of features that further facilitate the automatic selection of subsets for each split reducing the risk of overestimation of prediction values [41, 44].

3.1.2 Gradient Boosting Tree (GBT)

Gradient boosting tree (GBT) model is another ML technique for mechanical property prediction of materials using decision tree outputs [44]. Unlike RF, GBT utilizes shallow decision trees (known as weak learners) and incorporates the results during the training process. One of the versatile facts of GBT is that it uses a sequential approach in the sense that it builds one tree at a time before moving to others. As a consequence, each one of the trees corrects or minimizes the error that occurred in previous trees, thereby improving the mistakes/errors iteratively. This led GBT to construct a strong predictive model. This method of learning from previous productions (i.e., minimizing error) will help in refining predictions, thereby enabling GBT to capture complex relationship within the data. Furthermore, in comparison with RF, GBT is less prone to overfitting of data, thus minimizing the overestimation and making it suitable for various regression tasks, the computation resources required are high though.

3.1.3 Gaussian Process Regression (GPR)

Gaussian process regression (GPR) facilitates the deviation from the conventional output approach associated with other ML models. GPR provides probability distributions for all outputs rather than providing precise value for a parameter [41, 45]. GPR encompasses complex probability distributions that further allow them for predicting discernable confidence levels associated with each prediction parameters, thus making them suitable for informed decision-making in uncertain situations. All the above observations corroborate with the fact that GPR reduces the noise as a part of their inherent ability to capture the uncertainties in data. This further enhances reliability with limited or incomplete data. Note that with all the above features, GPR induces computational complexities, as the dataset sizes are large. Nevertheless, GPR is still recognized as an important machine learning domain.

3.1.4 Neural Network (NN)

Neural network (NN) represents the foundation of deep learning methodologies [41, 46]. NN mainly consists of interconnected neurons that further communicate with each other. NN comprises of input layers, hidden layers, and output layers during the AI model predictions. The dataset training in NN mainly entails integrative process of identifying and adjusting the weights via backprop agitation. With continuous iterations, NN facilitates minimization in loss function, thus the predictive model converges to a global minimum. With this it is possible to handle range of large and diverse datasets as well as their capacity to predict the hierarchical representation of data and accurate model. One of the intriguing features of NN is that they can also perform the task such as image analysis (e.g., fracture image analysis in case of AM parts) and speech recognitions with precise accuracy. Like GBT and GPR, training the datasets in NN requires high computational resources with substantial amounts of labeled data. However, NN is still attracting the considerable attention in many applications, thereby running on the forefronts of the technological advancements in the field.

3.1.5 Extreme Gradient Boosting (XGBoost)

Extreme gradient boosting (XGBoost) is another variant in gradient boosting algorithms that further features regularization technique for the improved accuracy of training datasets. Compared to traditional GBT, XGBoost offers superior speed, high performance, and the ability to conduct the parallel tree boosting. Notably, XGBoost offers high efficiency, therefore suitable for large datasets and intense computational tasks. Akbari et al. [41] have employed the XGBoost for various datasets obtained from different AM-processed materials and demonstrated its feature importance in SHapely Additive exPlanation (SHAP). The versatility of XGBoost lies in the fact that it simplifies the decision-making process even with large datasets, of course with the computation cost on bit higher side.

3.2 Mechanical Property Prediction of MAM Parts Using AI

In this section, we will leverage the techniques of ML that can analyze the dataset, and generate the correlation between processing parameters, microstructure, and desired mechanical properties. In addition, ML techniques can also predict the combined effect of multiple processing routes and extrapolate beyond existing experimental data. The identification of optimal processing parameters for desired properties provides better accuracy for the prediction of mechanical properties [41]. The ML techniques also allow the real-time auto detection of defects and process optimization.

Recognizing the potential of ML, many studies are now exploring the possibility of mechanical property prediction of MAM parts. Zhan et al. [47], in case of laser DED-processed SS 316L stainless steel, have demonstrated the fatigue life prediction model using data-driven modeling considering the parameters such as laser power, scan speed, and hatch spacing. The model prediction closely aligned with the experimental data. Furthermore, Xie and co-workers [48] have devised a framework for the prediction of location-dependent properties in Inconel 718 alloy using CNN. The predicted framework reflects the $R^2 \sim 0.7$, thereby indicating decent accuracy of ML models in predicting the mechanical properties. Compared to the accuracy of CNN model explored by Xie and co-workers [48], the σ_{ys} prediction of laser DED-processed SS 316L stainless steel with $R^2 \sim 0.84$ using XGBoost and ridge regression reflects in better accuracy [49]. The surface roughness of wire arc additive manufactured (WAAM) mild carbon steel ER70S – 6 was successfully predicted using different ML models by Xie at al. [50] with $R^2 \sim 0.93516$. Although the above studies have effectively leveraged ML models in predicting mechanical properties of MAM parts, the experimental datasets used to train and test the data are limited. The review paper by Akbari and co-workers [41] includes the studies that have benchmarked the uses of more extensive datasets collected from the individual MAM-processed materials with around 230 sources of experimental data. This eventually helps in creating a large repository (approximately with 1600 experimental data points that contain various MAM process and processing parameters) for predicting the mechanical properties that are governed by the wide spectrum of process parameters. The features that are obtained after benchmarking the ML models in MAM process are presented in Table 2.

Table 2 Summary of the parameters for featurization for benchmarking ML model [41].

Processing Parameters	*Materials Properties*	*Categorical Features*
Beam power	Density	Material
Layer thickness	Melting point Specific heat Thermal conductivity Coefficient of thermal expansion	Machine type Specimen orientation MAM process MAM subprocess Post-processing conditions

The crucial step while implementing the ML model in mechanical property prediction of MAM parts is the selection of set of features as input parameters. The first set of featurization parameters includes process variables such as beam power and layer thickness. Thus far the majority of the ML models are implemented over PBF and DED along with the respective physical properties. The extent to this is the categorical features that include material, machine types, MAM process details, and post-processing conditions. Considering the above features, some existing review articles focus on the application of ML algorithms in microstructure and mechanical property prediction of AM parts that further include properties and topology prediction [51, 52], materials design and processing [53, 54], AM process optimization [55], and effect of intrinsic defects on microstructure evolution [56, 57]. For instance, a review article by Toprak and co-workers [58] highlights the mechanical properties prediction of PBF-processed metals using different ML models, while another article by Abdelhamid et al. [59] shed light on applications of ML models in process–structure–property relationship of material extrusion AM metals. The application of ML models in mechanical property prediction is endless considering the combinations of materials used, AM processes adopted, its relevant parameters optimization, microstructure evolution, and defect detection. In the following section, we highlight such studies in detail according to the AM process adopted.

3.2.1 Mechanical Properties Prediction Using AI–ML of PBF-Processed Metals

In this section, we highlight some of the important case studies in mechanical property prediction using ML algorithms in PBF-processed metals. The study of Lesko and co-workers [60] shed light on the regression-based ML models such as ordinary least squares and partial least squares to predict the Vickers hardness in Ni-based superalloy 718. They used the above ML models for establishing a relation between process parameters laser power, scan speed, and layer thickness with the Vickers hardness, indicating the correlation between experimental and trained data from the models. Eshikabilov et al. [61] in case of laser-PBF-processed AM parts using K-nearest neighbor (KNN) and decision tree (DT) algorithms have demonstrated that laser power, laser energy, and scan speed have marked influence on mechanical properties of MAM parts. The mechanical property predictions such as local strains, plastic anisotropy, and tensile properties (mainly failure load during tensile deformation) of selective laser melted light weight high strength AlSi10Mg alloy have been demonstrated by Ali et al. [62], demonstrating that the process parameters such as print direction, porosity volume fraction, and size of pores have a marked influence on above mechanical properties. The application of multiple linear regression (MLR) has been demonstrated by Kusano et al. [63] for characterizing and predicting the microstructural defects and tensile properties of heat-treated SLM Ti – 6Al – 4V alloys. Having microstructural data collected from scanning electron microscope (SEM) and defects characterized from X-ray computed tomography (XCT), their findings mainly evolved in correlating α grain size and micro colonies with σ_{ys} and σ_{UTS} with an error less than 2%. The role of ML models is not only limited to predicting the accurate mechanical properties but also to refine the predictions (or minimize the errors). For instance, Wang and co-workers [64] have implemented KNN and DT techniques to optimize the process parameters in EBP-processed Ti – 6Al – 4V alloy for the accurate prediction of mechanical properties. It was found that there was a 5% improvement in σ_{UTS} after optimizing the process parameters through above ML models.

Post-processing fabrication treatments have pronounced influence on microstructure evolution that subsequently affects the mechanical properties of material. Therefore, it is of vital importance to study, analyze, train, and test the datasets related to microstructure evolution after post-processing treatments and their effect on mechanical properties. For instance, Wang et al. [65] have demonstrated the effect of annealing temperatures on the microstructure development of AM Ti – 6Al – 4V alloy and subsequently prediction of their mechanical properties. The accuracy of the predicted mechanical properties has a marked influence on microstructural data. They obtained a $R^2 \sim 0.8$ without incorporating microstructural data, while it was $R^2 \sim 0.9$ in case of microstructural data as one of the inputs during the implementation of ML models. The implementation of CNN-based deep learning model by Johnson et al. [66] with porosities (modeled from finite element simulations) as inputs was found useful in predicting critical peak load (accuracy ~ 84.2%), critical displacement (accuracy ~ 84.2%), equivalent plastic strain with critical peak load (accuracy ~ 89.2%), and equivalent plastic strain with critical peak displacement (accuracy ~ 91.2%). Besides this, an article by Baturynska [67] reports the effect of sample orientation and positioning during AM on mechanical properties of MAM parts, indicating the efficacy of DT and GBR models.

All the above studies corroborate that the efficacy of the ML models depends on the model, type of input data, and material used to fabricate the part. There is a growing need for the common data repository to streamline the mechanical properties prediction in PBF-processed parts. Table 3 summarizes the application of each ML model for the mechanical property prediction of PBF-processed parts along with the type of used input data.

3.2.2 Mechanical Properties Prediction Using AI–ML of DED-Processed Metals

The effect of thermal histories obtained from computation models on the mechanical properties of Ni-based Inconel 718 alloy processed via DED using CNN ML algorithm has been systematically studied by Fang and co-authors [69], demonstrating the training and testing score of 0.96 and 0.67, respectively for the prediction of σ_{UTS}. Chingilipalli and co-workers [70] have demonstrated the prediction of tensile deformation behavior of WAAM-processed Inconel 825 alloy using DT, KNN, and RF models and applied load, build orientation, and material positioning as input parameters. All the models exhibited excellent accuracy in prediction of tensile properties. The process parameters in DED process such as layer height, scan speed, laser power, and energy density markedly affect the tensile properties of SS 316L stainless steel [71], also demonstrating that the accuracy of XGBoost is better than the RF models for the prediction of σ_{ys} (~ 76%) and σ_{UTS} (~ 71%). The optimization of input process parameters will always lead to an increase in accuracy of training and testing the data. Barik et al. [72] in case of WAAM-processed low carbon steel showed that optimizing the input process parameters before using them as inputs support vector regression mode increased the accuracy marginally after optimizing the process parameters.

Although the above studies indicate the efficacy of ML models for predicting the mechanical properties of AM parts with various process parameters as input, the combined effect of process parameters and thermal histories as input data is yet to be investigated in great details. Therefore, in future it would be prudent to focus on predicting the mechanical properties based on process parameters and thermal histories.

3.2.3 Mechanical Properties Prediction Using AI–ML of Metals Fabricated via Other AM Processes

In this section, we mention some studies that have used different ML models to predict the microstructure evolution and mechanical properties of various AM processes other than PBF and DED (mentioned in Section 3.2.1. and 3.2.2.).

Table 3 Summary of implementation of ML models for the mechanical property prediction of PBF-processed parts [68].

ML Model	*Predicted Properties*	*AM Process*	*Material*	*ML Input Data*	*Ref.*
Multiple linear regression	YS, UTS, fracture elongation, elasticity Hardness	SLM	Ti – 6Al – 4V	Microstructural features and defect features	[63]
Lasso regression	Hardness	LPBF	Inconel 718	Process parameters	[60]
KNN, DT, SVM	Relative density, hardness, YS, tensile strength	LPBF	SS 316L	Process parameters	[61]
MLP, SVM, KNN, DT	Tensile strength	EBM	Ti – 6Al – 4V	Process parameters	[64]
DTR and its derivative, ETR	Density, tensile strength, hardness	LPBF	AlSi10Mg	Printing parameters	[77]
MLP	Tensile strength	LPBF	Ti – 6Al – 4V	Annealing temperatures, microstructural features	[65]
TL-ANN	UTS, YS, and relative density	SLM	Aluminum	Process parameters, properties of powder bed	[78]
GPR	Tensile properties	LPBF	TICN/AlSi10Mg	Process parameters	[79]
CNN	Tensile strength	LPBF	AlSi10Mg	Load and displacement data	[66]
CNN	UTS, yield stress, failure stress, modulus	DED	Inconel 718	Thermal history data	[69]
CNN	Tensile strength	DED	Inconel 718	Thermal history data	[80]
LR, SVM, KNN, DT, RF	Tensile strain	WAAM	Inconel 825	Location and orientation of samples, applied stress	[70]
XGBoost, RF	YS, UTS, and elongation	DED	SS 316L	Process parameters	[71]
SVR	Tensile strength	WAAM	ER70S6 steel	Process parameters	[72]

The study by Guetal et al. [73] reports on using microstructural unit cells as an input for predicting the mechanical properties of AM parts via CNN. Interestingly, they have also validated the model design via 3D printing. In the same vein, Herriott and Spear [74] used morphological and microstructural images as input to predict the tensile properties of MAM-processed SS 316 L stainless steel. On this front, they have used CNN model on VGGNet. In the studies reported by Yan et al. [75], the Gaussian process metamodels are used to train the datasets with composition and process parameters as inputs, while the work of Hu et al. [76] mainly investigates the effect of microstructural features such as micro defects and internal voids (porosities), and surface topographies on the mechanical properties of AM parts. Table 4 summarizes the implementation of ML models in the mechanical property prediction of AM parts using the processes other than PBF and DED, thereby adding the information to readers knowledge.

Table 4 Summary of implementation of ML models for the mechanical property prediction of DED and other AM-processed parts [68].

ML Model	*Predicted Properties*	*AM Process*	*Material*	*ML Input Data*	*Ref.*
ANN	Stress–strain behavior	Multiple processes	Multiple materials	Unit cell size, relative density, the topology of lattice structures, etc.	[62]
GBR	Stiffness and strength	Multiple processes	Multiple materials	Variables related to geometries of the beam, defects	[76]
Ridge, XGBoost, CNN	Yield strength	Metal AM	SS 316L stainlesss steel	Microstructural features	[74]
GPR	Yield strength	Metal AM	PH48S	Alloy composition and process parameters	[75]

4. Present Day Challenges and Futuristic Roadmaps

All the studies mentioned above hint toward the growing applications of AI tools and ML algorithms in AM, especially MAM parts for optimization of process parameters, prediction of microstructure, and mechanical properties [81, 82]. It is also notable here that ML-based algorithms are being used in all MAM techniques that typically include PBF, DED, materials extrusion, and BJ. In this section, we highlight some of the open important issues related to mechanical property prediction of MAM parts using AI–ML. This will eventually help readers to identify the unanswered key questions in the field that will further open up ample avenues of futuristic research work.

First and foremost, several studies are focused on the application of AI–ML in mechanical property prediction of MAM parts that are fabricated from a particular group of material such as steel (ER70S6), SS316, Ti – 6Al – 4V alloys, Inconel 718 super alloy, and TiCN; however, other engineering materials such as Mg and Mg alloys, Al and Al alloys, and other class of Ni-based superalloys that are fabricated via AM need further exploration. This ultimately opens up future prospects in thorough research in process parameter optimization, microstructure, and mechanical property prediction in above alloys using AI–ML. Furthermore, AI–ML approach implemented in above studies accurately predicts the different process parameters such as scan speed, laser power, hatch spacing, layer thickness, and laser spot size. The extent to this is the microstructure prediction, topography of the surface, and design properties. Following this, the above-mentioned studies also include prediction of mechanical properties of MAM parts such as compressive strength, σ_{ys}, ultimate tensile strength (σ_{UTS}), H, impact strength, and stiffness. The thermal histories of the sample during its processing and post-fabrication treatments play a crucial role in tailoring microstructural features and consequently the mechanical properties. Therefore, keeping this in view, the role of thermal histories and defect formation on the microstructure prediction using AI–ML should be addressed carefully in future, thereby posing a great demand for their subsequent effects on mechanical properties.

There is also a pressing need of the comparison of mechanical properties prediction of the same MAM parts using different AI–ML algorithms. Most of the ML models predict tensile strength of the MAM parts to a greater extent; however, there is an increasing demand of σ_{ys} and σ_{UTS} prediction of the same parts using these ML algorithms. In addition to this, the present literature focuses mostly on mechanical property prediction of MAM parts fabricated using PBF and BJ, thereby demanding the applicability of ML models for microstructure and mechanical property prediction of parts fabricated via DED. In the same vein, most of the studies are centered on using ML algorithms to predict the correlation between process parameters and mechanical properties, while the scarcity in research studies correlating microstructure with mechanical properties using ML tools pins toward the pressing demand in current scenario, as the microstructural features have pronounced influence on mechanical properties. As on date, only a single study attempted correlating process parameters, microstructure, and mechanical property prediction using AI–ML; however, the authors have used surface data instead of microstructural data due to the limitations on acquiring microstructures and high computational cost. Furthermore, the applications of AI–ML algorithms are limited by high computational cost and expensive data-acquiring systems. This poses a great challenge for the development of algorithms that are comparatively cheaper than the existing ones. The data required for training and testing the model are acquired from the experiments such as universal testing machines, hardness testers, infrared cameras, and different microscopy techniques, which are time-consuming and expensive. Computational and numerical simulation may help up to some extent that further has potential to generate the data; however, the high computational cost limits their uses. Alternative data-generation techniques such as data augmentation, transfer learning, few-shot, and one-shot learning are being explored nowadays to alleviate these problems.

5. Summary

This chapter explores the integration of AI and ML techniques in the mechanical property prediction of MAM parts. In recent years, MAM has gained a significant attention in various industrial sectors such as aerospace, healthcare, and defense due to rapid fabrication of complex, customized, and lightweight metal component through layer-by-layer deposition. On this front, various AM processes typically include PBF, DED, BJ, and materials extrusion utilizing different laser and energy sources. With this, a variety of metals ranging from steels, stainless steels, Ti, and Ni superalloys to high entropy alloys offer distinct advantages for high throughput applications.

Despite the above benefits, MAM parts are inherently prone to defects such as porosity, voids, and micro cracking that further affects the mechanical properties, thereby raising the operational reliability issues. To date, although the efforts are going on to tailor the mechanical properties of MAM parts via altering or optimizing the process parameters and post-fabrication treatments, the nonlinear relationship between process parameters, several combinations, and complexities associated with each process limits the experimental techniques, thus demanding more advanced prediction approaches.

The integration of AI and ML tools in microstructure evolution and mechanical property prediction seems to be promising and one of the key solutions to alleviate the above challenges, thereby enabling the data-driven analysis of extensive datasets obtained from previous experiments and simulations. On this front, the ML algorithms such as RF, GBT, GPT, NN, and XGBoost have been successfully implemented to predict the microstructure evolution and mechanical properties such as H, σ_{ys}, σ_{UTS}, and fatigue life. The input data for training and testing the datasets in the above are main AM process parameters followed by post-processing (treatment) parameters. One of the advantages of this ML-based approach is that it also facilitates the real-time defect detection and process control, thereby allowing dynamic optimization during the manufacturing process.

In view of this, the present chapter presents comprehensive outlook on the studies that report the efficacy of AI–ML tools in mechanical properties prediction of various MAM parts. It was observed that mechanical properties of MAM parts have pronounced influence on process parameters and all the predicted values through ML models closely corroborate with experimental data within the nominal error limit. For instance, fatigue life prediction in PBF-processed SS 316L stainless steel exhibited strong correlation with process parameters such as scan speed, layer thickness, and hatch spacing. Furthermore, it is worth mentioning here that CNN and XGBoost exhibit the higher accuracy than other ML models for forecasting the location-dependent mechanical properties in WAAM process. One of the key contributions in the literature sheds light on development of comprehensive dataset composing over 1600 data points that mainly cover various materials, processing methods, and parameters that subsequently enhance model training and reliability.

Besides this, this chapter also addresses the customization of ML models to specific MAM process in predicting the mechanical properties such as stress–strain behavior, fatigue life prediction, and failure modes. It also highlights the current limitations that can be further viewed as future research opportunities. Thus far most of AI–ML models have been implemented on MAM parts such as steels, Ni, and Ti alloys while research on other metals remain scarce. Furthermore, tensile strength prediction dominates the field with all other properties such as impact resistance and fracture behaviour remain undisclosed. There is also a notable gap in incorporating microstructural features as input data though they have pronounced influence on mechanical properties.

In summary, this chapter underscores the transformative potential of AI and ML in enhancing process optimization, defect prediction, and mechanical property prediction in MAM parts. Further advancement such as wide material coverage, integration of microstructural data, and development of computationally efficient models are indeed required. Improved data acquisition, augmentation techniques, and transfer learning approaches are expected to address the current limitations, thus creating ample pathways to drive the future generation innovations in MAM.

References

[1] Milewski, J. (2017). A book on "Additive Manufacturing of Metals From Fundamental Technology To Rocket Nozzles, Medical Implants, and Custom Jewellery". Springer, Cham, Switzerland.

[2] Toyserkani, E., Sarker, D., Ibhadode, O., Liravi, F., Russo, P. and Taherkhani, K. (2021). A book on "Metal Additive Manufacturing". John Wiley and Sons, Hoboken, NJ, USA.

[3] ASTM F2792-12a. (2012). ASTM standard for Standard Terminology for Additive Manufacturing Technologies.

[4] Alifui-Segbaya, F., Bowman, J., White, A., George, R., Fidan, I. and Love, R. (2020). Chemical characterization of additively manufactured methacrylate for dental devices. Additive Manufacturing, 31, 100944.

[5] Hasanov, S., Gupta, A., Nasirov, A. and Fidan, I. (2020). Mechanical characterization of functionally graded materials produced by the fused filament fabrication process. Journal of Manufacturing Processes, 58, 923–935.

[6] Putra, N., Mirzaali, J., Apachitei, I., Zhou, J. and Zadpoor, A. (2020). Multi-material additive manufacturing technologies for Ti-, Mg-, and Fe-based biomaterials for bone substitution. Acta Biomaterialia, 109, 1–20.

[7] Lewandowski, J. and Seifi, M. (2016). Metal additive manufacturing: A review of mechanical properties. Annual Review of Materials Research, 46, 151–186.

[8] Frazier, W. (2014). Metal additive manufacturing: A review. Journal of Materials Engineering and Performance, 23, 1917–1928.

[9] Dutta, B. and Froes, F. (2014). Additive manufacturing of titanium alloys. Advanced Materials Research, 1019, 19–25.

[10] Seifi, M., Salem, A., Beuth, J., Harrysson, O. and Lewandowski, J. (2016). Overview of materials qualification need for metal additive manufacturing. Journal of Metals, 68, 747–764.

[11] Rometsch, P., Zhu, Y., Wu, X. and Huang, A. (2022). Review of high-strength aluminium alloys for additive manufacturing by laser powder bed fusion. Materials & Design, 219, 110779.

[12] Gu, D., Wang, H., Dai, D., Yuan, P., Meiners, W. and Poprawe, R. (2015). Rapid fabrication of Al-based bulk-form nanocomposites with novel reinforcement and enhanced performance by selective laser melting. Scripta Materialia, 96, 25–28.

[13] Taminger, K. and Hafley, R. (2003). Electron beam freeform fabrication: A rapid metal deposition process. In Proceedings of Annual Automotive Composite Conference, 3rd Ed., 9–10.

[14] Li, X., Reynolds, A., Cong, B., Ding, J. and Williams, S. (2015). Production and properties of a wire-arc additive manufacturing part made with friction extruded wire. In TMS Proceedings, 445–452.

[15] Buican, G., Oancea, G., Lancea, C. and Pop, M. A. (2015). Some considerations regarding micro hardness of parts manufactured from 316-L steel using SLM technology. Applied Mechanics and Materials, 760, 515–520.

[16] Sun, G., Zhou, R., Lu, J. and Mazumder, J. (2015). Evaluation of defect density, microstructure, residual stress, elastic modulus, hardness and strength of laser-deposited AISI 4340 steel. Acta Materialia, 84, 172–189.

[17] Colegrove, P., Coules, H., Fairman, J., Martina, F. and Kashoob, T. (2013). Microstructure and residual stress improvement in wire and arc additively manufactured parts through high-pressure rolling. Journal of Materials Processing Technology, 213, 1782–1791.

[18] Murr, L., Martinez, E., Pan, X. and Gaytan, S. (2013). Microstructures of Rene 142 nickel-based superalloy fabricated by electron beam melting. Acta Materialia, 61, 4289–4296.

[19] List, F., Dehoff, R., Lowe, L. and Sames, W. (2014). Properties of Inconel 625 mesh structures grown by electron beam additive manufacturing. Materials Science and Engineering: A, 615, 191–197.

[20] Harrison, N., Todd, I. and Mumtaz, K. (2015). Reduction of micro-cracking in nickel superalloys processed by selective laser melting: A fundamental alloy design approach. Acta Materialia, 94, 59–68.

[21] Zhang, Y., Cao, X., Wanjara, P. and Medraj, M. (2014). Tensile properties of laser additive manufactured Inconel 718 using filler wire. Journal of Materials Research, 29, 2006–2020.

[22] Facchini, L., Magalini, E., Robotti, P. and Molinari, A. (2009). Microstructure and mechanical properties of Ti-6Al-4V produced by electron beam melting of pre-alloyed powders. Rapid Prototyping Journal, 15, 171–178.

[23] Gong, H., Rafi, K., Gu, H., Starr, T. and Stucker, B. (2014). Analysis of defect generation in Ti-6Al-4V parts made using powder bed fusion additive manufacturing processes. Additive Manufacturing, 1, 87–98.

[24] Challis, V., Xu, X., Zhang, L., Roberts, A., Grotowski, J. and Sercombe, T. (2014). High specific strength and stiffness structures produced using selective laser melting. Materials & Design, 63, 783–788.

[25] Qiu, C., Ravi, G., Dance, C., Ranson, A., Dilworth, S. and Attallah, M. (2015). Fabrication of large Ti-6Al-4V structures by direct laser deposition. Journal of Alloys and Compounds, 629, 351–361.

[26] Wang, F., Williams, S., Colegrove, P. and Antonysamy, A. (2013). Microstructure and mechanical properties of wire and arc additive manufactured Ti-6Al-4V. Metallurgical and Materials Transactions A, 44, 968–977.

[27] Yao, J., Suo, T., Zhang, S., Zhao, F., Wang, H., Liu, J., Chen, Y. and Li, Y. (2016). Influence of heat-treatment on the dynamic behavior of 3D laser-deposited Ti-6Al-4V alloy. Materials Science and Engineering: A, 677, 153–162.

[28] Dixit, T., Sahu, P., Jonnalagadda, K. and Prasad, K. (2021). Effect of powder layer thickness and scan orientation on the deformation and failure of selectively laser melted Ti-6Al-4V alloy over six decades of strain rates. Materials Science and Engineering: A, 822, 141656.

[29] Carlton, H., Haboub, A., Gallegos, G. and Parkinson, D. (2016). Damage evolution and failure mechanisms in additively manufactured stainless steel. Materials Science and Engineering: A, 651, 406–414.

[30] LeBrun, T., Nakamoto, T., Horikawa, K. and Kobayashi, H. (2015). Effect of retained austenite on subsequent thermal processing and resultant mechanical properties of selective laser melted 17-4 PH stainless steel. Materials & Design, 81, 44–53.

[31] Kempen, K., Yasa, E., Thijs, L., Kruth, J. and van Humbeeck, J. (2011). Microstructure and mechanical properties of selective laser melted 18Ni-300 steel. Physics Procedia, 12, 255–263.

[32] Gorsse, S., Hutchinson, C., Gouné, M. and Banerjee, R. (2017). Additive manufacturing of metals: A brief review of the characteristic microstructures and properties of steels, Ti-6Al-4V and high entropy alloys. Science and Technology of Advanced Materials, 18, 584–610.

[33] Attar, H., Calin, M., Zhang, L., Scudino, S. and Eckert, J. (2014). Manufacture by selective laser melting and mechanical behavior of commercially pure titanium. Materials Science and Engineering: A, 593, 170–177.

[34] Barbas, A., Bonnet, A., Lipinski, P., Pesci, R. and Dubois, G. (2012). Development and mechanical characterization of porous titanium bone substitutes. Journal of the Mechanical Behavior of Biomedical Materials, 9, 34–44.

[35] Campbell, F. (2012). Lightweight materials. ASM International, Materials Park, p. 231.

[36] Boyer, R., Welsch, G. and Collings, E. (1994). Materials Properties Handbook- Titanium alloys. ASM International, Materials Park, 1051–1057.

[37] ASTM F1472-14. (2014). Standard specification for wrought titanium-6aluminum-4vanadium alloy for surgical implant applications (UNS R56400). ASTM International, West Conshohocken, PA.

[38] ASTM F1108-14. (2014). Standard specification for titanium-6aluminum-4vanadium alloy castings for surgical implants (UNS R56406). ASTM International, West Conshohocken, PA.

[39] Murr, L., Esquivel, E., Quinones, S., Gaytan, S., Lopez, M., Martinez, E., Medina, F., Hernandez, D., Martinez, E., Martinez, J., Stafford, S., Brown, D., Hoppe, T., Meyers, W., Lindhe, U. and Wicker, R. (2009). Microstructures and mechanical properties of electron beam-rapid manufactured Ti-6Al-4V biomedical prototypes compared to wrought Ti-6Al-4V. Materials Characterization, 60, 96–105.

[40] Xu, W., Brandt, M., Sun, S., Elambasseril, J., Liu, Q., Latham, K., Xia, K. and Qian, M. (2015). Additive manufacturing of strong and ductile Ti-6Al-4V by selective laser melting via in situ martensite decomposition. Acta Materialia, 85, 74–84.

[41] Akbari, P., Zamani, M. and Mostafaei, A. (2024). Machine learning prediction of mechanical properties in metal additive manufacturing. Additive Manufacturing, 91, 104320.

[42] Watring, D., Benzing, J., Hrabe, N. and Spear, A. (2020). Effects of laser-energy density and build orientation on the structure-property relationship in as-built Inconel 718 manufactured by laser powder bed fusion. Additive Manufacturing, 36, 101425.

[43] Mehrabi, A., Zurob, H. and McDermid, J. (2024). Process maps for prediction austenite fraction (vol. %) in medium Mn-third generation advanced high strength steels. Materials, 17, 993.

[44] Friedman, J. (2001). Greedy function approximation: A gradient boosting machine. Annals of Statistics, 29, 1189–1232.

[45] Rasmussen, C. (2003). Gaussian processes in machine learning. In Summer School on Machine Learning. Springer, 63–71.

[46] Hinton, G., Osindero, S. and Teh, Y. (2006). A fast learning algorithm for deep belief nets. Neural Computation, 18, 1527–1554.

[47] Zhan, Z. and Li, H. (2021). Machine learning based fatigue life prediction with effects of additive manufacturing process parameters for printed SS 316L. International Journal of Fatigue, 142, 105941.

[48] Xie, X., Bennett, J., Saha, S., Lu, Y., Cao, J., Liu, W. and Gan, Z. (2021). Mechanistic data-driven prediction of as-built mechanical properties in metal additive manufacturing. Computational Materials, 7, 1–12.

[49] Herriott, C. and Spear, A. (2020). Predicting microstructure-dependent mechanical properties in additively manufactured metals with machine- and deep-learning methods. Computational Materials Science, 175, 109599.

[50] Xia, C., Pan, Z., Polden, J., Li, H., Xu, Y. and Chen, S. (2021). Modelling and prediction of surface roughness in wire arc additive manufacturing using machine learning. Journal of Intelligent Manufacturing, 1–16.

[51] Era, I. (2021). Prediction of tensile behaviors of L-DED 316 stainless steel parts using machine learning [Graduate Theses, Dissertations, and Problem Reports].

[52] Hurtado-Pérez, A., Pablo-Sotelo, A. de J., Ramírez-López, F., Hernández Gómez, J. and Mata-Rivera, M. (2023). On topology pptimisation methods and additive manufacture for satellite structures: A Review. Aerospace, 10, 1025.

[53] Zhou, H., Yang, H., Yu, S., Shi, J., Gao, P. and Yu, B. (2024). Advancements in machine learning for material design and process optimization in the field of additive manufacturing. China Foundry, 21, 1–15.

[54] Mattera, G., Nele, L. and Paolella, D. (2024). Monitoring and control the wire arc additive manufacturing process using artificial intelligence techniques: A Review. Journal of Intelligent Manufacturing, 35, 467–497.

[55] Zhang, Y. and Yan, W. (2022). Applications of machine learning in metal powder-bed fusion in-process monitoring and control: status and challenges. Journal of Intelligent Manufacturing, 34, 2557–2580.

[56] Selot, A. and Dwivedi, R. (2023). Machine learning and sensor-based approach for defect detection in MEX additive manufacturing process—A review. Journal of the Brazilian Society of Mechanical Sciences and Engineering, 45, 1–25.

[57] Gunasegaram, D., Barnard, A., Matthews, M., Jared, B., Reaco, A., Bartsch, K. and Murphy, A. (2024). Machine learning-assisted in-situ adaptive strategies for the control of defects and anomalies in metal additive manufacturing. Additive Manufacturing, 81, 104013.

[58] Toprak, C. and Dogruer, C. (2023). A critical review of machine learning methods used in metal powder bed fusion process to predict part properties. International Journal of Precision Engineering and Manufacturing, 25, 429–452.

[59] Abdelhamid, Z., Mohamed, H. and Kelouwani, S. (2024). The use of machine learning in process–structure–property modeling for material extrusion additive manufacturing: A state-of-the-art review. Journal of the Brazilian Society of Mechanical Sciences and Engineering, 46, 1–24.

[60] Lesko, C., Sheridan, L. and Gockel, J. (2021). Microhardness as a function of process parameters in additively manufactured alloy 718. Journal of Materials Engineering and Performance, 30, 6630–6639.

[61] Eshkabilov, S., Ara, I. and Azarmi, F. (2022). A comprehensive investigation on application of machine learning for optimization of process parameters of laser powder bed fusion-processed 316L stainless steel. The International Journal of Advanced Manufacturing Technology, 123, 2733–2756.

[62] Ali, M., Sajjad, U., Hussain, I., Abbas, N., Ali, H., Yan, W. and Wang, C. (2022). On the assessment of the mechanical properties of additively manufactured lattice structures. Engineering Analysis with Boundary Elements, 142, 93–116.

[63] Kusano, M., Miyazaki, S., Watanabe, M., Kishimoto, S., Bulgarevich, D., Ono, Y. and Yumoto, A. (2021). Tensile properties prediction by multiple linear regression analysis for selective laser melted and post heat-treated Ti-6Al-4V with microstructural quantification. Materials Science and Engineering: A, 787, 139549.

[64] Wang, C., Chandra, S., Huang, S., Tor, S. and Tan, X. (2023). Unraveling process-microstructure-property correlations in powder-bed fusion additive manufacturing through information-rich surface features with deep learning. Journal of Materials Processing Technology, 311, 117804.

[65] Wang, H., Li, B., Zhang, W. and Xuan, F. (2024). Microstructural feature-driven machine learning for predicting mechanical tensile strength of laser powder bed fusion (L-PBF) additively manufactured Ti6Al4V alloy. Engineering Fracture Mechanics, 295, 109788.

[66] Johnson, K., Maestas, D., Emery, J., Grigoriu, M., Smith, M. and Martinez, C. (2022). Failure classification of porous additively manufactured parts using deep learning. Computational Materials Science, 204, 111098.

[67] Baturynska, I. (2019). Application of machine learning techniques to predict the mechanical properties of polyamide 2200 (PA12) in additive manufacturing. Applied Sciences, 9, 1060.

[68] Deshmankar, A., Challa, J., Singh, A. and Regalla, S. (2024). A review of the applications of machine learning for prediction and analysis of mechanical properties and microstructures in additive manufacturing. ASME Journal of Computational Science and Engineering, 24, 120801.

[69] Fang, L., Cheng, L., Glerum, J., Bennett, J., Cao, J. and Wagner, G. (2022). Data-driven analysis of process, structure, and properties of additively manufactured Inconel 718 thin walls. Computational Materials, 8, 126.

[70] Chigilipalli, B. and Veeramani, A. (2023). A machine learning approach for the prediction of tensile deformation behavior in wire arc additive manufacturing. International Journal of Interactive Design and Manufacturing, 1–13.

[71] Era, I., Grandhi, M. and Liu, Z. (2022). Prediction of mechanical behaviors of L-DED fabricated SS 316L parts via machine learning. International Journal of Advanced Manufacturing Technology, 121, 2445–2459.

[72] Barik, S., Bhandari, R. and Mondal, M. (2023). Optimization of wire arc additive manufacturing process parameters for low-carbon steel and properties prediction by support vector regression model. Steel Research International, 95, 2300369.

[73] Gu, G., Chen, C., Richmond, D. and Buehler, M. (2018). Bioinspired hierarchical composite design using machine learning: Simulation, additive manufacturing, and experiment. Materials Horizons, 5, 939–945.

[74] Herriott, C. and Spear, A. (2020). Predicting microstructure-dependent mechanical properties in additively manufactured metals with machine- and deep-learning methods. Computational Materials Science, 175, 109599.

[75] Yan, F., Chan, Y., Saboo, A., Shah, J., Olson, G. and Chen, W. (2018). Data-driven prediction of mechanical properties in support of rapid certification of additively manufactured alloys. Computer Modeling in Engineering & Sciences, 117, 343–366.

[76] Hu, E., Seetoh, I. and Lai, C. (2022). Machine learning assisted investigation of defect influence on the mechanical properties of additively manufactured architected materials. International Journal of Mechanical Sciences, 221, 107190.

[77] Minkowitz, L., Arneitz, S., Effertz, P. and Amancio-Filho, S. (2023). Laser-powder bed fusion process optimisation of AlSi10Mg using extra trees regression. Materials & Design, 227, 111718.

[78] Pashmforoush, F. and Seyedzavvar, M. (2023). A transfer learning-based machine learning approach to predict mechanical properties of different material types fabricated by selective laser melting process. Proceedings of the Institution of Mechanical Engineers, Part E: Journal of Process Mechanical Engineering, 09544089231215683.

[79] He, P., Liu, Q., Kruzic, J. and Li, X. (2022). Machine-learning assisted additive manufacturing of a TiCN reinforced AlSi10Mg composite with tailorable mechanical properties. Materials Letters, 307, 131018.

[80] Xie, X., Bennett, J., Saha, S., Lu, Y., Cao, J., Liu, W. and Gan, Z. (2021). Mechanistic data-driven prediction of as-built mechanical properties in metal additive manufacturing. Computational Materials, 7, 86.

[81] Kathavate, V., Eswara Prasad, K., (2025) A chapter on *"Microstructure evolution and mechanical properties of an alloys fabricated via additive manufacturing"* in the book *"Metal*

Additive Manufacturing: Methods, Materials and Applications" Springer, Singapore, 131–162. https://doi.org/10.1007/978-981-96-8162-4_8

[82] Kathavate, V., Wu, S., Zhu, Y., Deshpande, P., (2025) A chapter on *"Additively Manufactured Stimuli Responsive Materials and Structures"* in the book *"Additively Manufactured Smart Materials and Structures: Design, Processing and Applications"* Elsevier, Netherland. https://doi.org/10.1016/B978-0-443-16144-5.00017-X

5 Artificial Neural Network-Based Mechanical Property Prediction for Additively Manufactured Parts

R. Venkatesh[1*], K. Premnath[2] and M. Pradeep[3]

Additive manufacturing (AM), often known as three-dimensional printing, is gaining increasing interest from academia and business because of its clear advantages over traditional subtractive production. Adjusting AM processing settings can be difficult since they can have a significant impact on the printed microstructure and the performance of the final products. Using traditional numerical and analytical models to construct a process–structure–property–performance (PSPP) relationship for AM is a difficult undertaking. Intricate pattern recognition and regression analysis may now be done legitimately using the machine learning (ML) approach without having to actively create and solve the underlying physical models. The neural network (NN) is the most often used model among machine learning algorithms because of its complex algorithm architecture, strong computational capabilities, and extensive dataset. An overview of the usage of artificial neural networks to forecast the mechanical characteristics of additively built parts is given in this chapter. The existing challenges in applying NNs to AM are then discussed, along with potential solutions. Finally, future developments are proposed to provide a broad picture of this multidisciplinary field.

1. Introduction

Ten years ago, the Rapid Prototyping sector gave rise to additive manufacturing (AM), which is now a popular alternative for a range of applications. Unlike traditional subtractive manufacturing technologies, additive manufacturing (AM) is a digital approach that has garnered international interest and has enormous potential for the modern industrial paradigm [1–4]. By fabricating products layer by layer from three-dimensional (3D) computer-aided design (CAD) models, additive manufacturing (AM) provides a number of benefits. It can be used to create products with complex shapes, such as topologically optimal structures, and unique material properties, such as dislocation networks, that are difficult to create with traditional casting or forging methods [5], which academics find very

[1] Assistant Professor, Karpaga Vinayaga College of Engineering and Technology, Chengalpattu, India.
[2] Assistant Professor, Coimbatore Institute of Engineering and technology, Coimbatore. India.
[3] Assistant Professor, Karpagam Institute of Technology, Coimbatore, India.
* Corresponding author: mechanical.ve@gmail.com

DOI: 10.1201/9781003635673-5

intriguing; it also reduces material waste, which decreases industry expenses. AM parts also contain dozens of different defects compared to their cast and wrought equivalents. These include a highly anisotropic microstructure parallel to and perpendicular to the printing direction, porosity from a lack of fusion and gas entrapment, and distortion from large residual stress caused by a high cooling rate and abrupt temperature gradient [6]. Understanding the complex interrelationships among a powder's metallurgical qualities, the printing process, and the mechanical and microstructural properties of AM parts is therefore essential. Several important criteria that are always present in the AM process can be used to determine the performance of the final products. For example, processing parameters such as laser power, scan speed, hatch spacing, and layer thickness have a significant impact on the quality of components made with selective laser melting (SLM). Unfortunately, the relationship between these parameters and the output quality is too complicated to fully understand because SLM is a multi-physics and multi-scale process that involves powder laser interaction at the microscale, melt pool dynamics and columnar grain growth at the mesoscale, and thermal–mechanical coupling at the macroscale. Researchers have tried to develop a number of physical models in an effort to more accurately and clearly classify this relationship. Fergani et al. [8] proposed an analytical model to assess residual stress in the AM process of metallic materials; Chen et al. [9] employed a finite-element model to investigate the shape of the beads and melt pool profiles; and Acharya et al. [7] developed a phase-field framework and computational fluid dynamics (CFD) to model the evolution of grain structure in the as-deposited state for the laser powder-bed fusion (PBF) process. As can be seen, the simulations listed above range from the powder scale to the component scale and only concentrate on one or two parts of the overall process because of a lack of a comprehensive understanding of AM [10]. Many forecasting techniques and models are available to ascertain the ideal limitations in additive manufacturing. The most popular approach for forecasting the best cycle boundaries across a range of materials and manufacturing processes is the use of artificial neural network (ANN) models. This methodology has a great deal of experience in forecasting network borders when the input cycle boundary relationships are very dynamic [11–13].

2. Artificial Neural Network

In recent years, research on artificial intelligence (AI) has been used more and more to represent complicated issues in the physical sciences, engineering, computing, and statistics. Pattern recognition, data categorization, and application-based challenges that are difficult for conventional approaches to address can be resolved in the environment with the help of an ANN, a learning information processing tool [14–16]. One of the best tools for making predictions is an ANN. After learning from the input data, the ANN model made accurate predictions about the future. Similar to human intelligence, the ANN approach works on the same basis. ANN establishes a link between input and output parameters based on

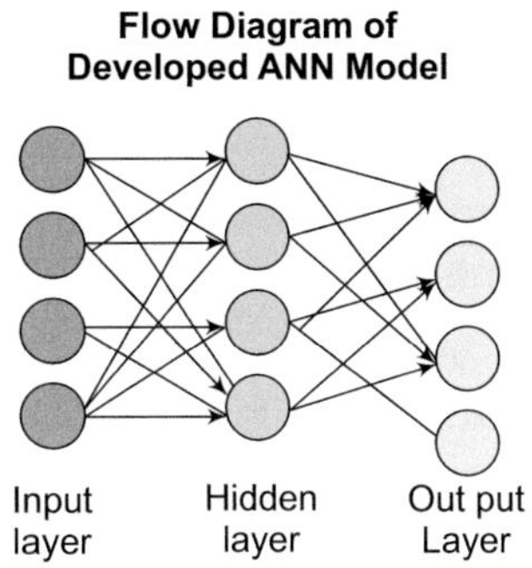

Fig. 1 ANN model diagram [25].

historical patterns and experience [17]. Moreover, depending on the complexity of the parameters, an ANN is composed of several hidden layers and neurons that are connected [18]. Figure 1 shows a vector flow diagram of the developed ANN model. The general structure of an artificial neural network includes the input layer, hidden layers, neurons, weights, activation function, and output layer [19]. To create a relationship between the input parameters and neurons with different activation functions and weights, the ANN algorithms start by choosing random data. The ANN algorithm continuously builds links between interlink neurons of different weights to extract output values from training data. The weight in the ANN algorithm regulates several input parameters on neurons [17]. Weights are compounded by each neuron's input, and the activation function determines the neurons' output [20]. Information is sent to subsequent neurons through a mathematical process known as the activation function. A single neuron cannot function unless the activation function is triggered. Saturated linear, hyperbolic tangent, sigmoid, linear, Gaussian, and other activation functions are used by ANN algorithms [21]. The main issue with ANN models is over-fitting, when the model keeps trying to match the outputs of the training dataset. Over-fitting can happen when more neurons learn relationships from data over an extended period of time. The overfit model worked effectively when used on a known dataset. However, the unidentified dataset cannot be reliably predicted by the over-fitted model [22]. The main application of ANN is in scenarios where the input and output data have complex interactions. The ANN algorithm learns complex associations and helps anticipate data without offering any useful interpretation. It is often referred to as the "Black box" approach because of this. These days, ANN is mostly used in sophisticated applications such as self-driving cars, speech and handwriting recognition, and weather forecasting [17]. In an effort to save money, time, and materials in complicated manufacturing processes, academics have lately begun researching the ANN model. This is especially true for additive manufacturing, since the building component is influenced by a wide range of factors, either directly or indirectly. Researchers have recently reported using the ANN model for the selective laser sintering (SLS) method [23, 24]. The ANN helps optimize process parameters to attain the necessary mechanical properties for newly created materials.

3. Estimating the Mechanical Characteristics of Additively Produced Components

An ANN-based predictive model was created utilizing empirical tensile strength data from test pieces that were additively built using PLA and the fused deposition modelling (FDM) technique. The empirical data from the tensile test results were used to train the neural network that will anticipate future parameter combinations. The results show a good correlation between tensile strength and process parameters, and the neural network can approximate the complicated function needed to forecast tensile strength values for a given set of parameters, with a prediction accuracy of about 93% [26].

The tensile strength of fused deposition models with varying process parameters is predicted using artificial neural networks. The process parameters that were chosen are the infill speed, layer thickness, and nozzle temperature. The Taguchi approach is used to create an L9 orthogonal array. The tensile strength of the polylactic acid specimens made via fused deposition modeling is being experimentally investigated. This dataset is used to train the artificial neural network, which predicts the tensile strength at various degrees of infill speed, layer thickness, and nozzle temperature. Artificial neural network modeling is used to forecast the ultimate tensile strength based on the experimental runs. The following input parameters are used to build the ANN model: layer thickness, nozzle temperature, and infill speed. The ANN module in the MATLAB software is used to forecast tensile strength. A number of confirmatory experiments were carried out in order to validate the ultimate tensile strength model results. The experiment values were in agreement with the expected values when rounded to the closest whole number [27].

The pieces made by the FDM process are measured using coordinate measuring and three-dimensional scanning. To evaluate the influence of different process parameters, a test part shape is designed and manufactured using PETG and PLA materials under varying printing rates, nozzle temperatures, and build orientations. The components' mechanical characteristics and tensile strength are measured. The mean absolute error (MAE) of the training predictions is plotted against the MAE of the predictions generated on the validation data in order to evaluate the efficacy of the tested ANN designs. An objective performance measurement pertinent to real-world scenarios is made possible by the validation data being invisible to the neural network during training, in addition to representing real-world data obtained from separating the dataset between training and validation data. Performance charts comparing the two designs for predicting tensile strength of distinct architectures 3 × 64 and 3 × 32 are presented in the two graphs. The data tend to be overfit by the 3 × 64 model. On the available dataset, the model with the 3 × 32 architecture performs the best. On the available dataset, the model with the 3 × 32 architecture performs the best. The model may develop a predictive and correlative function with the provided input and output parameters, as evidenced by the performance plot of the 3 × 32 architecture, which shows good model generalization and a trend for the MAE to drop [28].

We developed a molecular dynamics simulation-based artificial neural network to model the mechanical behavior of crystalline polymers. Initial configurations are created to simulate PA12 in γ crystal form. We chose ReaxFF for the force field and performed MD simulations under uniaxial and biaxial tensile deformations to generate a dataset for the learning model. The ANN architecture was selected using hyperparameter tuning methods and meticulously calibrated using the training set. The training and validation process revealed that ANN-4 was the best model; it was accurate, dependable, and showed no signs of overfitting. A range of strain rates that exceeded the loading rate range of the training set were used to assess the best model. The results demonstrated the predictive power and generalization capacity of our technique, even when strain rates were much slower or faster than the original data [29]. A Flash Forge Guider II 3D printer was utilized to create the specimens used in this investigation. With a 280 × 250 × 300 mm^3 build envelope, the printer can create items with a 0.2 mm precision level. ABS was chosen in the study since it is a common FDM material and a robust thermoplastic. ABS performs best on components that require strength and flexibility, such as those found in automobiles or appliances. When the error, or the difference between the expected and predicted output, is smaller than a predefined upper bound or when the number of epochs exceeds a predefined threshold, the ANN stops training. A score around 1 denotes a strong link, whereas a value near 0 denotes a random relationship. Regression plots indicated that the optimal fitness was recommended after multiple training sessions. Training, testing, validation, and total data had respective values of 0.98228, 0.85711, 0.92749, and 0.917. This implies that the ANN model's predicted outcomes appear to align with the experimental data. The ANN model demonstrated potential for future use with an average percentage error of 1.6437 based on the trial findings. The ANN model's accuracy ranges from 1% to 5%. This indicates that the experimental and projected results are in good accord [30].

In order to anticipate the surface roughness of specimens of Ti6Al4V alloy that are additively created following electro polishing and blasting, this research designs an ANN. This ANN is then used to determine the parameters of both finishing procedures to achieve a softer surface. The following conclusions are drawn from the results above: The ANN is a useful tool for virtually conducting blasting and electro polishing experiments, with a correlation factor greater than 0.9. The parameters needed to raise the surface roughness by around 60% are given by the optimization algorithm. By analyzing the difficulties, the ANN makes it possible to determine the most crucial parameters and suggests further research to enhance the results. These testing could also improve the ANN itself [31].

The laser additive manufacturing-direct energy deposition (LAM-DED) technology uses an ANN based on MATLAB software to precisely estimate and forecast the track dilution produced in single track deposition of Ti-15Mo alloy. The experimental calibration can be reduced by continuously observing the ANN model's response (dilution) for various sets of input variables. This study will help regulate the dilution for single track deposition, which results in porosity and ultimately distortion in the final multilayer deposited part. All things considered, our work will save a great deal of time and effort in precisely estimating the dilution in DED of Ti-15Mo alloy when compared to FEM or iteration-based simulation models [32].

Lattice structures are very versatile for applications needing specialized performance because they can be designed to maximize particular mechanical qualities. By customizing each unit cell and strut inside the lattice structure, this special ability enables more control over mechanical properties—a degree of flexibility that is not possible with foams and honeycombs. The mechanical performance of 3D-printed lattice structures has been the subject of numerous investigations [33–35]. The orthotropic mechanical response of 3D-printed polylactide (PLA) was examined by Song et al. [36]. They discovered that the direction of loading in relation to the printing orientation greatly affects the mechanical properties, such as toughness and tensile strength. The work highlighted the anisotropic nature of 3D-printed materials by showing that PLA that has been 3D-printed exhibits greater toughness in the extrusion direction as opposed to the transverse direction. These findings demonstrate that in order to effectively forecast the mechanical behavior of lattice structures—particularly those made using additive manufacturing techniques—it is imperative to take into account anisotropic and transversely isotropic models. The mechanical response of metallic pyramidal lattice truss core sandwich structures made with direct metal laser sintering (DMLS) technology was examined by Qi et al. [37]. To assess the stiffness and strength characteristics, the study included shear tests in addition to in-plane and out-of-plane compression tests. The results showed that strut buckling determines the out-of-plane compressive strength, whereas in-plane tests showed plastic yielding of face sheets. Node failure dominated shear performance. The experimental findings were supported by numerical simulations, highlighting DMLS's capacity to create high-performance lattice systems. The analysis of earlier research revealed that expensive and time-consuming methods should be used in order to comprehend and maximize the special mechanical characteristics of these structures. The use of artificial intelligence in engineering has grown in popularity in recent years due to technological advancements. In recent years, researchers have also looked into using artificial intelligence to build lattice structures and forecast their mechanical behavior [38]. Koeppe et al. investigated the use of neural networks for the efficient numerical modeling of 3D-printed lattice-cell structures [39]. In contrast to conventional finite element models, the study showed how to anticipate maximum stresses in lattice structures using neural networks and deep learning, which drastically cuts down on computing time. Alejandrino et al. [40] improved material efficiency by optimizing lattice infill patterns for additive manufacturing using a machine learning technique. The study improved the precision of 3D-printed lattice structures by modeling geometric corrections using ANNs. In comparison to traditional infill designs, the results demonstrated a significant improvement in material efficiency, lowering filament usage by as much as 61.3%. In order to maximize structural design, increase predictive accuracy, and reduce manufacturing defects, recent developments have placed a strong emphasis on combining ANN and finite element methods (FEM) inside AM processes. Through data-driven learning, ANN delivers strong prediction capabilities, whereas FEM enables the exacting mechanical and physical models required for precise simulations. Such integrated approaches were highlighted in a number of recent research. Without explicitly optimizing the hidden layer structure, Vasundara et al. [41] used ANN to analyze elastic deformation. They came to the conclusion that utilizing a single hidden layer was adequate and suggested that the number of input parameters might be equal to the number of neurons in this layer. Their findings provided compelling evidence for the accuracy of ANN techniques. In order to optimize the structural design of agricultural backhoes, Robles et al. [42] used ANN in conjunction with theoretical analysis (TA), finite element analysis (FEA), and response surface methodology (RSM). Their neural network, which had three layers for practical purposes—one for input, one for hidden, and one for output—proved to be quite successful and showed ANN to be an excellent modeling tool. In order to anticipate blast-induced damage on reinforced concrete walls reinforced with fiber-reinforced polymer (FRP), Tu et al. [43] first carried out experimental research. Based on the experimentally collected data, they then developed an ANN architecture. Four input parameters were taken into consideration in the study: wall thickness, FRP layer thickness, explosive weight, and standoff distance. Crater diameter, breaching, spalling, and FRP fracture length were the output metrics utilized to characterize damage in the reinforced concrete and FRP. The created ANN architecture was shown to have a strong prediction capability. Drieschner et al. [44] used an integrated ANN and FEM approach to investigate the mechanical behavior of a three-dimensional carbon fiber-reinforced plastic construction. Experimental results confirmed that the proposed hierarchical FEM–ANN combined method produced satisfactory factory results; nonetheless, it was recommended that the accuracy and efficiency of this method be compared with other numerical approaches. Aydin [45] used both FEM and ML techniques to examine the best machining parameters for Monel 400 superalloy while taking carbon emissions into account. The most efficient machining settings in terms of cutting force were first identified using ANN after FEM analysis. The ANN model demonstrated a 98.081% compatibility with FEM results for cutting force and a 99.978% agreement with carbon emission. In order to optimize the basic machining parameters and protrusion dimensions that impact tool shape degeneration in die-sinking electric discharge machining (EDM), Cogun and Ayli [46] looked into the application of machine learning (ML) approaches. Their study included three distinct machine learning techniques: support vector machines (SVM), adaptive-network-based fuzzy inference systems (ANFIS), and ANN. The ANN approach demonstrated improved prediction capability with a mean relative error (MRE) of 0.854% and a determination coefficient (R2) of 0.99985 when compared with experimental results. In order to minimize tool shape degeneration, four different EDM machining scenarios were provided, and the ANN technique was used to optimize the machining parameters and protrusion dimensions. This study shows how well ANN may improve tool performance and machining parameters in EDM operations. To forecast the mechanical behavior of pyramidal lattice truss core sandwich constructions under bending loads, an ANN model was created. In a much shorter amount of time, the model may be used to estimate the mechanical behavior of such structures during the design phase. The investigation began with the use of FDM technology to fabricate pyramidal lattice truss core sandwiches. Four-point bending tests were then conducted, yielding vital information on failure modes and deformation behavior. FEM was created to incorporate parameters and mimic bending behavior. Parametric models were created when the testing showed comparable behavior. The outcomes of this parametric model were used to construct the ANN model. An effective method for forecasting the mechanical behavior of pyramidal lattice truss core sandwiches is the ANN model. Its usefulness for design and optimization procedures is highlighted by its capacity to precisely replicate FEM results within the training data range. To enhance performance, particularly in interpolated and extrapolated predictions, future research should concentrate on growing the training dataset and improving the ANN architecture. Improved model accuracy and generalization may also result from investigating cutting-edge ANN structures and training methods such as deep learning and ensemble learning. It is also critical to note that the bending behavior can be significantly impacted by the printing direction and material composition. Therefore, this ANN can be enhanced to take these factors into account. The following are some of the findings: The accuracy of the model was confirmed by developed FEM findings, which closely matched experimental data with an error rate of 4.5%. In line with experimental findings, FEM analysis revealed substantial shear stresses at the facesheet/strut contact and bending-dominated behavior. The created ANN showed low errors for response forces and stresses and great accuracy, particularly within the training data range. For smaller networks, Levenberg–Marquardt (LM) offered the quickest convergence and the lowest mean squared error (MSE) of the three algorithms; nevertheless, it came with a high memory cost. For robustness and generalization, Bayesian regularization (BR) is recommended. Although it did not perform better than LM or BR, scaled conjugate gradient (SCG) provided moderate convergence and reduced memory utilization. The ANN model's capacity to extrapolate is still a major drawback, despite its great accuracy within the specified training dataset. When applying the model to unknown conditions, care must be taken because, like with other data-driven models, predictions made outside of the training range may result in notable variances. The model should only be used within the parameter ranges that were covered during training in order to guarantee dependable performance. By increasing the dataset, using hybrid modeling techniques, or including physics-informed constraints to improve generalizability, future research can overcome

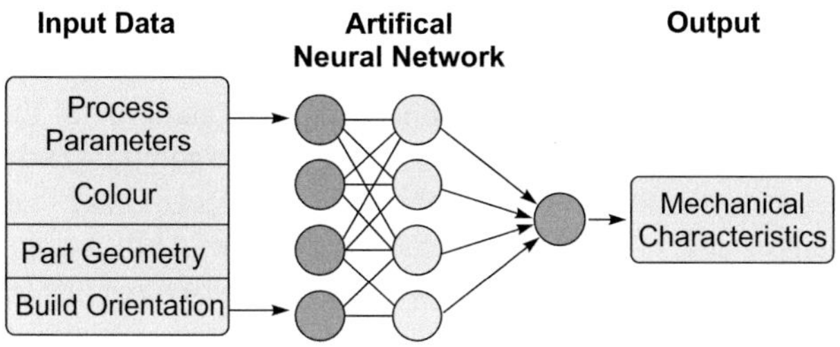

Fig. 2 Estimating the mechanical characteristics of additively produced components by ANN model [26].

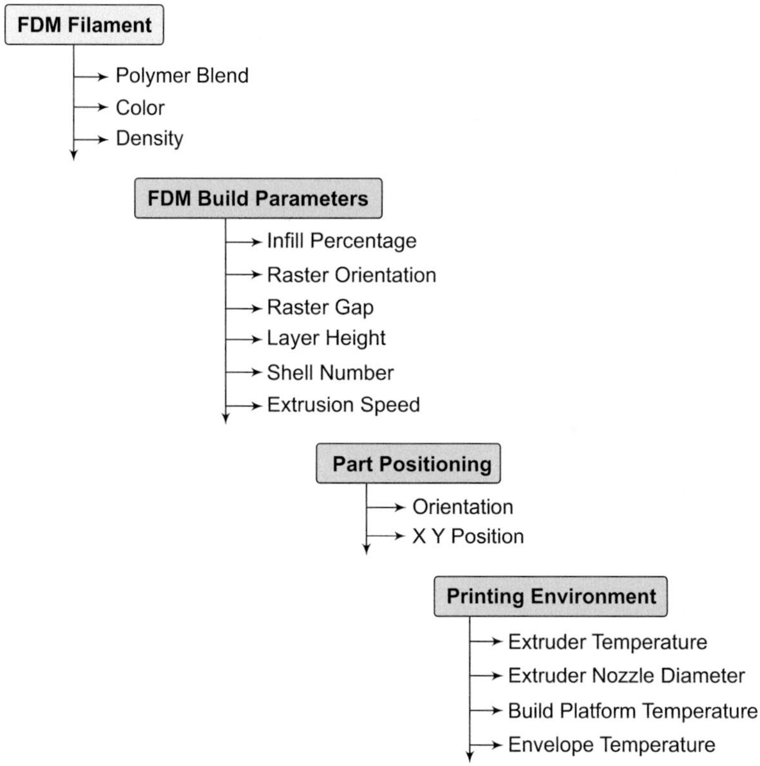

Fig. 3 Factors influencing the qualities of the printed material in the FFF process [47].

this constraint. To help engineers and researchers in real-world applications, clear instructions on how to use the model safely have been provided [47]. A variety of materials are employed in the FFF process as standard-sized filaments. Typically, filaments are made via the extrusion method, which involves melting polymers, pushing them through a nozzle, and then cooling them to create filaments with the desired sizes and shapes. Fillers can be combined with the base material and extruded as a single filament during the extrusion process. Based on the necessary qualities, additives such as fibers and fillers are chosen. Numerous elements, such as the percentage addition of fillers or fibers, the size and type of the fibers and fillers, the pre-processing of the fibers and fillers prior to addition, and many more, affect the final results of the 3D printed object. The impact of laser scanning speed on the mechanical and surface quality characteristics of Al/PLA composites was investigated during the FDM printing process. It was discovered that laser-induced polishing increases tensile strength by 25.1% and decreases surface roughness by 86.6% [48].

Table 1 Findings from adjusting various input parameters on a range of materials [49-55].

Material Type	*Experimental Variables*	*Observed Outcomes*
PLA	Raster angle	45° raster angle yields highest elongation and fracture toughness [49].
ABS	Print speed	Peak fracture toughness achieved at 50 mm/sec print speed [50].
ABS-CF	Raster angle	Raster angles between 0° and 60° enhance fracture toughness [51].
PLA	Raster angle	0° raster angle maintains superior mechanical properties post thermal aging [52].
PLA	Infill density, print speed, nozzle temperature	Optimal tensile strength at 210°C, 124.77 mm/sec, and 100% infill [53].
ABS-Al & PA6-Al	Infill density, nozzle angle	Maximum tensile strength at 100% infill and 60° nozzle angle [54].
ABS	Slice height, fill style, build orientation	Best machining time, surface finish, and hardness at 0.013" slice height, solid fill, and 45° orientation [55].

The impact of printing parameters on delamination, surface roughness, and surface roughness surrounding drilled holes was investigated and confirmed using neural network models. The experimental values were found to be exactly the same as the expected values. The extruder nozzle temperature must be kept at 210°C, the layer heights must be 100 μm, the raster angle must be 90°, and the filling pattern must be hexagonal in order to achieve the lowest possible surface roughness on the normal surface. A line-type filling pattern, a 45° raster angle, 200 μm layer heights, and an extruder nozzle temperature of 210°C are required to achieve the least amount of surface roughness near the drilled hole. A triangle filling pattern, a 45° raster angle, 100 μm layer heights, and an extruder nozzle temperature of 200°C are required to achieve a low delamination factor. When predicting surface roughness, the suggested neural network models have an overall prediction accuracy of 99.80%. The suggested neural network models predict the surface roughness near the hole with an overall prediction accuracy of 99.76%.

When predicting the delamination factor, the suggested neural network models have an overall prediction accuracy of 99.96% [56].

The creation of an ANN model was to assess the correctness of the model and forecast the flexural and compression characteristics of a 3D-printed PLA/CF sample. A 70% training dataset and a 15% testing and validation dataset were used to create the LM and SCG algorithms for the investigation. Although the LM algorithm showed minor overfitting as training went on, it performed better on small to medium datasets, reaching its best validation MSE at epoch 6. Furthermore, optimal results were obtained via the SCG method, which is renowned for its stability and applicability to large-scale issues [57]. Numerous scholars defend the precision of ANN modeling for mechanical behavior and tribology for a range of materials. Here, the neural network's versatility and ability to uncover hidden links come in handy. ANNs are essential for many applications because they reliably produce outcomes that closely match actual observations. ANNs are a convenient solution to problems such as regressions and classifications. In the former case, ANNs show proficiency in modeling and forecasting the values, while in the latter case, they separate data points into distinct groups. The dynamic character of the mating materials, their reactions to operational circumstances, and their interactions are all captured by these networks. Because of their accuracy, ANNs are essential for creating materials that work well in tribological settings. Mechanical, structural, and tribological characteristics all affect how easily the polymers wear. Additionally, loading and operating circumstances have an impact on tribological performance [58, 59].

Processing temperature and cooling rate have an impact on the mechanical and structural properties of the polymer during the fabrication process, such as injection molding, additive manufacturing, etc. The creation of any polymer sample in the context of additive manufacturing is contingent upon a set of printing parameters that can be altered for any given material. With the advent of ANN, it is now possible to estimate the impact of several input parameters and achieve accuracy with nonlinear relationships. The features of the additively manufactured samples differ from those of the standard method of producing polymeric samples or polymer composites [60, 61]. Any part's tribological characteristics are crucial when it is exposed to high loading circumstances that result in significant wear. A spinning shaft in the case of a journal bearing in the system can be supported by the part without producing catastrophic failure. The part's material can be metal, polymer, or polymer composite. To forecast the wear loss of the Al alloy under tribological conditions, the ANN model was created [62–64]. An ANN with back propagation is used in a hybrid regression model to forecast an Al-based composite's wear rate. The wear rate uses the hybrid regression model to compare the predictions of ANN and nonlinear regression with limited experimental data [62]. By adding zircon and carbon, the aluminum sample's wear loss, friction coefficient, and wear rate were predicted using the ANN regression model. It was discovered that the model's output prediction was very accurate [64].

Surface texture and loading circumstances are two of the many variables that affect the prediction of the polymers' tribological and mechanical characteristics. The amount of neurons in the ANN model and the training data gathered from the experiments have an impact on the prediction accuracy. Applications for polymers and their composites are numerous, and bearing applications can benefit from wear prediction [65]. In order to anticipate erosive wear, three polymers were examined using various input parameters. The fraction of the dataset appears to have an impact on the prediction quality or coefficient of determination [66]. In order to forecast the wear rate and friction of polyphenylene sulfide (PPS) composite sliding on a pin-on-disc test setup, Gyurova and Friedrich [67] used ANN. The mean relative error for both wear rate and friction coefficient decreases as the number of datasets increases. Additionally, without the combination of tests, the neural network was able to anticipate the tribological testing parameters. According to the prediction, a good proportion of a solid lubricant combination should be added to the PPS in order to improve the composite's resistance to wear. In other research, a PPS-based composite was created utilizing carbon fiber, graphite, polytetrafluoroethylene, and micro titanium dioxide (TiO2) powder. The sliding wear was examined and forecasted using artificial neural networks. For the composite improved wear rate, they forecasted the ideal filler percentages [68, 69]. Backpropagation learning algorithms are frequently used. With an accuracy of 94.4% in relation to the experimental data, it is discovered that the sliding speed affects the prediction of erosive wear for nylon 12 under four input parameters [70]. By minimizing the number of tests, wear rate or wear loss research under feed-forward neural networks (FFNN) and loss prediction can help save time. The polymers may perform better mechanically and tribologically as a result of parameter optimization. An efficient method of studying various loading conditions with fewer experiments is to forecast the sliding wear loss of polyamide [71]. The input parameters determine how well an ANN predicts mechanical qualities and wear loss. To improve prediction accuracy and provide a better match, more experimental data can be used to train the model [72]. With enough training data, the number of neurons in the ANN model can improve the prediction even more; nevertheless, a rise in the number of neurons eventually affects the projected values [69]. Machine learning optimization is proven to be helpful for wear prediction in material selection and procedure [73]. In order to influence the surface structure and attributes based on the printing parameters, additive manufacturing is essential [74]. Polyetheretherketone (PEEK) and acrylonitrile butadiene styrene (ABS) samples' wear rate and wear mechanism are influenced by the printed sample's orientation with regard to the sliding direction [75, 76]. According to the sliding direction with respect to the sample surface, the build orientation of the 3D-printed ABS and polylactic acid (PLA) samples produced using FDM exhibits a notable difference in wear volume and wear rate [77]. The journal bearing was created utilizing three distinct techniques and tested for self-lubrication properties under various lubricating circumstances using three-dimensional printing. The study examined the wear loss, friction coefficient, and temperature increase during the journal-bearing test [78]. For the journal-bearing application tested under dry sliding conditions, the PEEK samples were made via an extrusion technique with varying infill densities. Because it created room for the wear debris, the surface porosity was essential to deformation [79]. Porosity can store and supply lubricant during times of scarcity, according to samples tested under grease lubrication [80]. When evaluated in dry sliding, the 3D-printed ABS sample's surface inhomogeneity takes longer to reach steady-state conditions. The applied loads and the direction of the raster angle affect the friction behavior [81]. According to the experimental data, the artificial neural network's prediction of wear loss is consistent, indicating that it can be used to forecast wear loss. The activation function, learning rate, training ratio, and datasets utilized to train the model are the primary determinants of the prediction quality when employing ANN. Neural network optimization can improve prediction and performance. The measurement error structure is strongly impacted by the mean absolute and square errors with weights and biases. With a minimum accuracy of 94%, the design represented 3-5-5-1 using three input neurons, five neurons for each of the two hidden layers, and one for the output neuron. The model's predictions are reliable in forecasting wear loss and, consequently, weight reduction, and they closely match the experimental results. The best-fitting model that can forecast wear loss under abrasion is obtained by employing the ReLU activation function for wear prediction. The wear loss of the 3D-printed PEEK samples can be computed using the same ANN model that is used here to forecast the wear loss in microns under abrasion [82]. In order to overcome the difficulties involved in forecasting the characteristics of auxetic materials, machine learning has shown itself to be a revolutionary technique. Auxetic material research and development has benefited greatly from ML models due to their capacity to handle complicated, nonlinear behavior, lower computational costs, support inverse design and optimization, and enhance generalization and adaptability. They assist engineers and researchers realize the full potential of auxetic materials for cutting-edge applications by facilitating quick discovery, accurate property prediction, and effective optimization. Key challenges including high computational costs, complex nonlinear behavior, and the intricate relationships on geometric factors that define auxetic structures have been addressed by ML models through the use of data-driven surrogates. Real-time forecasts and multi-objective optimization capabilities of ML have sped up auxetics design and prototyping, enabling the investigation of new configurations and customized properties with previously unheard-of speed. Furthermore, by shedding light on the links between structure and property, machine learning models have helped to better understand auxetic behavior and direct the creation of novel materials for cutting-edge applications. Nevertheless, there is a dearth of research on the machine learning-guided response prediction of FEM-modeled and 3D-printed auxetics. Algorithms for machine learning used: One of the most popular algorithms for forecasting the auxetics' reaction to various loading scenarios is the neural network. However, a lot of datasets are needed to train the algorithms in neural networking, especially deep learning models. Getting a large dataset is a laborious stage in the surrogate creation process. Additionally, it takes a lot of computing power to train the model using neural networks. The black-box aspect of neural networks, particularly deep neural networks, makes it challenging for users or researchers to comprehend how the algorithm makes judgments. Neural networking has other drawbacks, such as handling outliers, hyperparameter adjustment, and sensitivity to data noise. Simpler machine learning methods, such as support vector machines, random forests, reinforcement learning algorithms, hybrid algorithms, and Gaussian process regression, can be used in place of neural networking to get around the problems with neural networking techniques. The literature only contains a small number of works that use these algorithms. Analysis of stochasticity and uncertainty: The impact of input parameter uncertainty on the auxetic structures' strength and behavior is not covered in any research. Investigations on the stochastic mechanical properties of the auxetic structures are still needed in this subject. The stochastic properties and behavior of the auxetics can be quantified by applying sophisticated techniques such as Monte Carlo or advanced Monte Carlo simulations such as Quasi-Monte Carlo or Latin Hypercube sampling techniques, stochastic finite element method, stochastic optimization techniques, Markov Chain Monte Carlo methods, stochastic Kriging, etc. Auxetic structure: The most extensively researched structure employing machine learning algorithms is the re-entrant honeycomb. Other structures, such as foam-based auxetics, spinning squares, double arrowheads, etc., require more thorough research utilizing machine learning algorithms in order to comprehend and measure the impact of the geometric features on the mechanical behavior of the auxetics. Some auxetics, such as bio-inspired auxetics and origami-based auxetics, are still awaiting investigation in related fields. Functionally graded auxetics: Researchers have generally suggested a graded distribution of the auxetic structure to improve auxetic performance. However, consideration must be given to the use of machine learning in the research of functionally graded auxetics. It is possible to study the functionally graded auxetics with less computational effort by using machine learning techniques. Multi-shape surrogate: Every study that is currently accessible in the literature focuses on examining the behavior of a single auxetic type through the use of surrogates. Nonetheless, a surrogate that has been trained to forecast the characteristics and potency of two or more distinct auxetic types may be used. In this instance, the qualities or strength of the auxetics can be regarded as the output variables, and the geometric properties as the input variables. Utilizing cutting-edge methods to track auxetics behavior in real time: It is necessary to track the behavior of 3D printed auxetics during the in-service period using real-time sophisticated monitoring techniques, such as machine learning and the Internet of Things. This will assist in tracking the auxetics' behavior in real time and offer valuable information on how the auxetics behave under service settings. There are a number of difficulties in using ML to investigate the mechanical characteristics of auxetic materials, such as the requirement for diverse, high-quality datasets, managing intricate and nonlinear behaviors, guaranteeing model interpretability, and verifying predictions. These problems show how crucial thorough data collection, feature engineering, model optimization, and validation are to creating trustworthy and broadly applicable machine learning models for auxetic materials. By overcoming these obstacles, researchers will be able to use ML more efficiently, which will make it easier to design and optimize auxetic materials for cutting-edge applications. An innovative strategy that strikes a good balance between computing efforts and dataset production is the use of machine learning to examine auxetic behavior and predict its features [83]. A novel deep learning technique for modeling AM processes' geometric deviations has been developed. CNN has opened up new avenues to examine the impact of numerous influencing factors on the resulting geometric deviations by leveraging the power of data. By varying the layer boundary points based on the anticipated deviations, the deformed part surface may be created from the nominal STL representation, aiding DfAM in assessing geometrical quality. A compensated STL model can also be obtained by point-wise reverse compensation, which will bring the surface of the manufactured item closer to the design purpose. More deviation data from a wider range of forms might be used to train this technique, perhaps increasing its power. Without a doubt, the predictive performance in real-world industrial applications will be enhanced with the addition of more realistic measurement data. However, in order to obtain layer-wise data, more sophisticated measurement methods such as computed tomography (CT) are required. Furthermore, by extending the input channel and running more simulations under such parameters, new process parameters can be considered. It is also important to note that this work can take into account other design limitations and requirements that might be included as new parameters if simulation techniques that go beyond basic FEA are available, such as generative design. The digital twin [84] for additive manufacturing, which uses data analytics as its foundation to create high-fidelity digital models powered by multi-domain data from the product life cycle [85], is another exciting area of research. The effects on dimensional accuracy, surface quality, and tensile strength of additively manufactured objects are examined under varying layer thicknesses, extrusion temperatures, printing speeds, build directions, and deposition directions. To capture the link between these process factors and part properties, artificial neural network models were created using experimental data. The training, validation, and testing outcomes for the suggested ANN models roughly match the experimental data, according to the root mean square error (RMSE) analysis. In terms of the RMSE and correlation coefficient for the training data, the ANN models for the individual part characteristics fared better than the model for the combination of three output part attributes. The main effect plot result showed that in order to produce goods with higher surface quality, layer thickness, printing speed, build orientation, and deposition direction required to be kept to a minimum. Higher extruder temperatures, however, produced high-quality parts. Therefore, the ideal combination to produce the best surface finish items was determined to be 0.1 mm layer thickness, 220 degrees Celsius extruder temperature, 40 mm/s printing speed, 0 degree build orientation, and 0 degree deposition direction. Although the 210-degree extruder temperature of our manufactured specimen produced the maximum tensile strength, the main effect plot result indicated that greater printing speed, build orientation, and deposition direction produced high tensile strength products. It was discovered that the primary cause of the materials' maximum tensile strength was their thinner layers. According to our research, the mechanical properties of the constructed specimen are improved by 0.1 mm layer thickness, an extruder temperature of 210 degrees Celsius, a printing speed of 80 mm/s, a 90-degree construction orientation, and a 45-degree deposition direction. The main effect plot results showed that dimensional accuracy was significantly impacted by extruder temperature and deposition orientation. After constructing an actual part, our investigation revealed that a 45-degree deposition direction and an extruder temperature of 210 degrees Celsius were required to obtain an accurate CAD model of the specimen's dimensions. The specimen's dimensional accuracy was less affected by construction orientation, printing speed, and layer thickness. In terms of training data and all datasets, it was discovered that individual ANN models performed better than the combined model based on RMSE [86].

The strength of the printed pieces, which results from the anisotropic stacking characteristic of AM techniques, is one of the primary limitations of extrusion-based AM technologies. Mathematical modeling of part strength variations has been used to improve important process parameters prior to printing, resulting in a more accurate additive fabrication process strategy and increased strength for components printed using extrusion-based AM processes. To increase the component strength, the design of experiment technique of part strength prediction has been used. However, when combined with various processing variables such as process parameters, materials, machines, etc., such empirical modeling methodologies may frequently fail to accurately estimate the surface roughness component strength. Therefore, using experimental data from the body of existing literature, the ANN technique has been employed in this research work. The created model has been trained several times to obtain better results because the ANN parameters also affect the accuracy of the prediction output. The parametric effect on the tensile property is then estimated using the trained ANN model. For both techniques (i.e., RSM and ANN), the variation of tensile strength with each process variable (i.e., layer thickness, orientation, rater angle, nozzle temperature, bed temperature, room temperature, air gap, and barrel temperature) shows almost the same data trends; however, the RSM model's prediction result has a higher error. It is discovered from the constructed models that, for the most part, the ANN's forecast is true. Additionally, it was discovered that infill density, barrel temperature, nozzle temperature, and layer thickness are important process variables. While the ANN is not appropriate for smaller datasets, the RMSE for the RSM occasionally performs better in situations with less data. In most cases, the ANN results have a lower RMSE, indicating the suitability of this model. For the RSM and ANN models, the average RMSE across all cases is 0.90 and 0.49, respectively. It is advised that ANN be used primarily for larger datasets and situations requiring predictions with lower error rates because of its time-consuming nature. This work demonstrated how to optimize MEAM process settings, prevent difficulties, and minimize the number of numerous experiments that are typical in many tests by utilizing an RSM and an ANN [87].

To develop an innovative and adaptable technique that integrates fuzzy logic with artificial neural networks for forecasting the mechanical properties of additively manufactured objects, two ANFIS models were created: a specialized model designed specifically to predict the mechanical behavior of PLA specimens produced via FFF under bending tests, and a generalized model that requires additional experimental data across diverse AM technologies, materials, and mechanical testing methods. According to the goal of improving the AM manufacturing process, most pertinent research usually concentrates on and looks at a small number of factors within the rigorous confines of the trials. The creation of a predictive model that can be used to any AM method and material while taking into account the several process parameters that influence the ultimate product's Young's modulus and maximum stress and strain is the novel part of this study. The increase in performance was obtained by employing nine input parameters instead of 21. This is because two different models were developed: one with nine inputs specifically focused on bending, and another with 21 inputs intended to be broader. The general model with 21 inputs is relevant to a wider range of AM technologies and materials since it attempts to capture a wide range of physical phenomena. Nevertheless, this model's performance may be constrained by the less complete data that are currently accessible. On the other hand, the nine-input bending model is tailored for particular characteristics that have a direct impact on the samples' bending behavior. A more thorough training procedure is made possible by the larger amount of data used for the bending model. We find areas that need improvement while also observing a high correlation in some aspects between the experimental results and the mechanical parameters that were numerically modeled. With RMSE values within 20% and 9.75%, respectively, the numerically predicted maximum stress and strain demonstrated a respectable degree of agreement with the experimental results, indicating that the model is quite accurate for these parameters. However, because of one outlier with a considerable departure, the forecast for Young's modulus had a higher RMSE of almost 40%. This suggests that although the model reflects the overall trend, further data and possible model improvement could increase the accuracy for Young's modulus. The outcomes of the built systems demonstrated the method's potential for successful prediction, underscoring the need for further input data. This would prevent the waste of important resources by enabling an accurate initial estimation of the applied AM process (including AM technologies, materials, structural settings, and process parameters) [88].

For the numerical simulation of an additively built test specimen under a uniaxial tensile load, determine the material parameters of the yield curve using an FFANN. The labeled data needed for the Feed-Forward Artificial Neural Network (FFANN's) training were produced by numerical simulations using MAT024 and LS-DYNA. A second-order polynomial and a root function were used to develop a custom yield curve formulation that would characterize the ABS's elastic-plastic material behavior without taking into account more intricate mechanisms such as damage, which would also require a Plasticity Integration (PI). The material parameters were obtained and the outcomes of the subsequent structural simulation of the direct inverse NN-based approach for PI and the iterative optimization-based method were compared. The findings demonstrate that the NN is appropriate for determining the yield curve parameter of the ABS under investigation. The NN-based approach outperformed the widely known iterative optimization method when appropriate boundary conditions and parameters were employed. The findings support the authors' hypothesis that parameter identifications of the complicated property spectrum for comprehensive material models of thermoplastics could be carried out in the future using such feedforward neural networks. Because the FFANN can be trained using data from numerical simulations, this would have the benefit of minimizing the amount of experimental research required for a material's PI. The trained FFANN can be reused for similar materials and the related PI in a matter of seconds, which is an even bigger benefit. Time spent developing new products could be shortened as a result. On the other hand, computationally demanding numerical simulations would be necessary if the material behavior differed only marginally from the commonly used iterative optimization strategy. However, the exact material behavior of the material to be characterized must be captured in the training input data for a trained FFANN to be employed repeatedly. The outcomes would be objectively comparable because the PI procedure only needs expert knowledge to create the FFANN—not for the actual identifying process. The effects of various materials (with comparable material qualities) on the structural-mechanical features of created components might therefore be assessed and contrasted, particularly during the early stages of development. A polymer that exhibits comparable material-mechanical behavior, such as a maximum tensile strength of 30% or a distinct nonlinear stress–strain curve contour, would be conceivable. The selection and processing of the FFANN's training data, as well as the selection of the hyperparameters, as described in this paper, are crucial for the accuracy of the predicted parameters and, consequently, the method's success. Future research in this area should find it easier to choose boundary conditions and settings thanks to the investigations. Additionally, various approaches for choosing and refining the FFANN's HPs were demonstrated and contrasted. HP optimization using the Bayesian optimization approach with the library Keras Tuner could result in the greatest improvement in the validation set's prediction accuracy [89].

4. Conclusion

This paper discusses the importance and applications of additively made parts. Additive manufacturing is one of the most widely used technologies in an increasing range of industries, including engineering, manufacturing, Industry 4.0 concepts, aerospace, and medical applications. It is a layer-by-layer production technique, which is particularly erratic since it depends on a number of factors, including the manufacturing settings used. Artificial neural networks are a new method for predicting the mechanical properties of additively built objects. It predicts the particular property based on the given set of parameters. Its disadvantages include its nature, higher computing burden, tendency to overfit, and the empirical nature of model construction. Future algorithms and computational approaches can easily show that our neural network predicts product attributes more correctly than existing models and simulations.

References

[1] Lu, B., Li, D. and Tian, X. (2015). Development trends in additive manufacturing and 3D printing. *Engineering*, *1*(1), pp.085–089.

[2] Derby, B. (2015). Additive manufacture of ceramics components by inkjet printing. *Engineering*, *1*(1), pp.113–123.

[3] Gu, D., Ma, C., Xia, M., Dai, D. and Shi, Q. (2017). A multiscale understanding of the thermodynamic and kinetic mechanisms of laser additive manufacturing. *Engineering*, *3*(5), pp. 675–684.

[4] Herzog, D., Seyda, V., Wycisk, E. and Emmelmann, C. (2016). Additive manufacturing of metals. *Acta Materialia*, 117, pp. 371–392.

[5] Liu, L., Ding, Q., Zhong, Y., Zou, J., Wu, J., Chiu, Y.L., Li, J., Zhang, Z., Yu, Q. and Shen, Z. (2018). Dislocation network in additive manufactured steel breaks strength–ductility trade-off. *Materials Today*, *21*(4), pp. 354–361.

[6] Gorsse, S., Hutchinson, C., Gouné, M. and Banerjee, R. (2017). Additive manufacturing of metals: a brief review of the characteristic microstructures and properties of steels, Ti-6Al-4V and high-entropy alloys. *Science and Technology of advanced MaTerialS*, *18*(1), pp. 584–610.

[7] Acharya, R., Sharon, J.A. and Staroselsky, A. (2017). Prediction of microstructure in laser powder bed fusion process. *Acta Materialia*, 124, pp. 360–371.

[8] Fergani, O., Berto, F., Welo, T. and Liang, S.Y. (2017). Analytical modelling of residual stress in additive manufacturing. *Fatigue & Fracture of Engineering Materials & Structures*, *40*(6), pp. 971–978.

[9] Chen, Q., Guillemot, G., Gandin, C.A. and Bellet, M. (2017). Three-dimensional finite element thermomechanical modeling of additive manufacturing by selective laser melting for ceramic materials. *Additive Manufacturing*, *16*, pp. 124–137.

[10] Qi, X., Chen, G., Li, Y., Cheng, X. and Li, C. (2019). Applying neural-network-based machine learning to additive manufacturing: current applications, challenges, and future perspectives. *Engineering*, *5*(4), pp. 721–729.

[11] Sun, J., Yang, Y. and Wang, D. (2013). Parametric optimization of selective laser melting for forming Ti6Al4V samples by Taguchi method. *Optics & Laser Technology*, 49, pp.118–124.

[12] Andrey Tyagunov, Oleg Milder, Dmitry Tarasov, "Deep Learning in Simulation of Nickel-based Superalloys Ultimate Tensile Strength: Accounting the Role of Alloying Elements", WSEAS Transactions on Environment and Development, vol. 15, pp. 340–345, 2019

[13] ZhengJ et al., (2009). *The International Journal of Advanced Manufacturing Technology* 44(7) pp. 667–674.

[14] Prieto, A., Prieto, B., Ortigosa, E.M., Ros, E., Pelayo, F., Ortega, J. and Rojas, I. (2016). Neural networks: An overview of early research, current frameworks and new challenges. *Neurocomputing*, *214*, pp.242–268.

[15] Firat, M. and Gungor, M. Generalized Regression Neural Networks and Feed Forward Neural Networks for Prediction of Scour Depth around Bridge Piers. *Advances in Engineering Software* 2009, 40(8), 731–737. https://doi.org/10.1016/j.advengsoft.2008.12.001.

[16] Akbari, M., Saedodin, S., Panjehpour, A., Hassani, M., Afrand, M. and Torkamany, M.J. (2016) Numerical Simulation and Designing Artificial Neural Network for Estimating Melt Pool Geometry and Temperature Distribution in Laser Welding of Ti6Al4V Alloy. *Optik-International Journal for Light and Electron Optics*, 127, 11161–11172. https://doi.org/10.1016/j.ijleo.2016.09.042

[17] Rathi, N.K. and Rathi, N. (2020). An application of ANN for modeling and optimisation of process parameters of manufacturing process: A review. *Int. J. Eng. Appl. Sci. Technol*, *4*, pp. 127–134.

[18] Stangierski, J., Weiss, D. and Kaczmarek, A. (2019). Multiple regression models and Artificial Neural Network (ANN) as prediction tools of changes in overall quality during the storage of spreadable processed Gouda cheese. *European Food Research and Technology*, *245*(11), pp. 2539–2547.

[19] Zhang, J., Wang, P. and Gao, R.X. (2019). Deep learning-based tensile strength prediction in fused deposition modeling. *Computers in industry*, *107*, pp.11–21.

[20] Ahmadloo, E. and Azizi, S., 2016. Prediction of thermal conductivity of various nanofluids using artificial neural network. *International Communications in Heat and Mass Transfer*, *74*, pp. 69–75.

[21] Lantz, B., 2013. *Machine learning with R: learn how to use R to apply powerful machine learning methods and gain an insight into real-world applications*. Packt Pub..

[22] Yao, X., Moon, S.K. and Bi, G. (2017). A hybrid machine learning approach for additive manufacturing design feature recommendation. *Rapid Prototyping Journal*, *23*(6), pp. 983–997.

[23] Desai, P.S. and Higgs III, C.F. (2019). Spreading process maps for powder-bed additive manufacturing derived from physics model-based machine learning. *Metals*, *9*(11), p. 1176.

[24] Amiri, M., Bakhshandeh Amnieh, H., Hasanipanah, M. and Mohammad Khanli, L. (2016). A new combination of artificial neural network and K-nearest neighbors models to predict blast-induced ground vibration and air-overpressure. *Engineering with Computers*, *32*(4), pp. 631–644.

[25] Gor, M., Dobriyal, A., Wankhede, V., Sahlot, P., Grzelak, K., Kluczyński, J. and Łuszczek, J. (2022). Density prediction in powder bed fusion additive manufacturing: machine learning-based techniques. *Applied Sciences*, *12*(14), p. 7271.

[26] Grozav, S.D., Sterca, A.D., Kočiško, M., Pollák, M. and Ceclan, V. (2023). Artificial neural network-based predictive model for finite element analysis of additive-manufactured components. *Machines*, *11*(5), p. 547.

[27] Pazhamannil, R.V., Govindan, P. and Sooraj, P. (2021). Prediction of the tensile strength of polylactic acid fused deposition models using artificial neural network technique. *Materials Today: Proceedings*, *46*, pp. 9187–9193.

[28] Grozav, S.D., Sterca, A.D., Kočiško, M., Pollák, M. and Ceclan, V. (2022). Feasibility of predictive models for the quality of additive manufactured components based on artificial neural networks. *Machines*, *10*(2), p. 128.

[29] Tamur, C., Li, S. and Zeng, D. (2023). Artificial neural networks for predicting mechanical properties of crystalline polyamide12 via molecular dynamics simulations. *Polymers*, *15*(21), p. 4254.

[30] Tura, A.D., Lemu, H.G. and Mamo, H.B. (2022). Experimental investigation and prediction of mechanical properties in a fused deposition modeling process. *Crystals*, *12*(6), p. 844.

[31] Soler, D., Telleria, M., García-Blanco, M.B., Espinosa, E., Cuesta, M. and Arrazola, P.J. (2022). Prediction of surface roughness of SLM built parts after finishing processes using an artificial neural network. *Journal of Manufacturing and Materials Processing*, *6*(4), p. 82.

[32] Bhardwaj, T. and Shukla, M. (2020). Laser additive manufacturing-direct energy deposition of Ti-15Mo biomedical alloy: artificial neural network based modeling of track dilution. *Lasers in Manufacturing and Materials Processing*, *7*(3), pp. 245–258

[33] Iranmanesh, N., Sarvestani, H.Y., Ashrafi, B. and Hojjati, M. (2023). Beyond honeycombs: Core topology's role in 3D-printed sandwich panels. *Materials Today Communications*, *37*, p. 107548.

[34] Deng, Y., Huang, Z., Liu, C. and Zhang, Y. (2024). Improved design of 3D printed star honeycomb pressure resistance and sandwich panel protective capability. *Materials Today Communications*, *38*, p. 108069.

[35] Ma, Q., Rejab, M.R.M., Song, Y., Zhang, X., Hanon, M.M., Abdullah, M.H. and Kumar, A.P. (2024). Effect of infill pattern of polylactide acid (PLA) 3D-printed integral sandwich panels under ballistic impact loading. *Materials Today Communications*, *38*, p. 107626.

[36] Song, Y., Li, Y., Song, W., Yee, K., Lee, K.Y. and Tagarielli, V.L. (2017). Measurements of the mechanical response of unidirectional 3D-printed PLA. *Materials & Design*, *123*, pp. 154–164.

[37] Qi, G., Ji, B. and Ma, L., 2019. Mechanical response of pyramidal lattice truss core sandwich structures by additive manufacturing. *Mechanics of Advanced Materials and Structures*, *26*(15), pp. 1298–1306.

[38] Sivakumar, N.K., Kaaviya, J., Palaniyappan, S., Nandhakumar, G.S., Prakash, C., Basavarajappa, S., Pandiaraj, S. and Hashem, M.I. (2024). A machine learning approach for predicting flexural strength of 3D printed hexagon lattice-cored sandwich structures. *Materials Today Communications*, *41*, p. 110230.

[39] Koeppe, A., Padilla, C.A.H., Voshage, M., Schleifenbaum, J.H. and Markert, B. (2018). Efficient numerical modeling of 3D-printed lattice-cell structures using neural networks. *Manufacturing Letters*, *15*, pp. 147–150.

[40] Alejandrino, J.D., Concepcion, R.S., Lauguico, S.C., Tobias, R.R., Venancio, L., Macasaet, D., Bandala, A.A. and Dadios, E.P. (2020). A machine learning approach of lattice infill pattern for increasing material efficiency in additive manufacturing processes. *Int. J. Mech. Eng. Robot. Res*, *9*(9), pp. 1253–1263.

[41] Vasundara, M., Padmanaban, K.P., Sabareeswaran, M. and RajGanesh, M. (2012). Machining fixture layout design for milling operation using FEA, ANN and RSM. *Procedia Engineering*, *38*, pp. 1693–1703.

[42] Saldaña-Robles, A.L., Bustos-Gaytán, A., Diosdado-De la Peña, J.A., Saldaña-Robles, A., Alcántar-Camarena, V., Balvantín-García, A. and Saldaña-Robles, N. (2020). Structural design of an agricultural backhoe using TA, FEA, RSM and ANN. *Computers and Electronics in Agriculture*, *172*, p. 105278.

[43] Tu, H., Yu, Q.J., Tan, K.H., Fung, T.C. and Riedel, W. (2024), March. FEM-and ANN-based design of CFRP-strengthened RC walls under close-in explosions. In *Structures* (Vol. 61, p. 105930). Elsevier.

[44] Drieschner, M., Wolf, C., Seiffarth, F. and Petryna, Y. (2023). Combined ANN-FEM approach for spatial–temporal structural response prediction: Method and experimental validation. *Thin-Walled Structures*, *189*, p.110800.

[45] Aydin, K. (2024). Investigation of optimal machining Monel 400 superalloy considering carbon emissions using FEM, regression and ANN methods. *Journal of Cleaner Production*, *447*, p. 141616.

[46] Cogun, C. and Ayli, E. (2025). Machine Learning-Driven Approach for Reducing Tool Wear in Die-Sinking Electrical Discharge Machining. *Arabian Journal for Science and Engineering*, pp. 1–21.

[47] Karagözlü, C.O., Ayli, E., Tanabi, H. and Sabuncuoglu, B. (2025). Artificial Neural Networks for Predicted Bending Properties of Additively Manufactured Pyramidal Lattice Truss Core Sandwich Structures. *Materials Today Communications*, p. 112926.

[48] Zhang, X. and Chen, L. (2020). Effects of laser scanning speed on surface roughness and mechanical properties of aluminum/Polylactic Acid (Al/PLA) composites parts fabricated by fused deposition modeling. *Polymer Testing*, *91*, p. 106785.

[49] Ayatollahi, M.R., Nabavi-Kivi, A., Bahrami, B., Yahya, M.Y. and Khosravani, M.R. (2020). The influence of in-plane raster angle on tensile and fracture strengths of 3D-printed PLA specimens. *Engineering Fracture Mechanics*, *237*, p. 107225.

[50] Nabavi-Kivi, A., Ayatollahi, M.R., Rezaeian, P. and Razavi, N. (2022). Investigating the effect of printing speed and mode mixity on the fracture behavior of FDM-ABS specimens. *Theoretical and Applied Fracture Mechanics*, *118*, p. 103223.

[51] Iyer, S.S.G. and Keles, O. (2022). Effect of raster angle on mechanical properties of 3D printed short carbon fiber reinforced acrylonitrile butadiene styrene. *Composites Communications*, *32*, p. 101163.

[52] Khosravani, M.R., Božić, Ž., Zolfagharian, A. and Reinicke, T. (2022). Failure analysis of 3D-printed PLA components: Impact of manufacturing defects and thermal ageing. *Engineering Failure Analysis*, *136*, p. 106214.

[53] Deshwal, S., Kumar, A. and Chhabra, D. (2020). Exercising hybrid statistical tools GA-RSM, GA-ANN and GA-ANFIS to optimize FDM process parameters for tensile strength improvement. *CIRP Journal of Manufacturing Science and Technology*, *31*, pp. 189–199.

[54] Kumar, R., Singh, R. and Ahuja, I.P.S. (2018). Investigations of mechanical, thermal and morphological properties of FDM fabricated parts for friction welding applications. *Measurement*, *120*, pp. 11–20.

[55] Vinodh, S. (2020). Parametric optimization of fused deposition modelling process using Grey based Taguchi and TOPSIS methods for an automotive component. *Rapid Prototyping Journal*, *27*(1), pp. 155–175.

[56] Sakthi Balan, G., Aravind Raj, S., Sridharan, M. and Adithya, R.N. (2025). A study of the effect of printing process parameters on the delamination and surface finish properties of aluminium-infused PLA printed via material extrusion. *Australian Journal of Mechanical Engineering*, *23*(1), pp. 17–37.

[57] Mishra, V. and Veeman, D. (2025). Artificial neural network-based predictive models for analyzing the flexural and compressive strength of PLA/carbon parts fabricated via material extrusion-based 3D printing. *Journal of Thermoplastic Composite Materials*, p. 08927057251314430.

[58] Paturi, U.M.R., Cheruku, S. and Reddy, N.S. (2022). The role of artificial neural networks in prediction of mechanical and tribological properties of composites—a comprehensive review. *Archives of Computational Methods in Engineering*, *29*(5), pp. 3109–3149.

[59] Nguyen, T.A., Ly, H.B., Mai, H.V.T. and Tran, V.Q. (2021). On the training algorithms for artificial neural network in predicting the shear strength of deep beams. *Complexity*, *2021*(1), p. 5548988.

[60] Hoskins, T.J., Dearn, K.D. and Kukureka, S.N. (2018). Mechanical performance of PEEK produced by additive manufacturing. *Polymer Testing*, *70*, pp. 511–519.

[61] Al Khatib, A., Le-Franc, R., Guin, J.P. and Coulon, J.F. (2023). Investigating the thermal effects of plasma surface treatment on crystallinity and mechanical behavior of PEEK. *Polymer Degradation and Stability*, *216*, p. 110500.

[62] Argatov, I.I. and Chai, Y.S. (2019). An artificial neural network supported regression model for wear rate. *Tribology International*, *138*, pp. 211–214.

[63] Argatov, I.I. and Chai, Y.S. (2021). Artificial neural network modeling of sliding wear. *Proceedings of the Institution of Mechanical Engineers, Part J: Journal of Engineering Tribology*, *235*(4), pp. 748–757.

[64] Ravitej, Y.P., Mohan, C.B. and Ananthaprasad, M.G. (2022). Dry sliding friction and wear behavior of LM13/Zircon/Carbon (HMMC's): an experimental, statistical and artificial neural network approach. *Tribology in Industry*, *44*(3), p. 374.

[65] Velten, K., Reinicke, R. and Friedrich, K. (2000). Wear volume prediction with artificial neural networks. *Tribology International*, *33*(10), pp. 731–736.

[66] Zhang, Z., Barkoula, N.M., Karger-Kocsis, J. and Friedrich, K. (2003). Artificial neural network predictions on erosive wear of polymers. *Wear*, *255*(1-6), pp. 708–713.
[67] Gyurova, L.A. and Friedrich, K. (2011). Artificial neural networks for predicting sliding friction and wear properties of polyphenylene sulfide composites. *Tribology International*, *44*(5), pp. 603–609.
[68] Gyurova, L.A., Miniño-Justel, P. and Schlarb, A.K. (2010). Modeling the sliding wear and friction properties of polyphenylene sulfide composites using artificial neural networks. *Wear*, *268*(5-6), pp. 708–714.
[69] Jiang, Z., Gyurova, L., Zhang, Z., Friedrich, K. and Schlarb, A.K. (2008). Neural network based prediction on mechanical and wear properties of short fibers reinforced polyamide composites. *Materials & Design*, *29*(3), pp. 628–637.
[70] Shuvho, B.A., Chowdhury, M.A. and Debnath, U.K. (2019). Analysis of artificial neural network for predicting erosive wear of nylon-12 polymer. *Materials Performance and Characterization*, *8*(1), pp. 288–300.
[71] Abdelbary, A.H.H.A., Abouelwafa, M.N., El Fahham, I.M. and Hamdy, A.H. (2012). Modeling the wear of Polyamide 66 using artificial neural network. *Materials & Design*, *41*, pp. 460–469.
[72] Jiang, Z., Zhang, Z. and Friedrich, K. (2007). Prediction on wear properties of polymer composites with artificial neural networks. *Composites Science and Technology*, *67*(2), pp. 168–176.
[73] Shah, R., Pai, N., Thomas, G., Jha, S., Mittal, V., Shirvni, K. and Liang, H. (2025). Machine learning in wear prediction. *Journal of Tribology*, *147*(4), p. 040801.
[74] Prajapati, S.K. and R, G. (2025). Mechanical and surface characterisation of additively manufactured polyetheretherketone for the tribo test. *Rapid Prototyping Journal*, *31*(3), pp. 584–598.
[75] Dangnan, F., Espejo, C., Liskiewicz, T., Gester, M. and Neville, A. (2020). Friction and wear of additive manufactured polymers in dry contact. *Journal of Manufacturing Processes*, *59*, pp. 238–247.
[76] Prajapati, S.K. and Gnanamoorthy, R. (2024). Effect of infill orientation on the wear behaviour of 3D printed polyetheretherketone. *Journal of Materials Science*, *59*(44), pp. 20840–20853.
[77] Mahmood, S., Guo, E., Stirling, A. and Schulze, K.D. (2023). Orientation Controls Tribological Performance of 3D-Printed PLA and ABS. *Tribology Online*, *18*(6), pp. 302–312.
[78] Lee, Y.J., Lee, K.H. and Lee, C.H. (2018). Self-lubricating and friction performance of a three-dimensional-printed journal bearing. *Journal of Tribology*, *140*(5), p. 054501.
[79] Prajapati, S.K. and Gnanamoorthy, R. (2024). Effect of the infill percentage of 3D printed Polyetheretherketone under the dry sliding condition. *Journal of Manufacturing Processes*, *131*, pp. 2073–2081.
[80] Prajapati, S.K. and Gnanamoorthy, R. (2025). Friction Performance of Three-Dimension Printed PEEK for Journal Bearing Application Under Grease Lubrication. *Journal of Tribology*, *147*(10), p. 102201.
[81] Dhakal, N., Wang, X., Espejo, C., Morina, A. and Emami, N. (2023). Impact of processing defects on microstructure, surface quality, and tribological performance in 3D printed polymers. *Journal of Materials Research and Technology*, *23*, pp. 1252–1272.
[82] Prajapati, S.K. (2025). Abrasive Wear Prediction of Three-Dimensional Printed PEEK Using Artificial Neural Network. *Journal of Tribology*, *147*(11), p. 114201.
[83] Garg, A., Sharma, A., Zheng, W. and Li, L. (2025). A review on artificial intelligence-enabled mechanical analysis of 3D printed and FEM-modelled auxetic metamaterials. *Virtual and Physical Prototyping*, *20*(1), p. e2445712.
[84] Schleich, B., Anwer, N., Mathieu, L. and Wartzack, S. (2017). Shaping the digital twin for design and production engineering. *CIRP annals*, *66*(1), pp. 141–144.

[85] Zhu, Z., Ferreira, K., Anwer, N., Mathieu, L., Guo, K. and Qiao, L. (2020). Convolutional neural network for geometric deviation prediction in additive manufacturing. *Procedia Cirp*, *91*, pp. 534–539.

[86] Deb, J., Ahsan, N. and Majumder, S. (2022). October. Modeling the interplay between process parameters and part attributes in additive manufacturing process with artificial neural network. In *ASME International Mechanical Engineering Congress and Exposition* (Vol. 86632, p. V02AT02A015). American Society of Mechanical Engineers.

[87] Gupta, A.K. and Taufik, M. (2023). Improvement of part strength prediction modelling by artificial neural networks for filament and pellet based additively manufactured parts. *Australian Journal of Mechanical Engineering*, *21*(5), pp. 1870–1887.

[88] Sagias, V.D., Zacharia, P., Tempeloudis, A. and Stergiou, C. (2024). Adaptive neuro-Fuzzy inference system-based predictive modeling of mechanical properties in additive manufacturing. *Machines*, *12*(8), p. 523.

[89] Meißner, P., Watschke, H., Winter, J. and Vietor, T. (2020). Artificial neural networks-based material parameter identification for numerical simulations of additively manufactured parts by material extrusion. *Polymers*, *12*(12), p. 2949.

6 Integrating Additive Manufacturing into Industrial Design Methods and Processes

Ahmet Atak[1*], Raheem Al-Sabur[2] and Hassanein I. Khalaf[3]

In early 1984, Chuck Hull revolutionized the manufacturing field with the creation of the first stereolithography-based 3D printer. This led to a significant acceleration in development, culminating in the late 1990s with the production of a scaffold for a bladder, followed by successful experiments that resulted in the creation and practical application of a human skull in 2014. 3D printers have continued to flourish steadily, becoming integrated into most products. This chapter aims to provide the reader with an in-depth overview of the evolution of this manufacturing process and the key aspects of its development.

The introduction begins by defining 3D printers, followed by a historical overview of this development. It then explores industrial design and its relationship to design for additive manufacturing (DFAM). The DFAM overview is comprehensive, including a definition of the concept and its fundamental principles. It also mentions some well-known brands that utilize 3D printing. The second section addresses 3D printing principles, starting with the main components of industrial design, moving through additive manufacturing technologies, and ending with material selection and properties. Each of these sections provides a detailed explanation. The third section focuses extensively on DFAM, including design analysis, design optimization, material selection, and prototype production. The fourth section discusses Challenges and Solutions in AMFD in detail, including the following subsections: Material and Machine Limitations, Software and Tools, and Standards and Certifications. The fifth section discusses future developments and opportunities related to DFAM. This section includes advancements in 3D printing technologies, innovations, and the expansion of DFAM applications. The final section contains a brief summary of DFAM, its applications, successes, and future developments.

Finally, this chapter analyzes the practical challenges facing DFAM, such as material limitations, software shortages, and the need for greater standardization, while also considering future innovation opportunities.

[1] Department of Industrial Product Design, OSTİM Technical University, OSTİM, 06374 Ankara/Türkiye.
[2,3] Mechanical Department, Engineering College, University of Basrah, Basrah 61004, Iraq.
* Corresponding author: ahmet.atak@ostimteknik.edu.tr

DOI: 10.1201/9781003635673-6

1. Introduction

The current role of 3D printing in production has changed due to several factors. 3D printers can now meet the needs of diverse and often disparate industries. They can be a significant solution in the automotive, aerospace, and even healthcare sectors. In the healthcare sector, for example, biocompatible materials enable the healthcare industry to manufacture implants and prosthetics tailored to individual patients.

Additive manufacturing relies primarily on 3D printing, and its quality depends on the printer's capabilities. The 3D printer creates the desired model layer by layer using pre-programmed digital data within specialized software packages [1, 2]. Therefore, each manufactured product is a reflection of a unique, pre-programmed code. However, this code programming often relies on supporting software such as 3D modeling kits. This step is essential for the rapid design of customized models, allowing companies to respond quickly and flexibly to customer requirements [3, 4].

Additive manufacturing adds material only to the areas necessary to create the product, reducing material waste and enhancing sustainability [5, 6]. As is well known, producing a product prototype using traditional methods is time-consuming and expensive. Therefore, additive manufacturing speeds up the prototyping process [7], reduces delivery times, and directly reduces logistics costs [8].

In 1984, Chuck Hull invented the first kind of 3D printing. The inventor called it "stereolithography" or SLA. Chuck Hull continued in the field of 3D printing until he became CEO of 3D Systems. The commercialization of 3D printing dates to the 1990s, when Stratasys produced some of the first 3D printers. Stratasys continued its innovations until it produced 3D printers based on fused deposition modeling (FDM). The basic idea behind this type of printing is to deposit the molten material in successive layers. This technology is the foundation of what is known today as 3D printing of houses.

In the medical field, in 1999, the Wake Forest Research Institute successfully 3D-printed the "scaffolding" for a bladder. This model is enhanced with real cells to grow a new bladder on top of this base. Moreover, in 2002, an animal kidney was produced by 3D printing. Later, the Organovo company offered the first 3D-printed blood vessel. After that, there were amazing developments in the medical field. For example, in 2014, doctors were able to use a 3D-printed human skull for the first time on a woman in Norway. The successes of 3D printers continued across various fields until the production of large structures became possible. A company in China was able to build an entire house using 3D printing technology. Figure 1 summarizes the 3D printing technology history.

Industrial design refers to the activities adopted by industrial companies to create competitive advantages in the market. This strategy controls consumer demand and anticipates future trends [9]. While industrial design has developed significantly in the past two decades, the emergence of 3D printers has been particularly pivotal in this field.

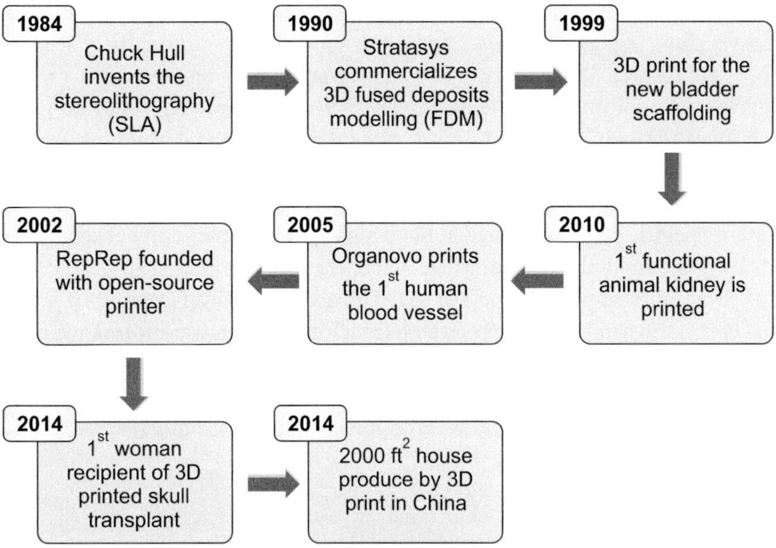

Fig. 1 3D printing technology history.

The 3D printers creating a technological leap led to the founding of a design for additive manufacturing (DFAM) in early 2000s. DFAM is a methodology for designing parts specifically to utilize 3D printing technologies. This approach has led to overcoming the limitations of traditional manufacturing. DFAM features include a greater capacity to manufacture complex parts and more sustainable raw material consumption. These features led to comprehensive improvements in production and faster turnaround times [10–12].

The production of 3D printers with exceptional capabilities has become a major objective for manufacturers to meet consumer demands. The consumer demands focused on lighter and more durable parts. Especially, manufacturers are focusing on complex geometric shapes, which traditional manufacturing processes cannot accommodate [13, 14]. Many companies now leverage DFAM as a core part of their brand identity. DFAM pays off in three ways: more attractive designs, lower manufacturing costs, and quicker time-to-market. There are many famous brands using DFAM production. Notable examples of products optimized through DFM include the Apple iPhone (reduced the component counts and simplified assembly), Tesla electric vehicles (increased production rates and lower costs compared to earlier models), IKEA flat-pack furniture, Dyson vacuum cleaners, the Boeing 787 Dreamliner, Whirlpool kitchen appliances, etc. Research also highlights that the intuitive nature of these processes makes implementation relatively straightforward [15–17].

The DFAM technique adds a layer of material on top of the previous layer. This technology results in less excess material and thus creates more complex shapes. So, it relies on traditional cutting or drilling, which depends on removing excess material. The DFAM techniques usually use photopolymerization, laser selective sintering, and molten deposition modeling to gradually add material to form the desired shape. It has given designers a broader vision for producing products using additive manufacturing [18, 19]. This approach relies on software that bridges the gap between a digital design and a physical object. It is guiding printers as they deposit layers of plastic, metal, or composites. It's becoming a staple in various sectors, and for good reasons. Not only does it make producing complex, specific parts much easier, but it also helps companies meet sustainability goals by reducing both material waste and energy consumption [20–22]. Polymers are of paramount importance in DFAM. The additive manufacturing of polymers opens up vast possibilities for producing functional and high-performance polymer parts. However, the fabrication of polymer parts requires an integrated approach based on real-world polymer material innovation and performance [23]. Figure 2 illustrates various design examples produced using 3D printing.

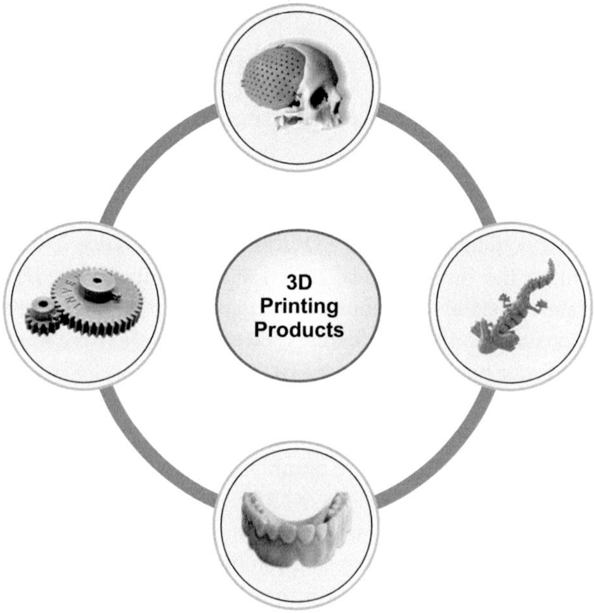

Fig. 2 Dfam Products Samples.

Figure 3 shows prototypes of design containers made using 3D printing.

2. 3D Printing Principles

2.1 Principles of Industrial Design

This section discusses the main components of industrial design. The main components often revolve around three aspects: the end-user approach, product

Fig. 3 3D-printed containers prototypes.

functionality, and ease of use. Moreover, it is preferred to be more aesthetic and visually appealing. To provide a comprehensive understanding of industrial design, this section explores some key points about it. Figure 4 indicates the industrial design elements.

The first stage, called the "Iterative Design Process," involves an iterative process of idea generation. This stage is followed by clear steps for building prototypes and then by refinement. Prototypes are often schematic or digital. The iteration and refinement process aims to create a product that meets its intended objectives. This stage is followed by user-centric design, which focuses on the needs and preferences of end users. Designers often concentrate on creating products that are visually appealing, simple to use, and highly comfortable. In addition, materials and manufacturing techniques should play a significant role in industrial design. Therefore, designers should have a thorough understanding of the materials used and their availability. Furthermore, they should also possess a strong grasp of the relevant manufacturing processes. Such a sound understanding enables manufacturers to consider manufacturing feasibility, improve the product, and enhance efficiency. Design for sustainability represents the fourth section that must be considered in industrial design. Sustainability has become an integral part of any manufacturing process. It is also mandated by international agreements binding on many countries and companies. Therefore, designers must work to minimize the environmental impact of their products throughout their lifecycle. The more recyclable and decomposable the products are, the more important they become. Industrial design is no longer limited to fulfilling requirements; it can also be an attractive and user-friendly product. Therefore, considering emotional and brand appeal has become essential for brand development and survival. A case study is the final step in the industrial prototyping and development process. Designers must have a case study for the model they intend to manufacture. For example, when designing a mobile phone, many factors must be considered. These include the phone's shape, screen size, battery capacity, visual appeal, user desire, and others. A case study can offer information about user feedback and market trends. These insights can significantly contribute to designers' understanding and ability to refine their prototypes.

So, the users' needs, understanding, preferences, and choice priorities should be considered in the DFAM [5]. Therefore, conducting surveys and opinion polls

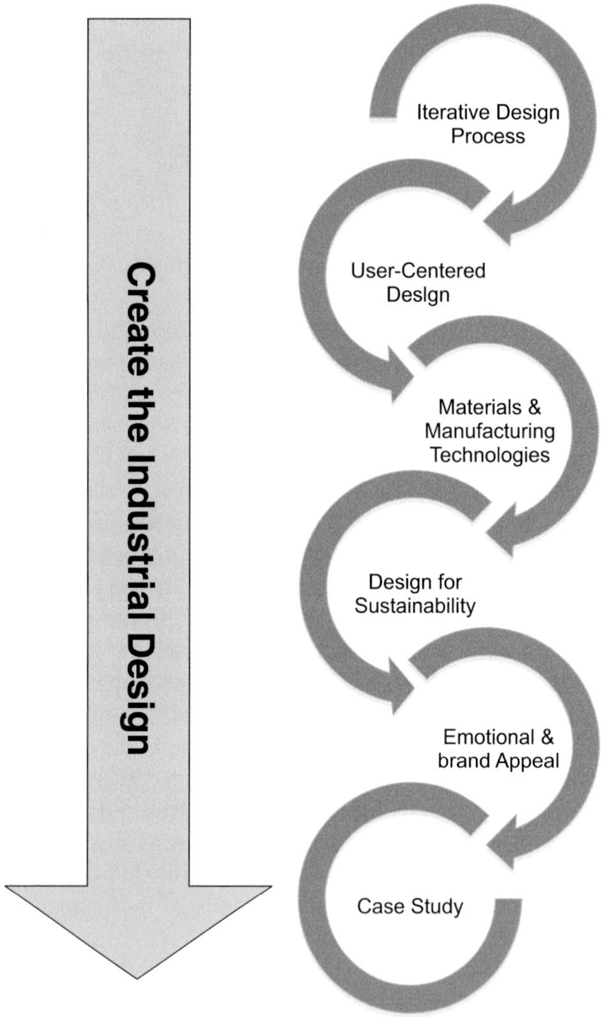

Fig. 4 Creation of the industrial design elements.

on user satisfaction or dissatisfaction has become essential and a given for every brand seeking to remain in the market longer and achieve industry leadership [24, 25]. DFAM designers have devoted more space (in later versions) to the concept of customizing products to users' specific needs based on actively collecting and evaluating user feedback [26, 27]. Therefore, product multi-functionality and ease of use are two primary goals of DFAM. Designers work with this approach, resulting in a product that is comfortable to use, with intuitive, smooth controls that enhance user interaction [28–30].

DFAM designers have also made aesthetics and visual appeal a significant and fundamental focus of their designs. Therefore, careful attention to form and proportion is essential, as they directly impact aesthetic value, enabling even complex geometric shapes to be more appealing to the user [31, 32]. High surfaces and great attention to surface finishing techniques have become a given in modern surface finishing designs, as smooth or textured finishes, along with the application of specific colors and patterns, can significantly enhance the visual impact of a product and increase user desirability [33, 34]. For DFAM designers, the ability to specialize is key. Optimal use and precise personalization can give traditional forms a unique character and greater appeal [35].

2.2 Advantages and Limitations of the Printing Process

The major benefit of the combination between 3D printers and DFAM is freedom to design. The combination offers many possibilities for making complex geometries. These possibilities are necessary for the parts manufacturing that fit together using 3D printing. In addition, these possibilities can enhance production time and improve cost reduction and reduce the need for extra material removal [36–38]. Moreover, 3D printing is very good for the environment and for using materials efficiently [39]. In 3D printing, parts are often made and put together in one step, so there is no need for extra joining or fixing processes [40]. The 3D part geometries can be optimized using DFAM to make the lightweight structures, holes, and complex surface shapes easy to manufacture [41]. These facilities are essential when compared with traditional manufacturing processes. Therefore, the cellular structures can be used to prepare strong solid or hollow parts [42]. Overall, DFAM makes it easy to make complex hollow shapes that cannot be made with traditional methods. Such performance improvement guided the parts to meet functional requirements [43]. Also, it costs less in terms of labor and tools than traditional manufacturing methods, which is a cost benefit [44].

Although 3D printers are faster now, they are still limited in use, especially for making a lot of things at once. Sometimes, there is an extra problem with related materials. Some materials don't have the same properties as those made with traditional methods [45]. Also, there are a lot of products that are hard to get due to 3D printing material availability [5].

Regarding the 3D printer's limitation, the removal of support structures is also classified as one of the main limitations. So many designers want to limit the use of support structures and make parts that can support themselves [46]. Moreover, the printer's speeds are still not quite suitable for mass production. There are limitations on the materials that can be printed; not all materials are printable. Products' accuracy/tolerance issues (layer lines) are also noticed. Finally, complex part design constraints like overhangs lead to weaker anisotropic parts compared to molding.

2.3 Additive Manufacturing Technologies

The principle behind 3D printers is to produce parts based on a feed of digital files through specialized software packages. The efficiency of the production process is measured by the quality of the product compared to the shortest time required for manufacturing. In DFAM, computer-aided design (CAD) software is often used to create a basic design and a digital model that can be run on a 3D printer. The additive manufacturing machine then understands the design model by slicing it into skinny layers. The additive manufacturing machine then deposits the material layer by layer using a specific type of material at predetermined time intervals.

There are several types of printers that can be used for AM, highlighted by FDM, SLA, selective laser sintering (SLS), digital light processing (DLP), and electron beam melting (EBM), as shown in Figure 5. Each of these printers operates differently, but the final product is similar, typically a 3D model based on a pre-designed computer code.

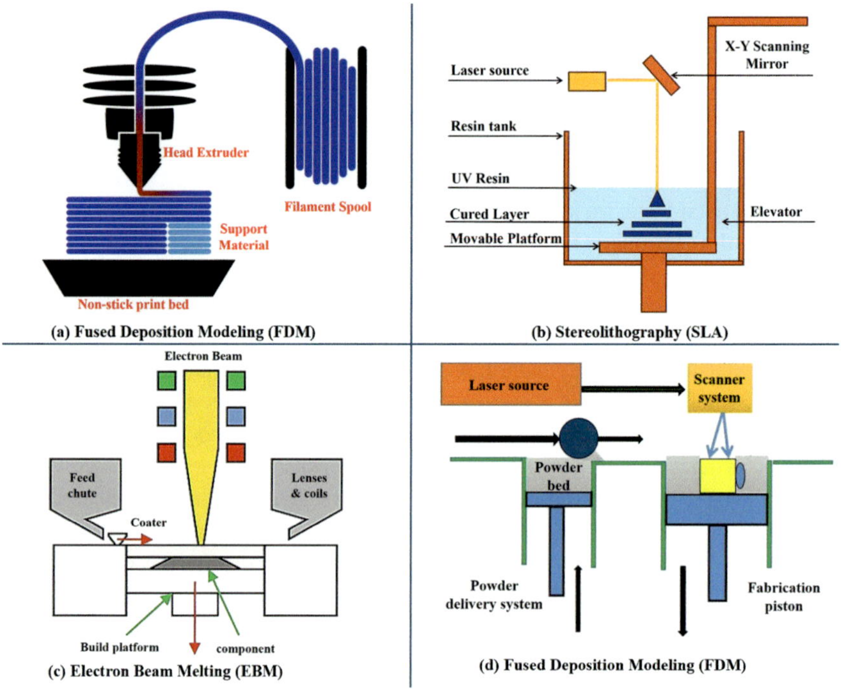

Fig. 5 Main 3D printer types.

FDM is the most popular DFAM method. In this method, a thermoplastic polymer is melted into a wire (called a filament) and pressed through a fine-diameter nozzle. In a semi-liquid case, it is noticed that the cured material is added layer by layer [4]. In SLA, the desired object is first programmed and interfaced with a 3D printer using special software. The software divides the object into cross-sections with thicknesses typically starting at 0.05 mm along the vertical axis.

In this method, laser beams are necessary to crystallize the liquid layers until the whole model is made [47]. The proper use of lasers is fundamental to the production of solid parts through SLS. Many polymer materials, particularly nylon, are used in this technique. Obtaining strong components with complex shapes and no supporting structures is often a desirable outcome. This method may require post-treatment to improve surface smoothness [48]. In 1987, DLP was first used. It is later categorized as a type of SLS technique. The processes are based on liquid resin additive, and a digital light processing (DLP) process the liquid photopolymer layer by layer to develop three-dimensional parts. The process is repeated frequently to produce successive layers until the final shape of the object is achieved. This method utilizes micromirrors that direct light onto a digital micromirror device (DMD) and selectively expose areas of resin for curing. Furthermore, DLP can produce larger quantities of objects in larger sizes at competitive production speeds [49]. EBM is an advanced manufacturing technique that differs from other methods. It uses an electron beam to selectively melt and combine metal powders. Then the melted powder deposits layer by layer to build a final product. The products of this technique are considered high-density metal parts. So, the EBM is suitable for precision applications, particularly those related to life, such as the fabrication of medical components [50]. Sheel Metal fabrication of the Laminated Object Manufacturing (LOM) is primarily used for rapid prototyping. Moreover, it is an impact process for the rapid, large-scale printing [51]. In this method, sheets of material are joined together and cut to the appropriate geometric shape according to a 3D model.

Figure 6 illustrates the main materials that are used in 3D printing. It can be found as liquids, sheets, filaments, and powders. DFAM typically requires post-production processes, which focus on removing support materials, smoothing surfaces, and coating [52, 53].

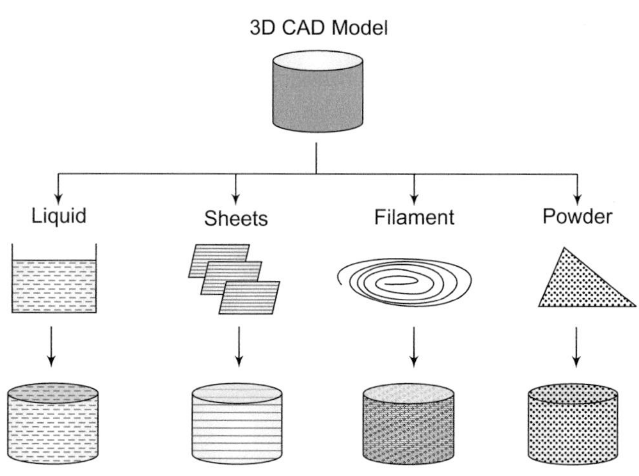

Fig. 6 The main materials that are used in 3D printing.

2.4 Material Selection and Properties

Flexibility, strength, and weight are key factors in selecting materials for DFAM processes, so plastics and metals are classified as the main materials used in these processes [54]. Hardness and thermal conductivity also play a role, especially when the material to be manufactured is metal. Although titanium is relatively expensive compared to other metals, it is widely accepted in DFAM technologies [55]. Furthermore, aluminum also possesses desirable properties for DFAM processes, and recently, the use of stainless steel in additive manufacturing has increased. On the other hand, polymers and plastics are crucial components of DFAM. Many products manufactured using 3D printers contain polyamide (PA), polycarbonate (PC), polyethylene terephthalate (PET), and polypropylene (PP) [33]. Figure 7 illustrates the main materials used in DFAM manufacturing.

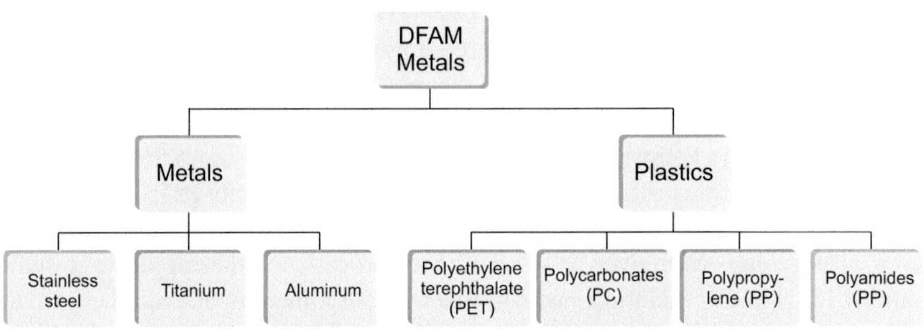

Fig. 7 Main materials used in the DFAM.

3. Design Process for DFAM

DFAM logic is based on a step-by-step method to improve how parts perform and how easily they can be made. It starts with a clear assessment of functional requirements. These functional requirements then lead into a design phase focused on efficiency. Efficiency is achieved through lattice structures and topology optimization. Success of mechanical needs at this stage usually depends on matching the suitable materials selection and the correct printing parameters. During this period, the design reaches the prototyping and testing phase, where the heaviest lifting is prepared. This process allows for final refinement before full-scale manufacturing begins. This workflow ensures that the unique benefits of AM are fully realized without sacrificing quality or budget.

3D modeling is essential for additive manufacturing as the foundation for precise and functional components. It enables designers to develop the required geometric shapes. Digitally controlled production processes are then guided to ensure the model conforms to the desired specifications. This conformity includes dimensions, tolerances, and material types. Furthermore, 3D modeling aids in rendering and evaluating the final product to identify issues that need to be addressed during manufacturing. The process can be included:

1. Design analysis, which refers to meeting customer expectations. This aspect is crucial in the DFAM process. Therefore, the intended design should be based on a careful study of these expectations and the nature of their needs [56]. However, careful analysis should not go over budget or make the production process more difficult. Consequently, a meticulous analysis must be predicated on the assessment of customer feedback, market research, and user experiences during the needs analysis [56, 57].
2. Design Optimization: CAD software is commonly employed to model and enhance designs in DFAM applications by generating a three-dimensional digital model utilizing parametric design methodologies, encompassing all specifications and dimensions of the intended product, including requisite structures [58, 59]. This is how simulations work, and they often use the finite element method (FEM). FEM allows the product design to have a mesh structure that shows detailed modeling for optimization and simulation [60, 61]. Figure 8 shows the FEM mesh form of lattice structures in relation to the structures.

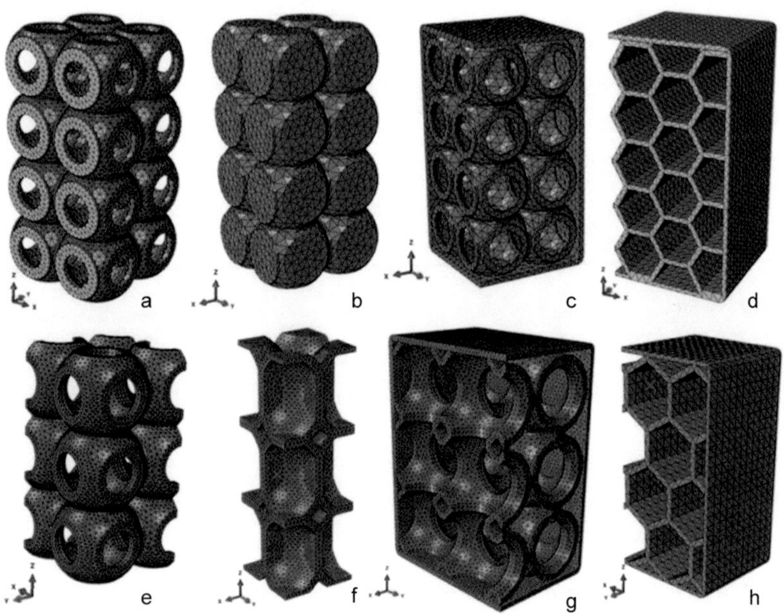

Fig. 8 FEM mesh form of lattice structures with reference to the structures a–h [61].

3. Material Selection: Polymers, metals, ceramics, and composites are the primary materials used in DFAM. The appropriate choice of materials can be a critical factor in the design's functionality, durability, sustainability, and cost. Materials are selected based on the nature of the product to be manufactured, and therefore, mechanical properties and thermal behavior are

the most important characteristics to consider. In addition, the availability of supply chains and cost also have an important impact [62, 63].

4. Prototype Production: The prototype represents an essential stage in the DFAM process chain, as it is always followed by manufacturing tests to improve the design and address defects. 3D printers have provided individual capabilities for producing and visualizing models at record speeds and lower costs [64, 65]. The prototype is tested according to the target environment or application. Tests can check performance characteristics, mostly mechanical properties, which are very important before going from the design stage to mass production [66, 67]. Figure 9 illustrates the importance of testing, as it gives a clear impression of the durability and its relationship to the shape and design of the prototype.

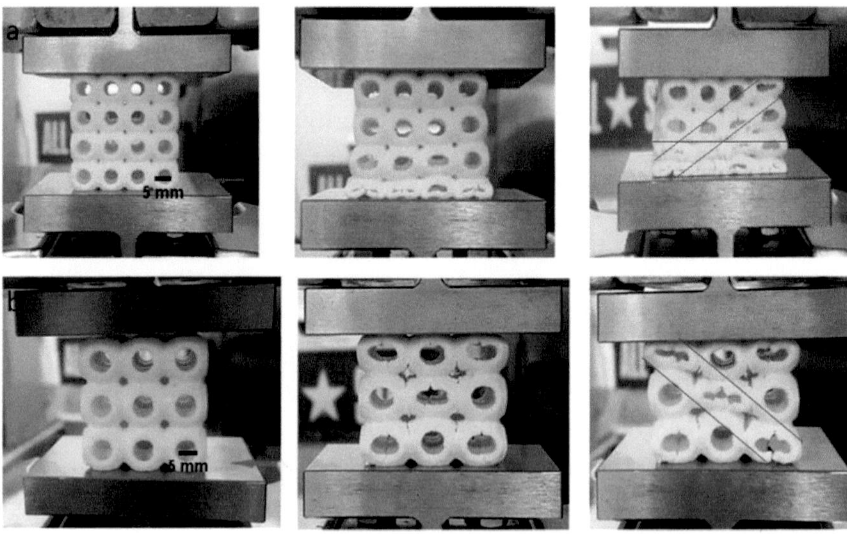

Fig. 9 Experimental compressive deformation for different prototypes [61].

4. Challenges and Solutions in AMFD

4.1 Material and Machine Limitations

The importance of durability in AMFD processes has already been mentioned, as it is a key factor in material selection. Durability varies depending on the material, being greater in metals than in plastics. Choosing the appropriate manufacturing process depends not only on the properties of the materials used but also on considering geometric parameters such as size and thickness [68, 69].

On the other hand, the speed of 3D printers and the quality of the product relative to the printer speed are the decisive factors in choosing a 3D printer, and they often determine the printer's price [70]. Despite significant advancements in

3D printer technology, their speeds are still below expectations. The low production speeds are the worst factor in the limitation of the production of materials using 3D printers. Another obstacle in manufacturing is the need for support when using 3D printers. Therefore, the need for supporting materials means additional work later in 3D printing [71]. This additional work influences the widespread use of 3D printing. So, it is clear that the right choice of 3D printer ensures a high-quality surface and prevents surface roughness [72].

4.2 Software and Tools

The growth of AMFD manufacturing is closely tied to different types of software. CAD software is where most manufacturing happens. The programming starts with the details and dimensions of the prototype. After that the process moves on to the modeling. Besides CAD, there are a lot of software packages, including SolidWorks, Autodesk Fusion 360, and CATIA [73]. Moreover, the topology optimization software is influential in the AMFD process. These software packages can make sure that the design space is used in the best way possible and that parts are lighter. Topology optimization often uses software like ANSYS and Altair OptiStruct [74]. The lattice design software is also used by manufacturers, especially Autodesk Within and Materialize 3-Matic [75]. Recently, simulation software has been used more in AMFD processes. The simulation software can do a lot of important analyses, like stress analysis and temperature distribution.

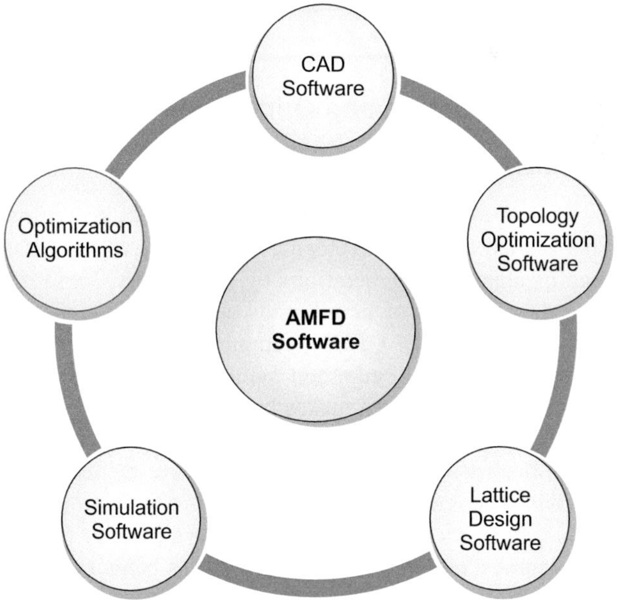

Fig. 10 Main software used in AMFD processes.

ANSYS, COMSOL, and Moldex3D are the most common simulation software packages in this field [76]. Moreover, AMFD processes often use algorithms. These algorithms are used to find the best design results by optimizing parameters. Many algorithms based on MATLAB and Python are used to write optimization algorithms [77]. The most common types of software used during the AMFD stages are shown in Figure 10.

4.3 Standards and Certifications

The perfect AMFD manufacturers follow specific standards and certifications during the production processes. These requirements let them make sure that the parts they make meet safety and quality standards. Many standards-focused organizations periodically issue updated standards. ISO standards decide if the materials, design, and production used in the AMFD process are safe and of good quality. Standards such as ISO 9001 Quality Management System, ISO 13485 Quality Management System for Medical Devices, and ISO 14001 Environmental Management System are widely recommended in the AMFD processes [78]. On the other hand, ASTM standards such as ASTM F2792-12 also set out requirements for materials, testing methods, and performance [79].

Regarding certificates, CE Certification is the certificate that is required in the European Economic Area (EEA). CE certification means that a product meets safety, health, and environmental standards in this specific region [80]. In the United States, the Food and Drug Administration (FDA) must approve any medical device made. The FDA requires all medical devices, including those made using AMFD processes [81].

5. Future Developments and Opportunities

5.1 Advancements in 3D Printing Technologies

3D printing has become one of the most active investments in new manufacturing processes. This has led to rapid developments. These developments have led to the appearance of many new generations of 3D printers. Various trends have emerged in the development of new brands and their corresponding models. Laser powder bed fusion (LPBF), EBM, ceramic printing, bioprinting (BP), and composite printing are the most common models that have appeared in the last ten years. In the LPBF field, which makes layered structures by directly melting or sintering metals, progress has been made in making strong, complicated metal parts [82]. In the field of EBM, where metal parts are melted with an electron beam, products have become very accurate, and ways to control production have improved, making them more efficient and effective [83]. In the last ten years, ceramics have become a part of 3D printing. The idea behind ceramic printing is to combine and sinter ceramic powders layer by layer to make ceramic parts. This technology has changed the ceramics industry and produced an unexpected jump

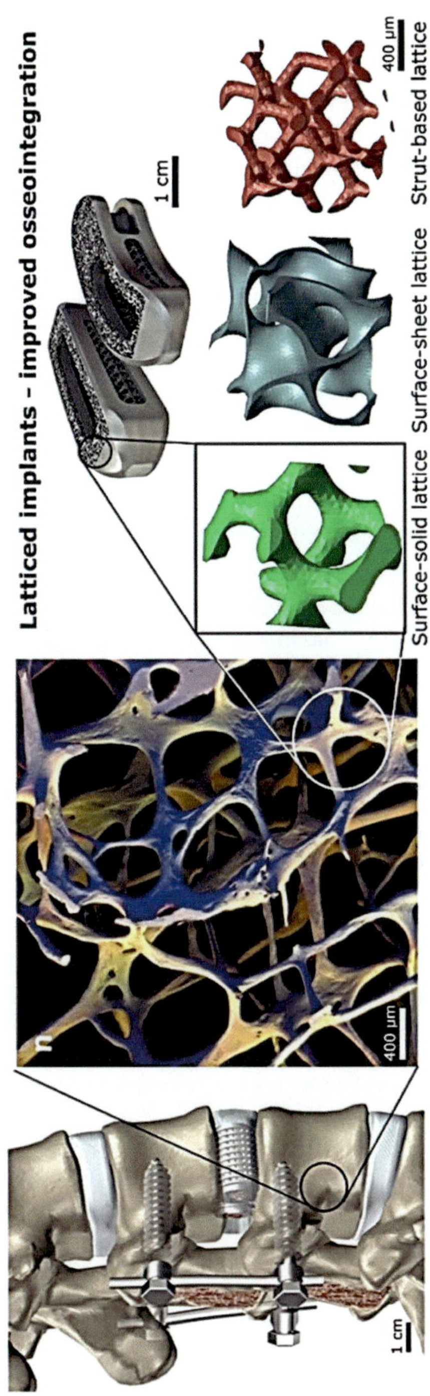

Fig. 11 (Left) Spine with implanted inter-disk metal device; (center) trabecular bone structure; (right) porous spinal implant device on top and three different types of lattice topologies at the bottom. The surface-based skeleton type lattice is selected for this work [87].

in this field. The new technology makes it possible to make products with shapes that are both complex and unusual [84]. Another remarkable development is that 3D printers are now capable of creating products from different materials, not just identical ones. This has paved the way for the ability to manufacture parts from composite materials. Composite parts manufactured using additive manufacturing techniques have the same mechanical and physical properties as those obtained using traditional methods or sometimes even surpass those of traditional products [85].

Perhaps the most attractive aspect of 3D printer development is what is known as bioprinting—a technology that enables the production of biological tissues and organs using 3D printing. Although it is still in the research and biomedical treatment phase, it is a promising technology despite its ethical complexities [86]. The operating principle of a bioprinter is similar to that of any other 3D printer: it assembles cells and biological materials layer by layer to create complex tissue structures. Figure 11 shows the use of 3D printing in the field of bioprinting.

5.2 Innovations and Expansion of DFAM Applications

DFAM is constantly undergoing innovation in the industry, and its application areas are expanding. Many sectors have witnessed increased use of DFAM-manufactured products. DFAM has increased the ability to produce metal materials directly through 3D printing, making it suitable for industrial applications [88]. Applications such as the production of high-strength parts include vital applications such as aerospace components and automotive parts, which have significantly contributed to producing more efficient products with reduced weight and fuel consumption [89, 90]. DFAM has also played an increasing role in bioprinting and medical applications. Applications such as organ and tissue engineering, the production of prosthetic limbs, and customized medical devices have become a broad field for designers and manufacturers of additive manufacturing technologies [91].

One of the best things about DFAM's growth is how important it has been for building capacity in schools and other educational settings. This has helped students become more creative and better at engineering. DFAM has made it possible to make models, and it has grown to include toy making and the arts [5]. Most architects who worked in this type of manufacturing found their calling because it let them speed up the design and manufacturing processes [92]. Table 1 shows the main topics and uses of DFAM in architectural design.

Table 1 Main applications of DFAM in architectural design.

Ref. No.	*Main Topics*	*Applications*
[92]	DFMA concepts in architecture, engineering, and construction; DFAM; ML integration	Enhancing process efficiency in building projects via digital and additive manufacturing

Contd.

Table 1 *Contd.*

Ref. No.	*Main Topics*	*Applications*
[93]	Curved concrete structures; CAD and structural analysis; 3D printing (3DP)	Case study on 3DP in creating a multipurpose pavilion with complex geometry
[94]	Thermal performance of 3D printed concrete buildings; infrared thermography	Identifying thermal flaws in 3DPC buildings; measuring R, U values and thermal conductivity
[95]	Review of 3DPC technology; printing parameters; sustainability and economics	Large-scale building construction using 3DPC; evaluation of current challenges and future prospects
[96]	Sustainable construction via 3D printing; mortar performance; insulation studies	Thermal and sound insulation of 3D printed wall panels; evaluating mortar and panel performance

6. Conclusion

The defining characteristic of today's competitive manufacturing landscape is its rapid pace of change. Additive manufacturing, particularly with regard to 3D printing, is undoubtedly a key element of this evolution. The assumption that additive manufacturing was merely a tool for prototyping has been proven false. Additive manufacturing, especially in the context of DFAM, has become a game changer, maximizing efficiency and minimizing waste. It has also opened up vast new design possibilities. This study examines the transformative impact of DFAM on industrial design and its potential to create new opportunities in product development. The results show some of the main benefits of DFAM, such as:

1. DFAM makes it now possible to make complex, high-performance shapes that were once impossible, so it can offer a better design.
2. The DFAM is active for the efficient use of materials. In this process, waste is minimized by using only what is necessary.
3. In the DFAM process, the designs can go straight to 3D printing, which cuts down on the time it takes to make them. Therefore, the DFAM process is an influential tool for faster production.
4. The DFAM can make products fit for required needs without having to spend a lot of money on new tools. Therefore, the personalization and customization are achieved in the DFAM process.
5. During the DFAM process, the products can be achieved as lighter, stronger, and more useful because of design freedom compared with traditional manufacturing methods.

DFAM takes away many design restrictions that are common in traditional manufacturing. This lets for new structures that are more efficient and last longer. Being able to 3D-print complex parts on demand means less waste and faster turnaround times.

References

[1] I. Gibson, D. Rosen, and B. Stucker, *Additive manufacturing technologies: 3D printing, rapid prototyping, and direct digital manufacturing, second edition*. 2015. doi: 10.1007/978-1-4939-2113-3.

[2] A. Nazir, K.M. Abate, A. Kumar, and J.Y. Jeng, "A state-of-the-art review on types, design, optimization, and additive manufacturing of cellular structures," *International Journal of Advanced Manufacturing Technology*, vol. 104, no. 9–12, 2019, doi: 10.1007/s00170-019-04085-3.

[3] T. Wohlers, "3D Printing and Additive Manufacturing State of the Industry. Annual Worldwide Progress Report.," 2017.

[4] I. Campbell, D. Bourell, and I. Gibson, "Additive manufacturing: rapid prototyping comes of age," *Rapid Prototyp J*, vol. 18, no. 4, 2012, doi: 10.1108/13552541211231563.

[5] B. Berman, "3-D printing: The new industrial revolution," *Bus Horiz*, vol. 55, no. 2, 2012, doi: 10.1016/j.bushor.2011.11.003.

[6] M. Kreiger and J.M. Pearce, "Environmental life cycle analysis of distributed three-dimensional printing and conventional manufacturing of polymer products," *ACS Sustain Chem Eng*, vol. 1, no. 12, 2013, doi: 10.1021/sc400093k.

[7] S.H. Huang, P. Liu, A. Mokasdar and L. Hou, "Additive manufacturing and its societal impact: A literature review," 2013. doi: 10.1007/s00170-012-4558-5.

[8] Y. Huang, M.C. Leu, J. Mazumder, and A. Donmez, "Additive manufacturing: Current state, future potential, gaps and needs, and recommendations," *Journal of Manufacturing Science and Engineering, Transactions of the ASME*, vol. 137, no. 1, 2015, doi: 10.1115/1.4028725.

[9] A. Alfaify, M. Saleh, F.M. Abdullah, and A.M. Al-Ahmari, "Design for additive manufacturing: A systematic review," 2020. doi: 10.3390/SU12197936.

[10] Y.S. Leung, T.H. Kwok, X. Li, Y. Yang, C.C.L. Wang, and Y. Chen, "Challenges and status on design and computation for emerging additive manufacturing technologies," *J Comput Inf Sci Eng*, vol. 19, no. 2, 2019, doi: 10.1115/1.4041913.

[11] Martin Leary, *Design for Additive Manufacturing*. Elsevier, 2020. doi: 10.1016/C2017-0-04238-6.

[12] B. Durakovic, "Design for additive manufacturing: Benefits, trends and challenges," *Periodicals of Engineering and Natural Sciences*, vol. 6, no. 2, 2018, doi: 10.21533/pen.v6i2.224.

[13] M.K. Thompson *et al.*, "Design for Additive Manufacturing: Trends, opportunities, considerations, and constraints," *CIRP Ann Manuf Technol*, vol. 65, no. 2, 2016, doi: 10.1016/j.cirp.2016.05.004.

[14] A. Jadhav and V.S. Jadhav, "A review on 3D printing: An additive manufacturing technology," in *Materials Today: Proceedings*, 2022. doi: 10.1016/j.matpr.2022.02.558.

[15] K. Ulrich, S.D. Eppinger, and M.C. Yang, *Product Design and Development - Seventh Edition*. 2020.

[16] B.B. De Mozota, "Design Management: Using Design to Build Brand Value and Corporate Innovation," *Technovation*, vol. 25, no. 3, 2005.

[17] J. HESKETT, *Design - Very Short Introductions*. 2005.

[18] V. Goodship, B. Middleton and R. Cherrington, *Design and Manufacture of Plastic Components for Multifunctionality: Structural Composites, Injection Molding, and 3D Printing*. 2015. doi: 10.1016/C2014-0-00223-7.

[19] G. Liu, Y. Xiong and D.W. Rosen, "Multidisciplinary design optimization in design for additive manufacturing," *J Comput Des Eng*, vol. 9, no. 1, 2022, doi: 10.1093/jcde/qwab073.

[20] S.A.S. Almtori, R.J. Jassim, D.C. Ali, E.H. Kadhim, and R. Al-Sabur, "Sustainable Manufacturing Process Applied to Produce Waste Polymer-Polymer Matrix Composites," *Environmental*

Research, Engineering and Management, vol. 79, no. 1, pp. 122–132, Apr. 2023, doi: 10.5755/j01.erem.79.1.32907–.

[21] R. Al-Sabur and M. Serier, "Material Sustainability During Friction Stir Joining," in *Sustainable Machining and Green Manufacturing*, 2024. doi: 10.1002/9781394197866.ch7.

[22] B. Almangour, *Additive manufacturing of emerging materials*. 2018. doi: 10.1007/978-3-319-91713-9.

[23] J.W. Booth, J. Alperovich, P. Chawla, J. Ma, T.N. Reid, and K. Ramani, "The design for additive manufacturing worksheet," *Journal of Mechanical Design*, vol. 139, no. 10, 2017, doi: 10.1115/1.4037251.

[24] B.A. Aljabali, S.K. Parupelli and S. Desai, "Generalized Design for Additive Manufacturing (DfAM) Expert System Using Compliance and Design Rules," *Machines*, vol. 13, no. 1, p. 29, Jan. 2025, doi: 10.3390/machines13010029.

[25] H. Ko, S. Enea, Z.Y. Chua, S.K. Moon and K.N. Otto, "User-centered design for additive manufacturing as a customization strategy," in *Proceedings of the International Conference on Progress in Additive Manufacturing*, 2016.

[26] "Printed and Hybrid Electronics Enabled by Digital Additive Manufacturing Technologies," in *Additive Manufacturing*, 2015. doi: 10.1201/b19360-10.

[27] C. Weller, R. Kleer and F.T. Piller, "Economic implications of 3D printing: Market structure models in light of additive manufacturing revisited," *Int J Prod Econ*, vol. 164, 2015, doi: 10.1016/j.ijpe.2015.02.020.

[28] C.K. Chua, K.F. Leong and C.S. Lim, *Rapid prototyping: Principles and applications, third edition*. 2010. doi: 10.1142/6665.

[29] N. Hopkinson, R.J.M. Hague and P.M. Dickens, *Rapid Manufacturing: An Industrial Revolution for the Digital Age*. 2006. doi: 10.1002/0470033991.

[30] C. Chu, G. Graf and D.W. Rosen, "Design for additive manufacturing of cellular structures," *Comput Aided Des Appl*, vol. 5, no. 5, 2008, doi: 10.3722/cadaps.2008.686-696.

[31] S.A.M. Tofail, E.P. Koumoulos, A. Bandyopadhyay, S. Bose, L. O'Donoghue and C. Charitidis, "Additive manufacturing: scientific and technological challenges, market uptake and opportunities," 2018. doi: 10.1016/j.mattod.2017.07.001.

[32] M. Gopal, H.G. Lemu and E.M. Gutema, "Sustainable Additive Manufacturing and Environmental Implications: Literature Review," 2023. doi: 10.3390/su15010504.

[33] J. Edgar and S. Tint, "Additive Manufacturing Technologies: 3D Printing, Rapid Prototyping, and Direct Digital Manufacturing', 2nd Edition," *Johnson Matthey Technology Review*, vol. 59, no. 3, 2015, doi: 10.1595/205651315x688406.

[34] J.M. Costa, E.W. Sequeiros, D. Figueiredo, A.R. Reis and M.F. Vieira, "Optimizing Metal AM Potential through DfAM: Design, Performance, and Industrial Impact," in *Additive Manufacturing - Present and Sustainable Future, Materials and Applications [Working Title]*, IntechOpen, 2024. doi: 10.5772/intechopen.1007309.

[35] P.F. Egan, "Design for Additive Manufacturing: Recent Innovations and Future Directions," 2023. doi: 10.3390/designs7040083.

[36] B. Kianian, "Wohlers Report 2017: 3D Printing and Additive Manufacturing State of the Industry, Annual Worldwide Progress Report: Chapters titles: The Middle East, and other countries," *Wohlers Report - 3D Printing and Additive Manufacturing State of the Industry*, vol. 22, 2017.

[37] T. Campbell, C. Williams, O. Ivanova and B. Garrett, "Could 3D Printing Change the World?," *Atlantic Council*, no. May 2021, 2012.

[38] K. Wong, "K.V. Wong, A. Hernandez, 'A Review of Additive Manufacturing,' ISRN Mechanical Engineering, Vol 2012 (2012), Article ID 208760, 10 pages.," *ISRN Mechanical Engineering*, 2012.

[39] J. Gardan, "Additive manufacturing technologies: State of the art and trends," *Int J Prod Res*, vol. 54, no. 10, 2016, doi: 10.1080/00207543.2015.1115909.

[40] J.P. Kruth, X. Wang, T. Laoui and L. Froyen, "Lasers and materials in selective laser sintering," *Assembly Automation*, vol. 23, no. 4, 2003, doi: 10.1108/01445150310698652.

[41] 3Dprinting scratch, "Types of 3D printers or 3D printing technologies overview," *3D Printing from scratch*, 2020.

[42] N. Khan and A. Riccio, "A systematic review of design for additive manufacturing of aerospace lattice structures: Current trends and future directions," *Progress in Aerospace Sciences*, vol. 149, p. 101021, Aug. 2024, doi: 10.1016/j.paerosci.2024.101021.

[43] Q. Zhang *et al.*, "Superior energy absorption characteristics of additively-manufactured hollow-walled lattices," *Int J Mech Sci*, vol. 264, 2024, doi: 10.1016/j.ijmecsci.2023.108834.

[44] M. Attaran, "The rise of 3-D printing: The advantages of additive manufacturing over traditional manufacturing," *Bus Horiz*, vol. 60, no. 5, 2017, doi: 10.1016/j.bushor.2017.05.011.

[45] S.C. Ligon, R. Liska, J. Stampfl, M. Gurr and R. Mülhaupt, "Polymers for 3D Printing and Customized Additive Manufacturing," 2017. doi: 10.1021/acs.chemrev.7b00074.

[46] Y.H. Kuo, C.C. Cheng, Y.S. Lin and C.H. San, "Support structure design in additive manufacturing based on topology optimization," *Structural and Multidisciplinary Optimization*, vol. 57, no. 1, 2018, doi: 10.1007/s00158-017-1743-z.

[47] S.S. Crump, "US5121329A - Apparatus and method for creating three-dimensional objects - Google Patents," 1989

[48] J.P. Kruth, P. Mercelis, J. Van Vaerenbergh, L. Froyen and M. Rombouts, "Binding mechanisms in selective laser sintering and selective laser melting," 2005. doi: 10.1108/13552540510573365.

[49] S. Swetha, T.J. Sahiti, G.S. Priya, K. Harshitha and A. Anil, "Review on digital light processing (DLP) and effect of printing parameters on quality of print," *Interactions*, vol. 245, no. 1, p. 178, Jul. 2024, doi: 10.1007/s10751-024-02018-5.

[50] T. DebRoy *et al.*, "Additive manufacturing of metallic components – Process, structure and properties," 2018. doi: 10.1016/j.pmatsci.2017.10.001.

[51] S.D. Nath and S. Nilufar, "An overview of additive manufacturing of polymers and associated composites," 2020. doi: 10.3390/polym12112719.

[52] Y. Zhai, D.A. Lados and J.L. Lagoy, "Additive Manufacturing: Making imagination the major Limitation," *JOM*, vol. 66, no. 5, 2014, doi: 10.1007/s11837-014-0886-2.

[53] S.K. Samal *et al.*, "3D-Printed Satellite Brackets: Materials, Manufacturing and Applications," 2022. doi: 10.3390/cryst12081148.

[54] A.S. Safaa, N.H.J. Al Hassan, D.C. Ali, M.A. Mohammed and R. Al-Sabur, "Assessing the compressive performance of PVC coating on steel wire mesh reinforced concrete," *Cogent Eng*, vol. 10, no. 1, 2023, doi: 10.1080/23311916.2023.2231700.

[55] T. Mukherjee, H.L. Wei, A. De, and T. DebRoy, "Heat and fluid flow in additive manufacturing – Part II: Powder bed fusion of stainless steel, and titanium, nickel and aluminum base alloys," *Comput Mater Sci*, vol. 150, 2018, doi: 10.1016/j.commatsci.2018.04.027.

[56] S. Yang and Y.F. Zhao, "Additive manufacturing-enabled design theory and methodology: a critical review," 2015. doi: 10.1007/s00170-015-6994-5.

[57] A. Wiberg, J. Persson and J. Ölvander, "Design for additive manufacturing – a review of available design methods and software," 2019. doi: 10.1108/RPJ-10-2018-0262.

[58] Y. Huang and M.C.L. March, "Frontiers of Additive Manufacturing Research and Education," *NSF workshop report*, no. March, 2014.

[59] L.E. Murr *et al.*, "Microstructure and mechanical behavior of Ti-6Al-4V produced by rapid-layer manufacturing, for biomedical applications," 2009. doi: 10.1016/j.jmbbm.2008.05.004.

[60] Mikell P. Groover, *Fundamentals of Modern Manufacturing Materials, Processes, and Systems Seventh Edition*, vol. 53, no. 9. 2020.

[61] A. Kumar, L. Collini, A. Daurel and J.Y. Jeng, "Design and additive manufacturing of closed cells from supportless lattice structure," *Addit Manuf*, vol. 33, 2020, doi: 10.1016/j.addma.2020.101168.

[62] J.O. Milewski, *Additive manufacturing of Metals - From Fundamental Technology to Rocket Nozzles, Medical Implants, and Custom Jewelry*, vol. 69, no. 12. 2018.

[63] M.A. de Pastre, Y. Quinsat and C. Lartigue, "Effects of additive manufacturing processes on part defects and properties: a classification review," *International Journal on Interactive Design and Manufacturing*, vol. 16, no. 4, 2022, doi: 10.1007/s12008-022-00839-8.

[64] C. Wang, X.P. Tan, S.B. Tor and C.S. Lim, "Machine learning in additive manufacturing: State-of-the-art and perspectives," 2020. doi: 10.1016/j.addma.2020.101538.

[65] M. Jiménez, L. Romero, I.A. Domínguez, M.D.M. Espinosa and M. Domínguez, "Additive Manufacturing Technologies: An Overview about 3D Printing Methods and Future Prospects," *Complexity*, vol. 2019, 2019, doi: 10.1155/2019/9656938.

[66] T. Vaneker, A. Bernard, G. Moroni, I. Gibson and Y. Zhang, "Design for additive manufacturing: Framework and methodology," *CIRP Annals*, vol. 69, no. 2, 2020, doi: 10.1016/j.cirp.2020.05.006.

[67] B. Ahuja, M. Karg, and M. Schmidt, "Additive manufacturing in production: challenges and opportunities," in *Laser 3D Manufacturing II*, 2015. doi: 10.1117/12.2082521.

[68] J.J. Lewandowski and M. Seifi, "Metal Additive Manufacturing: A Review of Mechanical Properties," 2016. doi: 10.1146/annurev-matsci-070115-032024.

[69] M. Manoharan and S. Kumaraguru, "Novel process planning approach for support-free additive manufacturing using multi-axis deposition systems," *Int J Comput Integr Manuf*, vol. 36, no. 6, 2023, doi: 10.1080/0951192X.2022.2145020.

[70] Z.C. Eckel, C. Zhou, J.H. Martin, A.J. Jacobsen, W.B. Carter and T.A. Schaedler, "Additive manufacturing of polymer-derived ceramics," *Science (1979)*, vol. 351, no. 6268, 2016, doi: 10.1126/science.aad2688.

[71] R. Spina and L. Morfini, "Material Extrusion Additive Manufacturing of Ceramics: A Review on Filament-Based Process," *Materials*, vol. 17, no. 11, p. 2779, Jun. 2024, doi: 10.3390/ma17112779.

[72] I. Buj-Corral, X. Sánchez-Casas and C.J. Luis-Pérez, "Analysis of am parameters on surface roughness obtained in pla parts printed with fff technology," *Polymers (Basel)*, vol. 13, no. 14, 2021, doi: 10.3390/polym13142384.

[73] A.H. Azman, F. Vignat, and F. Villeneuve, "Evaluating current cad tools performances in the context of design for additive manufacturing," *Joint Conference on Mechanical, Design Engineering & Advanced Manufacturing*, no. January 2016, 2014.

[74] W. Du *et al.*, "Innovative joint for cable dome structure based on topology optimization and additive manufacturing," *Materials*, vol. 14, no. 18, 2021, doi: 10.3390/ma14185158.

[75] Z.H. Neamah, L.A.H. Al-Kindi and G. Al-Kindi, "Designing and manufacturing a patient-specific implant for a part of the femur using additive manufacturing," 2024, p. 040011. doi: 10.1063/5.0236269.

[76] C.Y. Chung, "Integrated optimum layout of conformal cooling channels and optimal injection molding process parameters for optical lenses," *Applied Sciences (Switzerland)*, vol. 9, no. 20, 2019, doi: 10.3390/app9204341.

[77] D. Kim, Y. Ji, J. Lee, J. Yoo, S. Min and I.G. Jang, "A MATLAB code of node-based topology optimization in 3D arbitrary domain for additive manufacturing," *Structural and Multidisciplinary Optimization*, vol. 65, no. 11, 2022, doi: 10.1007/s00158-022-03339-1.

[78] A. Chiarini and A. Cherrafi, "Integrating ISO 9001 and Industry 4.0. An implementation guideline and PDCA model for manufacturing sector," *Total Quality Management and Business Excellence*, vol. 34, no. 13–14, 2023, doi: 10.1080/14783363.2023.2192916.

[79] C. Wang, H. Bai, C. Ren, X. Fang and B. Lu, "A comprehensive prediction model of bead geometry in wire and arc additive manufacturing," in *Journal of Physics: Conference Series*, 2020. doi: 10.1088/1742-6596/1624/2/022018.

[80] J. McDonald, "Regulatory Considerations for Devices Manufactured Using Additive Manufacturing Technologies," in *Polymer-Based Additive Manufacturing*, 2019. doi: 10.1007/978-3-030-24532-0_11.

[81] Z. Rahman, S.F. Barakh Ali, T. Ozkan, N.A. Charoo, I.K. Reddy and M.A. Khan, "Additive Manufacturing with 3D Printing: Progress from Bench to Bedside," 2018. doi: 10.1208/s12248-018-0225-6.

[82] J. Elambasseril, J. Rogers, C. Wallbrink, D. Munk, M. Leary and M. Qian, "Laser powder bed fusion additive manufacturing (LPBF-AM): the influence of design features and LPBF variables on surface topography and effect on fatigue properties," 2023. doi: 10.1080/10408436.2022.2041396.

[83] L.E. Murr, "Metallurgy of additive manufacturing: Examples from electron beam melting," *Addit Manuf*, vol. 5, 2015, doi: 10.1016/j.addma.2014.12.002.

[84] Z. Chen *et al.*, "3D printing of ceramics: A review," 2019. doi: 10.1016/j.jeurceramsoc.2018.11.013.

[85] S. Yuan, S. Li, J. Zhu and Y. Tang, "Additive manufacturing of polymeric composites from material processing to structural design," 2021. doi: 10.1016/j.compositesb.2021.108903.

[86] N. Cubo-Mateo and M. Gelinsky, "Wound and Skin Healing in Space: The 3D Bioprinting Perspective," *Front Bioeng Biotechnol*, vol. 9, 2021, doi: 10.3389/fbioe.2021.720217.

[87] D. Barba, E. Alabort and R.C. Reed, "Synthetic bone: Design by additive manufacturing," *Acta Biomater*, vol. 97, 2019, doi: 10.1016/j.actbio.2019.07.049.

[88] B.A. Praveena, N. Lokesh, A. Buradi, N. Santhosh, B.L. Praveena and R. Vignesh, "A comprehensive review of emerging additive manufacturing (3D printing technology): Methods, materials, applications, challenges, trends and future potential," in *Materials Today: Proceedings*, 2022. doi: 10.1016/j.matpr.2021.11.059.

[89] M. Trovato and P. Cicconi, "Design tools for metal additive manufacturing: a critical and perspective overview," in *Procedia CIRP*, 2023. doi: 10.1016/j.procir.2023.03.151.

[90] V. Mohanavel, K.S. Ashraff Ali, K. Ranganathan, J. Allen Jeffrey, M.M. Ravikumar, and S. Rajkumar, "The roles and applications of additive manufacturing in the aerospace and automobile sector," in *Materials Today: Proceedings*, 2021. doi: 10.1016/j.matpr.2021.04.596.

[91] A.J. Sheoran, H. Kumar, P.K. Arora and G. Moona, "Bio-medical applications of additive manufacturing: A review," in *Procedia Manufacturing*, 2020. doi: 10.1016/j.promfg.2020.10.093.

[92] W. Tuvayanond and L. Prasittisopin, "Design for Manufacture and Assembly of Digital Fabrication and Additive Manufacturing in Construction: A Review," 2023. doi: 10.3390/buildings13020429.

[93] L. Prasittisopin, T. Sakdanaraseth and V. Horayangkura, "Design and Construction Method of a 3D Concrete Printing Self-Supporting Curvilinear Pavilion," *Journal of Architectural Engineering*, vol. 27, no. 3, 2021, doi: 10.1061/(asce)ae.1943-5568.0000485.

[94] J. Sun, J. Xiao, Z. Li and X. Feng, "Experimental study on the thermal performance of a 3D printed concrete prototype building," *Energy Build*, vol. 241, 2021, doi: 10.1016/j.enbuild.2021.110965.

[95] M.T. Souza, I.M. Ferreira, E. Guzi de Moraes, L. Senff, and A.P. Novaes de Oliveira, "3D printed concrete for large-scale buildings: An overview of rheology, printing parameters, chemical admixtures, reinforcements, and economic and environmental prospects," 2020. doi: 10.1016/j.jobe.2020.101833.

[96] L. Prasittisopin, K. Pongpaisanseree, P. Jiramarootapong and C. Snguanyat, "Thermal and Sound Insulation of Large-Scale 3D Extrusion Printing Wall Panel," in *RILEM Bookseries*, vol. 28, 2020. doi: 10.1007/978-3-030-49916-7_111.

7 Convolution Neural Network Prediction of Surface Quality of Additively Manufactured Metal Parts

K.G. Anbarasu[1*], M. Satthiyaraju[2], A. Umapathy[3], K.G. Ashok[4] and R. Majusha[5]

Though additive manufacturing, or 3D printing, offers many advantages, the application of printed parts in certain fields is hindered by issues such as poor surface quality, geometrical inaccuracy, anisotropic properties, residual stress, and porosities. Among these, surface quality has a significant impact on the fatigue life, corrosion resistance, and wear resistance of a part. Generally, secondary post-processing methods—such as machining, coating, peening, and laser polishing—are used to improve the surface quality of printed parts. However, these post-processing steps consume additional time and increase the overall cost of production. Moreover, due to the complex nature of the additive manufacturing process, it is difficult to establish clear relationships between input and output parameters. In recent times, convolutional neural network (CNN) models have been used to predict the relationships between process inputs and resulting outputs. This approach is highly useful for understanding the process and reducing the time and cost associated with secondary post-processing. This chapter discusses the development of a convolutional neural network model for predicting the surface quality of additively manufactured metal parts based on process parameters.

1. Introduction

Surface quality has a significant impact on the functional performance and service life of components. The importance of surface quality for various engineering applications is elaborated in Section 2. Additive manufacturing (AM) processes offer many advantages over conventional manufacturing methods. However, AM also presents challenges in producing parts with smooth surface finishes. In some applications, due to stringent surface quality requirements, further post-

[1-3] Department of Mechanical Engineering, SRM Institute of Science and Technology, Vadapalani, Chennai, Tamil Nadu 600026, India.

[4] Department of Mechanical Engineering, Easwari Engineering College, Ramapuram, Chennai, Tamil Nadu 600089, India.

[5] Department of Electrical and Electronics Engineering, St Peter's College of Engineering and Technology, Avadi, Chennai, Tamil Nadu 600054, India.

Email: satthiym@srmist.edu.in[2]; umapatha@srmist.edu.in[3]; ashok6733@gmail.com[4], majushaash@gmail.com[5]

* Corresponding author: anbarask1@srmist.edu.in

DOI: 10.1201/9781003635673-7

processing becomes inevitable. Section 3 discusses the surface quality of additively manufactured metal parts. To access the surface quality, various in situ methods such as contact profilometer, non-contact profilometer, and microscope are used. However, these methods are time-consuming and costly. On the other hand, monitoring the melt pool characteristics gives the information about size, shape, temperature distribution, and stability of melt pool (Jincheng Wang et al. 2023; Zhao et al. 2021; Scime and Beuth 2019). This helps to identify various defects such as balling effect, lack of fusion, keyhole porosity, temperature distribution, spatter, and so on (Brennan, Keist, and Palmer 2021; Du, Mukherjee, and DebRoy 2021). Thus, in-process monitoring is a viable technique to reduce the defect and it helps to understand the physics behind the process.

Process monitoring can also help to achieve the desired surface finish by reducing defects such as spatter, balling, partially melted powders, and so on. In general, in-process monitoring is conducted by monitoring the melt pool characteristics using sensors (photodiode, thermocouple, acoustic emission sensor, and pyrometer) and cameras (thermal or visible light camera) (Ye et al. 2018; Bhandarkar et al. 2025; Grasso et al. 2017). Images captured by cameras contain detailed information about the process. In recent times, supervised machine learning (ML) algorithms such as support vector machine, random forest, decision tree, K-Nearest neighbor, and logistic regression are widely used to predict the defects and help for process modification to reduce the defects (Memari et al. 2025; Scime et al. 2020). Even unsupervised machine learning algorithms are also used for grouping the defects such as lack of fusion, gas porosity, and keyhole porosity. Though the above ML algorithm can predict the defects, their accuracy is relatively less than the deep learning algorithms. Among the various deep learning algorithms, the convolutional neural network (CNN) is very popular, since it can learn features deeply and can provide better accuracy, especially for complex phenomena (Asadi et al. 2024).

Furthermore, the aforementioned traditional methods face challenges to extract the deeper features from the images and need separate feature extraction steps to build a model. However, in the case of CNN, the image is the input. Thus, there is no need of separate feature extraction steps when compared with traditional ML models. Owing to this, CNN is widely used for image processing applications (Asadi et al. 2024). Especially, for surface quality prediction, CNN has been used (Wang et al. 2023; Asadi et al. 2024). CNNs can predict the surface quality in relation to changes in process condition or else it can be used for classification of surface quality (good or bad). This helps in identifying optimal process parameters to achieve better surface quality and in planning subsequent operations such as finishing (machining, grinding, burnishing, heat treatment and so on). More details about the CNN model are presented in Section 4, and the CNN model development and testing procedures are elaborated in Section 5. Finally, the conclusions of this chapter are provided in Section 6.

2. Significance of Surface Quality for Various Engineering Applications

Generally, surface quality is defined in terms of the level of surface imperfections, defects, and lay patterns present on a surface. The surface quality of manufactured parts varies depending on the manufacturing process, such as casting, machining, forming, and additive manufacturing. Among these, casting and additive manufacturing are considered primary manufacturing processes. In contrast, processes such as machining and finishing are typically used to improve surface quality. A typical machined surface is shown in Figure 1. The machined surface contains high-frequency, short-wavelength imperfections known as surface roughness, and low-frequency, long-wavelength imperfections referred to as waviness. Additionally, a dominant pattern in a specific direction, known as the lay, along with various defects, may also be present on the surface (Figure 1) (Hashmi et al. 2023).

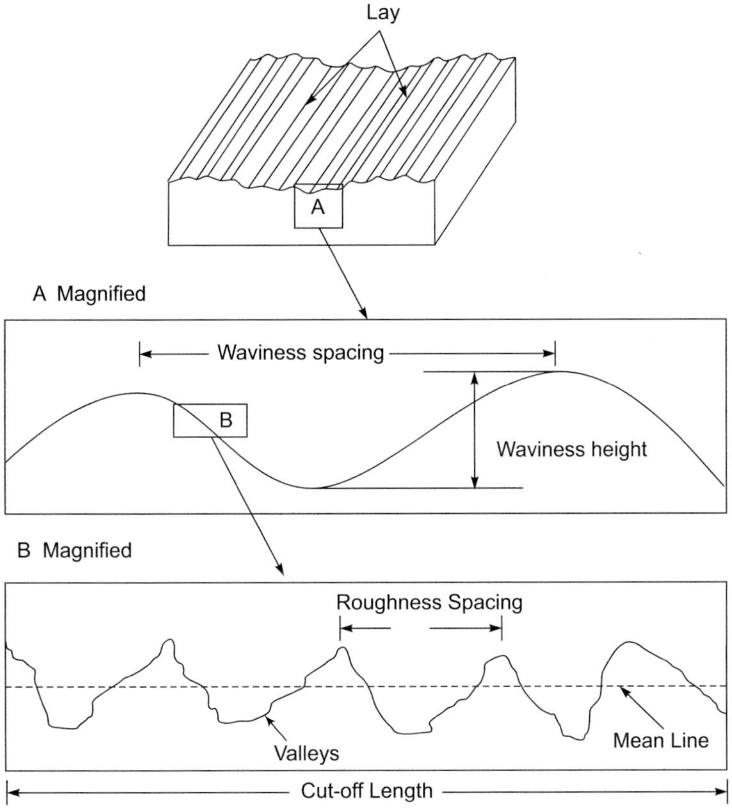

Fig. 1 Surface characteristics (Hashmi et al. 2023).

Surface quality has a significant impact on the functionality and service life of parts used in various engineering applications (Sherrington and Smith, 1986). Functional properties such as biocompatibility, self-cleaning ability, adhesion, wettability, reflectivity, heat transfer, fluid flow, friction, and sealing performance depend on the surface quality of the part (Sherrington and Smith, 1986). On the other hand, surface quality also influences the corrosion resistance and fatigue life of parts (Sherrington and Smith, 1986). The corrosion rate increases with an increase in surface roughness, whereas a polished surface exhibits a lower corrosion rate (Toloei, Stoilov, and Northwood 2013). Rough surfaces are more prone to localized corrosion, such as pitting, which results in a higher corrosion rate compared to polished surfaces (Sherrington and Smith, 1986; Toloei, Stoilov, and Northwood 2013). Similarly, the fatigue strength of parts decreases as surface roughness increases (Sherrington and Smith, 1986; Zeng, Ye, and Liu 2023). The rough surface acts as a stress concentration site, leading to crack initiation, nucleation, and growth (Sherrington and Smith, 1986; Zeng, Ye, and Liu 2023). Furthermore, an increase in roughness directly affects fluid flow, light scattering, skin friction, sealing, and overall friction (Sherrington and Smith, 1986). On the other hand, an increase in roughness can improve adhesion, wettability, heat transfer, and other similar properties. Applications such as bioimplants, dies, guideways, metal mirrors, bearings, microfluidics, sealants, shafts, gears, turbine blades, and others require smooth surfaces—i.e., low surface roughness—to ensure better lifespan and functional performance. However, higher roughness is beneficial for applications such as heat exchangers, surface coatings, clutch plates, brake pads, grippers, and similar components. Overall, surface quality plays a vital role in the functional performance and endurance of a part. In recent times, metal additively manufactured (MAM) parts have been used in a wide range of applications, from rocket engines to biomedical devices. This transformation is driven by several key factors:

- Design freedom
- Able to make complex parts
- Able to make multi-material parts
- Able to repair damages of parts such as turbine blades, gears, and so on
- Requires short time to introduce new product to the market
- Less wastage of material
- No need for tool and dies
- Mass customization

On the other hand, MAM parts are not directly used for certain applications where surface properties play a significant role in functional performance and lifespan. The following section elaborates on the surface quality of MAM parts (Lee, Nagalingam, and Yeo 2021).

3. Surface Quality of Additively Manufactured Metal Parts

The surfaces produced by MAM processes have poor surface quality. Moreover, the top, bottom, and side surfaces have different roughness as well as surface features (Tato et al. 2020; Nasab et al. 2018). The rough surface is due to the following surface features (Lee, Nagalingam, and Yeo 2021).

- Powders that are bonded loosely to the surface: These are raw material powders, which are attached to the surface during printing process (Figure 2(a)).
- Staircasing or layer by layer features: Occurs due to inherent nature of layer-by-layer building process (Figure 2(b)).
- Powders which are partially melted: During printing heat transfers from the melt pool to surroundings and results in partially melting of powders present near to edges of the melt pool (Figure 2(c)).
- Balling effect: Occurs due to high traverse speed of laser source and less power (Figure 2(d)).
- Semi-welded tracks: The fusion followed by subsequent solidification of melt pool results in welded tracks (Figure 2(e)).
- Spatters or ejection of spherical droplets from the melt pool: The usage of high energy density results in high thermal gradient. This in turn causes gradient in surface tension of melt pool eventually break down of melt pool and ejection of spherical droplets (Figure 2(f)).
- Porosities on the surface: Occurs due to high transverse speed of laser source and less amount of liquid front rates.

The surface morphology of MAM part is shown in Figure 2 (Lee, Nagalingam, and Yeo 2021), and the surface roughness of MAM part by different processes is given in Table 1.

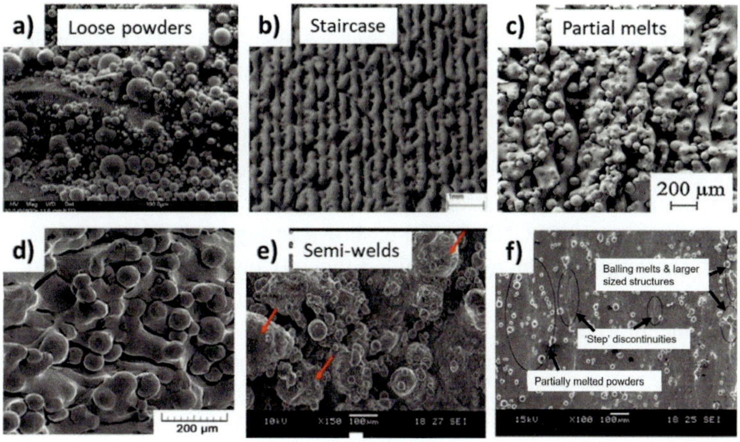

Fig. 2 (a) Loose powders (Gu and Shen 2009), (b) sintering layer steps (Simchi 2006), (c) partial melts (Karlsson et al. 2013), (d) electron beam melted balling effect (Fousová et al. 2018), (e) semi-welded structures (Nagalingam, Yuvaraj, and Yeo 2020), and (f) balling melts and step discontinuities (Tan and Yeo 2017).

Table 1 Average surface roughness of metal additive manufactured parts.

S. No	*Metal Additive Manufacturing Processes*	*Average Surface Roughness*
1	Selective laser melting (SLM)	14 µm (Wang, Fu, and Gao, 2021), 12–14 µm (Duval-Chaneac et al. 2018), 9.48 µm (Wang, Zhu, and Liew 2019), 7.6–10 µm (Hofele et al. 2017), 24 µm (Scherillo 2019)
2	Electron beam melting (EBM)	37.5 µm (Scherillo et al. 2020), 23–32µm (GUO et al. 2021) 13–17 µm (Spitaels et al. 2020), 31 µm (Iquebal et al. 2016), 27 µm (Genna and Rubino 2020)
3	Direct energy deposition (DED)	23.5 µm (Dadbakhsh, Hao, and Kong 2010), 21 µm (Rosa, Mognol, and Hascoët 2015), 56 µm (Careri et al. 2021)
5	Direct metal laser sintering (DMLS)	16.72 µm (Atzeni et al. 2016) , 6.5–7.5 µm (Tan and Yeo 2017), 15–25 µm (Nagalingam and Yeo 2018), 19 µm (Nagalingam, Yuvaraj, and Yeo 2020)

4. Introduction of Convolutional Neural Network Model

Convolutional neural networks (CNNs) are deep learning models that are specifically designed to handle structured grid data, such as images (Zhao et al. 2024; Taye 2023). The image may be RGB image or grayscale image. When processing input images, a CNN passes them through several layers that identify patterns and extract features. Traditional neural networks struggle to process high-dimensional data such as images. CNNs overcome these challenges by automatically learning spatial features, making them ideal for deep learning tasks like image recognition, object detection, and computer vision (Zhao et al. 2024; Taye 2023). Its ability to automatically learn spatial hierarchies of features makes it extremely useful for image recognition, object detection, and computer vision tasks. CNNs play an important role in deep learning for computer vision. They provide automated feature extraction, translation invariance, and high accuracy. The general structure of CNN model is shown in Figure 3 (Zhao et al. 2024). In Figure 3, the original data are convolved twice (Convolution 1, Convolution 2), pooled twice (Max Pooling 1, Max Pooling 2), output to the fully connected layer (fully connection), and finally the Softmax activation function compresses the output vectors of the full connection layer into (0, 1) and outputs them in the output layer. The Data Cost 1 represents the probability of belonging to the n categories; the larger the value, the greater the possibility of belonging to the category. The commonly used CNN architecture are LeNet, NexNet, ZfNet, VGG, GoogLe Net, Inception-V3, ResNet, and DenseNet (Zhao et al. 2024; Taye 2023).

Main components of CNN are as follows (Zhao et al. 2024; Taye 2023):

- Convolutional layers
- Rectified Linear Unit (ReLU)
- Pooling layers or sub-sampling layers
- Fully connected layers

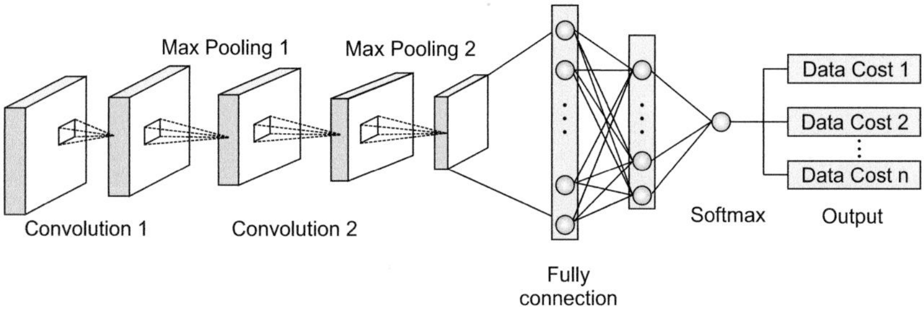

Fig. 3 Structure of CNN (Zhao et al. 2024).

Convolution layers: The convolutional layer is the central component of a CNN. It is in charge of detecting patterns such as edges, textures, and shapes in input images. Moreover, this layer is the first building block. In this layer, the mathematical operation such as convolution is used (Zhao et al. 2024; Taye 2023). In addition, in this layer, filters or kernel or feature detector are used to identify the specific features and textures from the image. For example, in convolution layer the pixel representation of an image (5×5 matrix) is multiplied with kernel (3×3 matrix). These kernels slide across the image performing element-wise multiplication with the part of the image they cover, extracting features such as edges, textures, and shapes. This generates a 2D representation of image called feature map (Zhao et al. 2024; Taye 2023) (shown in Figure 4). The size of sliding of the kernel is known as stride. The different kernels are used for identifying various features. The length of stride and padding used during convolution is chosen as one to preserve the spatial dimensions of data (Zhao et al. 2024; Taye 2023). The number of filters doubles in the succeeding layers. Hence this transforms to doubling of the depth of input volume in every layer (Zhao et al. 2024; Taye 2023).

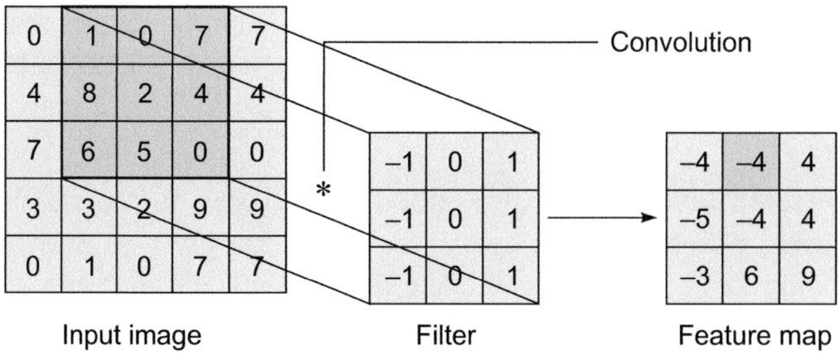

Fig. 4 Convolution procedure (Zhao et al. 2024).

Rectified Linear Unit (ReLU): This is the most commonly used unit. By applying ReLU activation, the ReLU layer introduces non-linearity to the network. In addition to that, ReLU activation sets all negative values in the feature map to zero and it does not alter the positive values (Figure 5) (Zhao et al. 2024; Taye 2023).

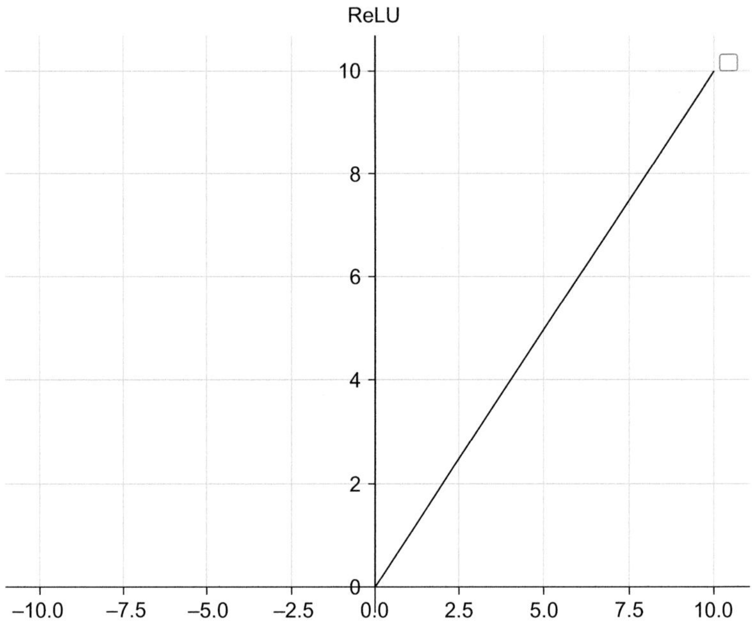

Fig. 5 Function curves of ReLU (Zhao et al. 2024).

Pooling or sub-sampling layer: This layer is also called down sampling layer. The main function of this layer is to extract the most important features from the convoluted matrix. The pooling layer reduces the spatial dimensions (height and width) of feature maps while maintaining important information. It improves the network's computational efficiency and generalization by reducing overfitting (Zhao et al. 2024; Taye 2023). The types of pooling are as follows:

1. Maximum pooling: It is the maximum value from the feature map and also chooses the significant features from the feature map (shown in Figure 6).
2. Average pooling: It is an average value from the feature map (shown in Figure 7).

Fully connected layer: It is also called dense layer. It is a standard neural network layer, which connects every neuron in one layer to every neuron in the succeeding layer. It is the final stage of CNN model, after convolution and pooling (Zhao et al. 2024; Taye 2023). Its function is to predict the output based on the extracted features. In addition, the extracted features are used to perform classification or regression tasks. At first, the extracted feature map from the pooling layer is flattened (i.e., converted the 2D matrix into 1D vectors) and then connected to fully connection layers, in which every input is connected to every output by a learnable weight (Zhao et al. 2024; Taye 2023). This layer executes a linear combination of input vector with a weight matrix tailed by nonlinear activation/Sigmoid (or) Softmax function. In general, the output from this layer is a vector with N dimension (N denotes the number of classes). Moreover, this layer correlates hierarchical features to feasible classes and provide the probabilities to each class being true (Zhao et al. 2024; Taye 2023).

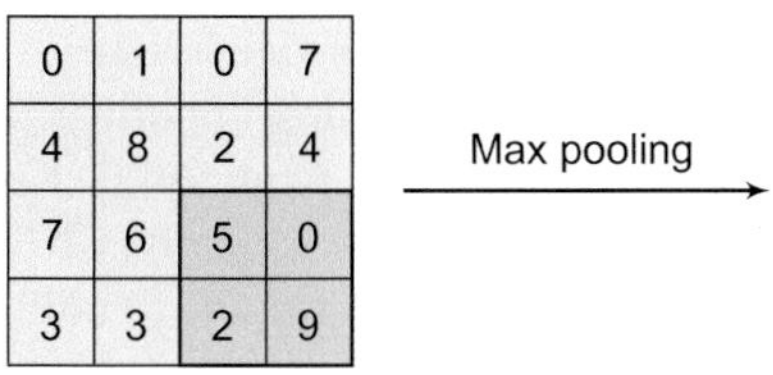

Fig. 6 Maximum pooling procedure (Zhao et al. 2024).

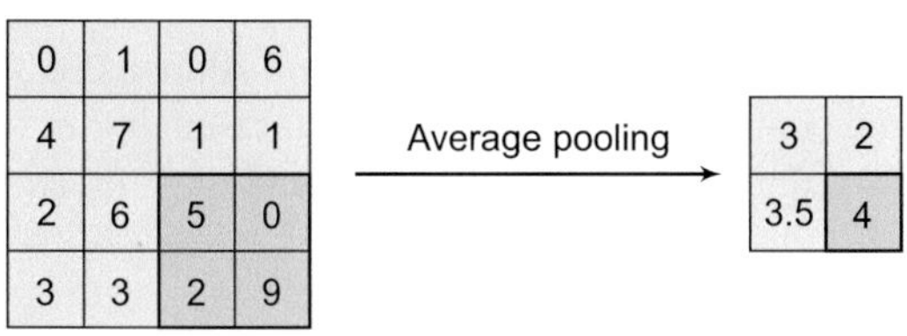

Fig. 7 Average pooling procedure (Zhao et al. 2024).

Softmax/Sigmoid activation Provides a probability distribution for multi-class classification (Abhilash and Ahmed 2023). The output layer of a CNN makes predictions using the Softmax and Sigmoid activation functions (Abhilash and Ahmed 2023; Knaak et al. 2021). Sigmoid is best for binary classification, and Softmax is best for multi-class problems. CNNs power practical applications such as self-driving cars and medical diagnostics. Furthermore, to evaluate the model performance various loss functions such as mean absolute error (MAE), mean square error (MSE), mean absolute percentage error (MAPE), LOG-COSH, cross entropy, and Huber loss are used (Rifai et al. 2020; Abhilash and Ahmed 2023).

5. Convolutional Neural Network Model for Predicting Surface Quality

In general, there are two ways to predict the surface roughness: (1) by using statistical models and (2) by using analytical models. The analytical model depends on physical relationship between input parameters and output, whereas in statistical models the experiments are conducted and then empirical relationship is developed. The developed relation is then used for prediction. Nowadays, the modern statistical methods like CNN are widely used to predict the surface quality (Zhang et al. 2019; Valizadeh and Wolff 2022; Rifai et al. 2020). As mentioned in the previous Section 4, CNNs are deep learning techniques that use interconnected neurons to classify images into different groups or categories. Owing to their superior efficiency, robustness, and generalization capabilities, CNNs are better suited for AM surface classification than other machine learning models and their ability to automatically learn hierarchical features makes them well suited for surface quality analysis (Rifai et al. 2020; Valizadeh and Wolff 2022; Zhang et al. 2019).

The typical CNN works by extraction of spatial features from images using filters, reducing the dimensionality while retaining important features to predict surface quality (Rifai et al. 2020; Tatzel and León 2020; Knaak et al. 2021; Abhilash and Ahmed 2023). As aforementioned, normally, CNN models are used to classify the surface images (smooth or rough) (Figure 8) (Abhilash and Ahmed 2023; Knaak, et al. 2021). On the other hand, to continuously predict the surface roughness the regression task is included in the CNN architecture (Figure 9) (Rifai et al., 2020). The surface quality evaluation by image classification using CNN model has the convolutional layer, pooling layer, fully connected layer or dense layer and then Softmax layer (Abhilash and Ahmed 2023) (Figure 8). On the other hand, the continuous prediction of surface quality in terms of average surface roughness requires regression layer instead of Softmax layer (Figure 9) (Rifai et al. 2020).

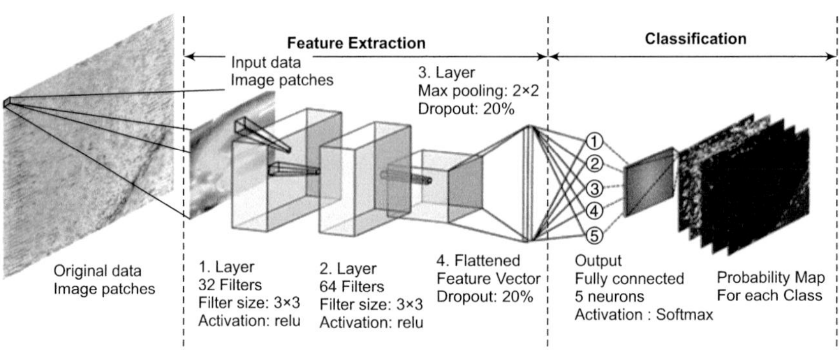

Fig. 8 CNN architecture used to classify the surface quality (Knaak et al. 2021).

The process starts with data acquisition and preprocessing. The surface images were captured using industrial cameras, microscopes, or 3D scanners (Zhang et al. 2019; Abhilash and Ahmed 2023; Gerdes et al. 2021; Knaak et al. 2021). Next step is preprocessing. In this process, the normalization, denoising, and augmentation of images are carried out to enhance the model prediction. Also, the preprocessing stage plays a crucial role in CNN model prediction, since the computation time and prediction accuracy depend on this stage. The preprocessed image is used as the input for the CNN model. The first part of CNN model (Convolution and Pooling layers) extracts the features from the surface images.

Then the second part of CNN model (dense layers or fully connected layers and output layers) receives the extracted features from the first part and then classifies the surface quality (rough, smooth) using Softmax layer (Abhilash and Ahmed 2023) or it predicts the surface roughness using regression layer as mentioned in the previous paragraph. In the case of surface roughness, prediction, using regression layer, works by using the learned relationship between the extracted features from the image and corresponding surface roughness (R_a or R_z) during the model training stage (Rifai et al. 2020). The hyperparameters (number of layers, size of layers, filter size, drop out, and epoch) optimization are done to obtain better prediction (Zhang et al. 2019; Rifai et al. 2020). CNNs provide an automated, efficient, and scalable solution for predicting surface quality in manufacturing. By leveraging deep learning, industries can enhance defect detection, optimize quality control processes, and reduce reliance on manual inspection methods for defective areas (Knaak et al. 2021; Zhang et al. 2019). The details about CNN model development and testing used for surface quality prediction of MAM components are elaborated in the subsequent sections.

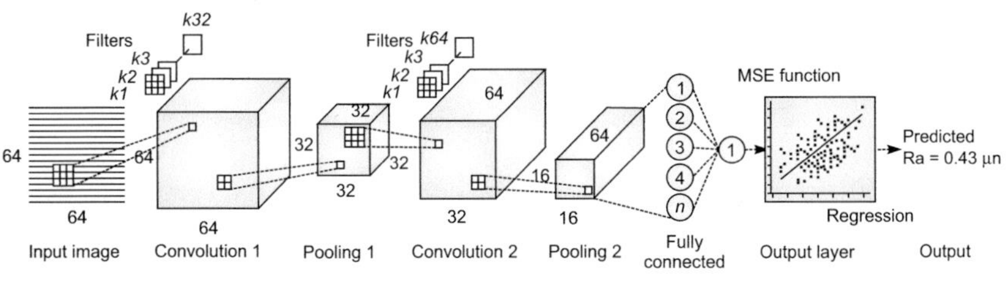

Fig. 9 CNN architecture used to predict the surface roughness (Rifai et al. 2020).

5.1 Data Collection and Preprocessing

The input for CNN model is image. The images are captured either during the process or after process using cameras, microscopes, and so on. Then the captured image is converted into grayscale images and normalized. In the preprocessing, the unwanted regions from the image are removed and split into the small images according to required sizes. Processing large size requires higher computation time (Knaak et al. 2021). Normalizing images is guaranteed in which pixel values stay within a standard range. The rotation, flipping, and adjustments to brightness enhances the diversity of the dataset which supports the model in generalization (Knaak et al. 2021). The images after pre-processing are labeled using the surface roughness value (R_a or R_z), which is measured by using contact or non-contact surface profilometers as well as confocal microscopes on the same location of the surface where the image is captured (Knaak et al. 2021; Zhang et al. 2019; Rifai et al. 2020). Furthermore, the surface images captured and processed are split into three categories: (1) Training images, (2) Validation images, and (3) Testing (Abhilash and Ahmed 2023; Zhang et al. 2019; Knaak et al. 2021). Generally, training set of images is used for learning purpose and the validation set of images is required for hyperparameters optimization. The images used for testing are required to evaluate the performance of the model. The images used in testing are untrained data. The typical image used for CNN model training is shown in Figure 10 (Abhilash and Ahmed 2023), and the corresponding surface roughness, which is used for label and correlation purpose, is given in Table 2 (Abhilash and Ahmed 2023).

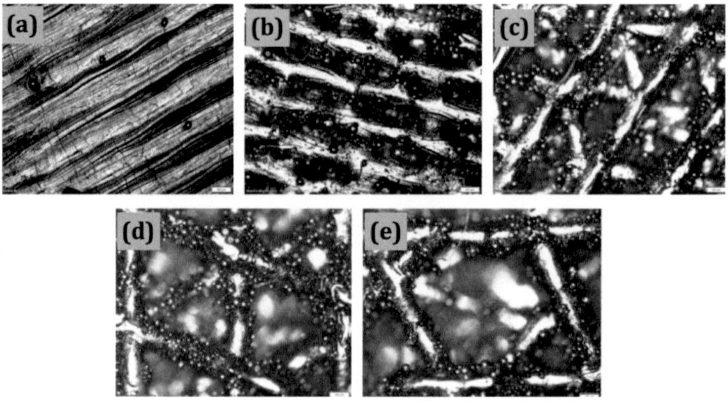

Fig. 10 Different metal AM surface images: (a) Level 1, (b) Level 2, (c) Level 3, (d) Level 4, (e) Level 5 (Abhilash and Ahmed 2023).

Table 2 Surface roughness for different surface label (Abhilash and Ahmed 2023).

Surface Label	*Sa (μm)*	*SD*	*Sz (μm)*	*SD*
Level 1	7.87	1.37	51.88	8.48
Level 2	13.01	2.85	73.76	9.06
Level 3	14.63	3.01	101.47	11.54
Level 4	15.49	4.52	105.74	13.85
Level 5	23.00	7.63	153.45	17.93

5.2 Model Architecture

The CNN architecture plays a crucial role in achieving better accuracy in prediction of surface quality. Moreover, the computation time also depends upon the type of architecture (Rifai et al. 2020; Knaak et al. 2021; Zhang et al. 2019). As mentioned in Section 5.3, depending upon the need, the CNN architecture is modified for either classification of surface quality in terms of rough or smooth surface or modified to predict the surface roughness values (Abhilash and Ahmed 2023;

Knaak et al. 2021; Rifai et al. 2020). A CNN architecture designed for predicting surface quality is organized in a systematic way. It starts with an input layer that handles either grayscale or color images of printed components. The convolutional layers identify features associated with surface imperfections, followed by pooling layers that lower computational demands and extract the most significant features (Gerdes et al. 2021; Abhilash and Ahmed 2023). Dropout layers are incorporated to mitigate overfitting by randomly turning off specific neurons during the training process. Then, the fully connected layers make predictions about surface quality which involve classifying it into groups such as smooth or rough (Abhilash and Ahmed 2023; Knaak et al. 2021; Gerdes et al. 2021).

The output layer then presents the prediction which helps manufacturers in estimating the quality in real time (Valizadeh and Wolff 2022). One of the commonly used CNN architecture for predicting the surface quality is ResNet 50 (Abhilash and Ahmed 2023; Knaak et al. 2021; Figure 11). The key benefits of using ResNet50 than other architecture is its ability to learn hierarchical features from the image using residual connection. Also, this particular architecture is suitable for classifying the similar-looking images with better accuracy (Abhilash and Ahmed 2023). This architecture contains 50 layers. It includes convolutional layer, pooling layer, fully connected layer, ReLU activation function, Softmax layer, loss functions, and dropouts (Abhilash and Ahmed 2023). In general, the hyperparameter tuning plays a significant role in design of CNN architecture (Gerdes et al. 2021; Abhilash and Ahmed 2023; Knaak et al. 2021; Zhang et al. 2019). The adams and nadam optimizer are used for optimizing the hyperparameters (Gerdes et al. 2021).

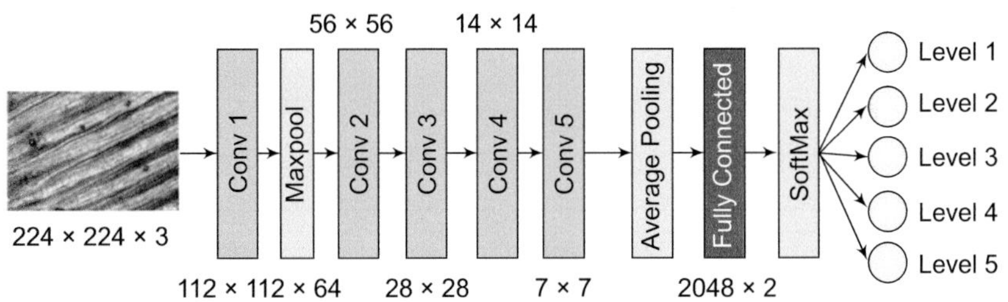

Fig. 11 The structure of the CNN classifier (Abhilash and Ahmed 2023).

5.3 Model Training and Validation

The training process for a convolutional neural network model involves several key steps. First, a dataset of 3D-printed surface images is collected using high resolution cameras or microscopes with each image labeled according to its surface quality. Next, the preprocessing techniques can be applied to enhance the quality of the images and prepare them for the model. Generally, for training the CNN architecture supervised learning procedure is followed (Gerdes et al. 2021; Abhilash and Ahmed 2023; Zhang et al. 2019; Liu et al. 2023). One of the crucial stages in CNN model development is training. The surface image after preprocessing is split into three categories (1) images for training purpose, (2) images for validation purpose, and (3) images for testing purpose. Among these three categories, most of the images were used for training purpose (70%) and then for validation purpose (30%). Only small number of images are used for testing purpose (10%) (Knaak et al. 2021; Gerdes et al. 2021). During training, the preprocessed surface images of AM part with label are given as input to the model to learn about features (first-order and second-order statistical features) present on the surface as well as to learn the correlation between the surface features and surface roughness of that particular surface (Zhang et al. 2019; Abhilash and Ahmed 2023; Gerdes et al. 2021; Knaak et al. 2021). The typical learning rate used is 0.001 (Gerdes et al. 2021; Zhang et al. 2019). Also, the hyperparameter tuning is carried out with the training and validation set of images. This is required to obtain better prediction accuracy. The change in loss function (MSE, MAE, and so on) with respect to epoch is shown in Figure 12 (Gerdes et al. 2021). Likewise, the effect of change in dropout rate on loss function is given in Figure 13 (Gerdes et al. 2021).

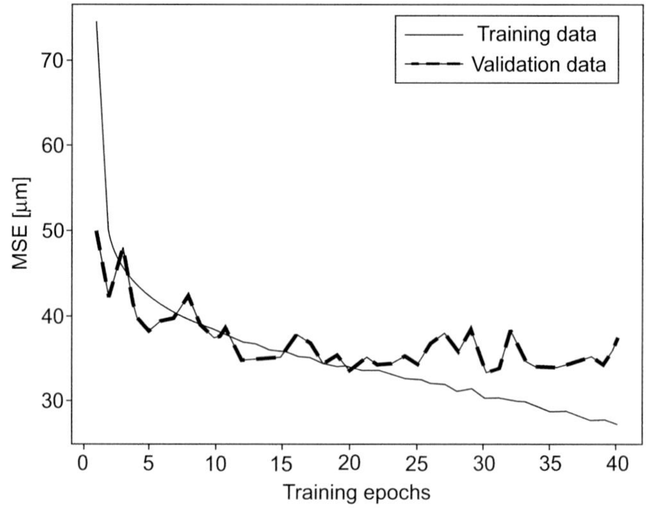

Fig. 12 Evolution of the loss function (mean squared error) for training and validation set during training of the CNN (Gerdes et al. 2021).

5.4 Performance Evaluation

The surface images used for testing are independent of the images used for training and validation (Gerdes et al. 2021; Abhilash and Ahmed 2023; Knaak et al. 2021). The testing is carried out with untrained images and then model performance or loss function is evaluated using the loss functions such as mean square error, mean absolute percentage error, mean absolute error, and so on (Rifai et al. 2020; Tatzel and León 2020; Abhilash and Ahmed 2023; Gerdes et al. 2021). The prediction

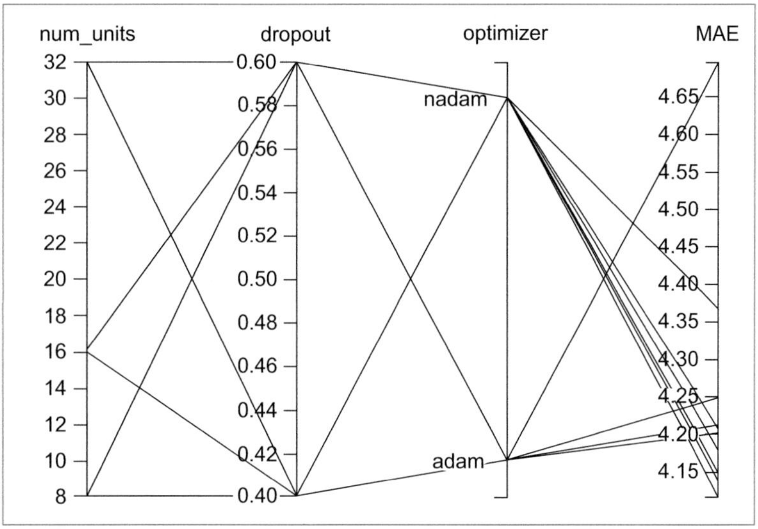

Fig. 13 Illustration of the hyperparameter tuning (Gerdes et al. 2021).

accuracy of CNN model is in the range of 82–96% (Abhilash and Ahmed 2023; Zhang et al. 2019). Typical loss function calculated for CNN prediction is given in Table 3 (Gerdes et al. 2021). Moreover, the output of classifier is given in confusion matrix (table used to evaluate the performance of model). The row represents the actual value obtained from the measurement, whereas column represents the predicted value. There are four categories in confusion matrix:

1. True positive (TP)
2. True negative (TN)
3. False positive (FP)
4. False Negative (FN)

Furthermore, the above categories of classification are used to find the accuracy, precision, and so on. A typical confusion matrix made out of surface quality classification is shown in Figure 14 (Abhilash and Ahmed 2023).

Table 3 Prediction accuracy of the convolutional neural network for different parameter sets compared to the inherent statistical spread within each set (Gerdes et al. 2021).

Process Conditions	*MAE R_z (μm) (Measured)*	*MAE R_z (μm) Predicted*
A (50 W; 75 mm/s)	2.2	2.3
B (50 W; 40 mm/s)	5.1	4.3
C (20 W; 75 mm/s)	8.4	9.6

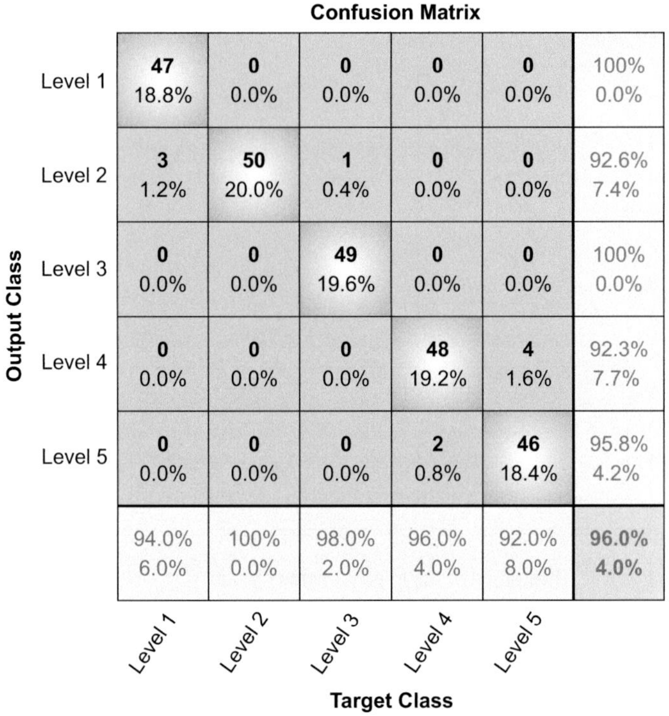

Fig. 14 Confusion matrix for CNN classifier (Abhilash and Ahmed 2023).

6. Conclusion

Understanding the physical phenomena occurring during the additive manufacturing of metal parts is crucial to achieving optimal surface quality. In general, sensors and cameras are used to comprehend the process physics and the relationship between input and output parameters. However, conducting numerous experimental trials to study the process is time-consuming and costly. Therefore, deep learning techniques such as CNNs can provide a cost-effective solution to this problem, as they are robust tools for image processing and prediction. In the case of CNNs, only a camera and computer are needed to process the captured images, and they also offer better prediction accuracy for unlabeled or test data. By using CNNs, process conditions can be simulated for different parameter settings. This helps identify the optimal conditions to achieve better surface quality at a lower cost and with shorter computation time. Moreover, in recent times, CNN models have been increasingly used for inspection purposes.

References

[1] Abhilash, P. M. and Afzaal Ahmed. (2023). Convolutional Neural Network–Based Classification for Improving the Surface Quality of Metal Additive Manufactured Components. *International Journal of Advanced Manufacturing Technology* 126, 3873–3885.

[2] Asadi, Reza, Antoine Queguineur, Olli Wiikinkoski, Hossein Mokhtarian, Tommi Aihkisalo, Alejandro Revuelta, and Iñigo Flores Ituarte. (2024). Process Monitoring by Deep Neural Networks in Directed Energy Deposition: CNN-Based Detection, Segmentation, and Statistical Analysis of Melt Pools. *Robotics and Computer-Integrated Manufacturing* 87, 102710-24.

[3] Atzeni, E., M. Barletta, F. Calignano, L. Iuliano, G. Rubino and V. Tagliaferri. (2016). Abrasive Fluidized Bed (AFB) Finishing of AlSi10Mg Substrates Manufactured by Direct Metal Laser Sintering (DMLS). *Additive Manufacturing* 10, 15–23.

[4] Careri, Francesco, Stano Imbrogno, Domenico Umbrello, Moataz M. Attallah, José Outeiro and António C. Batista. (2021). Machining and Heat Treatment as Post-Processing Strategies for Ni-Superalloys Structures Fabricated Using Direct Energy Deposition. *Journal of Manufacturing Processes* 61, 236–44.

[5] Dadbakhsh, Sasan, Liang Hao and Choon Yen Kong. (2010). Surface Finish Improvement of LMD Samples Using Laser Polishing. *Virtual and Physical Prototyping* 5 (4), 215–21.

[6] Duval-Chaneac, M. S., S. Han, C. Claudin, F. Salvatore, J. Bajolet and J. Rech. (2018). Experimental Study on Finishing of Internal Laser Melting (SLM) Surface with Abrasive Flow Machining (AFM). *Precision Engineering* 54,1–6.

[7] Fousová, Michaela, Dalibor Vojtěch, Karel Doubrava, Matěj Daniel and Chiu Feng Lin. (2018). Influence of Inherent Surface and Internal Defects on Mechanical Properties of Additively Manufactured Ti6Al4V Alloy: Comparison between Selective Laser Melting and Electron Beam Melting. *Materials* 11 (4), 357–75.

[8] Genna, Silvio and Gianluca Rubino. (2020). Laser Finishing of Ti6Al4V Additive Manufactured Parts by Electron Beam Melting. *Applied Sciences*. 10 (1). 183–96.

[9] Gerdes, Niklas, Christian Hoff, Jörg Hermsdorf, Stefan Kaierle and Ludger Overmeyer. (2021). Hyperspectral Imaging for Prediction of Surface Roughness in Laser Powder Bed Fusion. *International Journal of Advanced Manufacturing Technology*. 115 (4), 1249–58.

[10] Gu, Dongdong and Yifu Shen. (2009). Balling Phenomena in Direct Laser Sintering of Stainless-Steel Powder: Metallurgical Mechanisms and Control Methods. *Materials and Design*, 30 (8): 2903–10.

[11] GUO, Jiang, Min Hao GOH, Pan WANG, Rui HUANG, Xiaohua LEE, Bin WANG, Sharon Mui Ling NAI and Jun WEI. (2021). "Investigation on Surface Integrity of Electron Beam Melted Ti-6Al-4 V by Precision Grinding and Electropolishing." *Chinese Journal of Aeronautics*. 34 (12), 28–38.

[12] Hashmi, Abdul Wahab, Harlal Singh Mali, Anoj Meena, Mohammad Farukh Hashmi and Neeraj Dhanraj Bokde. (2023). "Surface Characteristics Measurement Using Computer Vision: A Review." CMES - Computer Modeling in Engineering and Sciences. *Tech Science Press*. 135(2), 917–1005.

[13] Hofele, M., J. Schanz, B. Burzic, S. Lutz, M. Merkel and H. Riegel. (2017). Laser Based Post Processing of Additive Manufactured Metal Parts. Lasers in Manufacturing Conference.

[14] Iquebal, Ashif Sikandar, Sanjay Shrestha, Zimo Wang, Guha P. Manogharan, A.S. Iquebal, S. Shrestha, Z. Wang, G.P. Manogharan, and S. Bukkapatnam. (2016). Influence of Milling and Non-Traditional Machining on Surface Properties of Ti6Al4V EBM Components. *Proceedings of the 2016 Industrial and Systems Engineering Research Conference*, 1–6.

[15] Karlsson, Joakim, Anders Snis, Håkan Engqvist and Jukka Lausmaa. (2013). "Characterization and Comparison of Materials Produced by Electron Beam Melting (EBM) of Two Different Ti-6Al-4V Powder Fractions." *Journal of Materials Processing Technology* 213 (12), 2109–18.

[16] Knaak, Christian, Lukas Masseling, Emil Duong, Peter Abels and Arnold Gillner. (2021). Improving Build Quality in Laser Powder Bed Fusion Using High Dynamic Range Imaging and Model-Based Reinforcement Learning. *IEEE Access* 9, 55214–31.

[17] Lee, Jian Yuan, Arun Prasanth Nagalingam and S. H. Yeo. (2021). A Review on the State-of-the-Art of Surface Finishing Processes and Related ISO/ASTM Standards for Metal Additive Manufactured Components. *Virtual and Physical Prototyping* 16 (1), 68–96.

[18] Liu, Chenang, Rongxuan Raphael Wang, Ian Ho, Zhenyu James Kong, Christopher Williams, Suresh Babu and Chase Joslin. (2023). Toward Online Layer-Wise Surface Morphology Measurement in Additive Manufacturing Using a Deep Learning-Based Approach. *Journal of Intelligent Manufacturing* 34 (6), 2673–89.

[19] Luke Scime, Derek Siddel, Seth Baird and Vincent Paquit (2020). Layer-wise anomaly detection and classification for powder bed additive manufacturing processes: A machine-agnostic algorithm for real-time pixel-wise semantic segmentation, *Additive Manufacturing*, 36, 101453, ISSN 2214-8604, https://doi.org/10.1016/j.addma.2020.101453. (https://www.sciencedirect.com/science/article/pii/S2214860420308253)

[19] Nagalingam, Arun Prasanth and S. H. Yeo. (2018). "Controlled Hydrodynamic Cavitation Erosion with Abrasive Particles for Internal Surface Modification of Additive Manufactured Components. Wear 414–415, 89–100.

[20] Nagalingam, Arun Prasanth, Hemanth Kumar Yuvaraj and S. H. Yeo. (2020). "Synergistic Effects in Hydrodynamic Cavitation Abrasive Finishing for Internal Surface-Finish Enhancement of Additive-Manufactured Components." *Additive Manufacturing* 33, 101110.

[21] Nasab, Milad Hamidi, Dario Gastaldi, Nora Francesca Lecis and Maurizio Vedani. (2018). On Morphological Surface Features of the Parts Printed by Selective Laser Melting (SLM). *Additive Manufacturing* 24, 373–77.

[22] Rifai, Achmad P., Hideki Aoyama, Nguyen Huu Tho, Siti Zawiah Md Dawal and Nur Aini Masruroh. (2020). Evaluation of Turned and Milled Surfaces Roughness Using Convolutional Neural Network.Measurement: *Journal of the International Measurement Confederation* 161, 107860-71.

[23] Rosa, Benoit, Pascal Mognol and Jean-yves Hascoët. (2015). Laser Polishing of Additive Laser Manufacturing Surfaces. *Journal of Laser Applications* 27, S29102-7.

[24] Scherillo, Fabio. (2019). Chemical Surface Finishing of AlSi10Mg Components Made by Additive Manufacturing. *Manufacturing Letters* 19, 5–9.

[25] Scherillo, Fabio, Emanuele Manco, Andrea El Hassanin, Stefania Franchitti, Carmine Pirozzi and Rosario Borrelli. (2020). Chemical Surface Finishing of Electron Beam Melted Ti6Al4V Using HF-HNO_3 Solutions. *Journal of Manufacturing Processes* 60, 400–409.

[26] Scime, Luke and Jack Beuth. (2019). Using Machine Learning to Identify In-Situ Melt Pool Signatures Indicative of Flaw Formation in a Laser Powder Bed Fusion Additive Manufacturing Process. *Additive Manufacturing* 25, 151–65.

[27] Sherrington, I and E H Smith. (1986). The Significance of Surface Topography in Engineering, *Precision Engineering*. 8(2),79–87.

[28] Simchi, A. (2006). Direct Laser Sintering of Metal Powders: Mechanism, Kinetics and Microstructural Features. *Materials Science and Engineering*: A 428 (1–2), 148–58.

[29] Spitaels, Laurent, François Ducobu, Anthonin Demarbaix, Edouard Rivière-Lorphèvre and Pierre Dehombreux. (2020). Influence of Conventional Machining on Chemical Finishing of Ti6Al4V Electron Beam Melting Parts. *In Procedia Manufacturing*, 47:1036–42.

[30] Tan, K. L. and S. H. Yeo. (2017). Surface Modification of Additive Manufactured Components by Ultrasonic Cavitation Abrasive Finishing. Wear. 378–379, 90–95.

[31] Tato, W., L. Blunt, I. Llavori, A. Aginagalde, A. Townsend and A. Zabala. (2020). Surface Integrity of Additive Manufacturing Parts: A Comparison between Optical Topography Measuring Techniques. In Procedia CIRP, 87, 403–8.

[32] Tatzel, Leonie and Fernando Puente León. (2020). Image-Based Roughness Estimation of Laser Cut Edges with a Convolutional Neural Network. In Procedia CIRP, 94, 469–73.

[33] Taye, Mohammad Mustafa. (2023). Theoretical Understanding of Convolutional Neural Network: Concepts, Architectures, Applications, Future Directions. Computation. 11(3), 52,

[34] Toloei, Alisina, Vesselin Stoilov and Derek Northwood. (2013). The Relationship between Surface Roughness and Corrosion. ASME 2013 International Mechanical Engineering Congress and Exposition. Vol. 2 B. *American Society of Mechanical Engineers* (ASME).

[35] Valizadeh, Mahsa and Sarah Jeannette Wolff. (2022). Convolutional Neural Network Applications in Additive Manufacturing: A Review. *Advances in Industrial and Manufacturing Engineering*. 4,100072–84.

[36] Wang, Jincheng, Rui Zhu, Yujing Liu and Laichang Zhang. (2023). Understanding Melt Pool Characteristics in Laser Powder Bed Fusion: An Overview of Single- and Multi-Track Melt Pools for Process Optimization. Advanced Powder Materials. KeAi Communications Co. 4(2), 100137–78.

[37] Wang, Jingsi, Jiaqi Zhu and Pay Jun Liew. (2019). Material Removal in Ultrasonic Abrasive Polishing of Additive Manufactured Components. *Applied Sciences* (Switzerland) 9 (24), 5359

[38] Wang, Xuanping, Youzhi Fu and Hang Gao. (2021) Finishing of Additively Manufactured Metal Parts by Abrasive Flow Machining. Solid Freeform Fabrication 2016: Proceedings of the 26th Annual International Solid Freeform Fabrication Symposium – An Additive Manufacturing Conference

[39] Zeng, Guowei, Wanrong Ye and Lei Liu. (2023). Effects of Surface Roughness on Fatigue Behavior of Additively Manufactured Stainless Steel 316L. *In Journal of Physics: Conference Series*. 2459, 012059–66.

[40] Zhang, Binbin, Prakhar Jaiswal, Rahul Rai, Paul Guerrier and George Baggs. (2019). Convolutional Neural Network-Based Inspection of Metal Additive Manufacturing Parts. *Rapid Prototyping Journal* 25 (3), 530–40.

[41] Zhao, Xia, Limin Wang, Yufei Zhang, Xuming Han, Muhammet Deveci and Milan Parmar. (2024). A Review of Convolutional Neural Networks in Computer Vision. *Artificial Intelligence Review* 57 (4), 1–44.

[42] Zhao, Xiyue, Aidin Imandoust, Mojtaba Khanzadeh, Farhad Imani and Linkan Bian. (2021). Solid Freeform Fabrication: Proceedings of the 32nd Annual International Solid Freeform Fabrication Symposium – An Additive Manufacturing Conference, 377–392.

8 Supervised Machine Learning Algorithms for Prediction of Ultimate Tensile Strength and Yield Strength of Metal Additive Manufactured Parts

M Arunadevi[1*], Vinayakumar R[2]

The reliable predictive models that predict the mechanical properties, such as yield strength and ultimate tensile strength, are in high demand for the growth of metal additive manufacturing. Nontraditional methods often struggle to find the influence of process parameters of additive manufacturing on its material behavior. So, this chapter explores the various supervised regression algorithms, including linear regression (LR), K-nearest neighbors (KNN), and artificial neural networks (ANN) for the prediction of yield strength and ultimate tensile strength. Prediction of strength levels is also explored using supervised classification algorithms, including logistic regression, decision trees, and support vector machines. Algorithms are evaluated using confusion matrices and accuracy, while the performance of regression algorithms is assessed through metrics like root mean squared error (RMSE) and R^2. This chapter demonstrates the accurate prediction of regression and classification algorithms and also classifies mechanical properties that emphasize the role of machine learning in material selection, process management, and enhancing quality assurance in additive manufacturing and leads to smarter manufacturing practices.

1. Introduction

Metal additive manufacturing (MAM) is a nontraditional manufacturing process that builds metal parts from a drawing. It builds the part layer by layer by adding the molten metal to produce precise components for aerospace applications [1]. More researchers are attracted toward metal additive manufacturing due to its ability to create complex shapes, minimal material waste, and provide design flexibility [2]. These printed parts ensured consistent performance and suitability for the intended application, which mainly depends on the mechanical properties of MAM. Evaluation of mechanical properties in metal additive manufacturing is expensive and time-consuming, often available only for a few materials and

[1] Associate Professor, Department of Mechanical Engineering, Dayananda Sagar College of Engineering, Bangalore 560111, India.

[2] Assistant Professor, Department of Mechanical Engineering, Dayananda Sagar College of Engineering, Bangalore 560111, India

* Corresponding author: arunadevi.dsce@gmail.com

DOI: 10.1201/9781003635673-8

techniques. To overcome the above issue, machine learning algorithms can be applied to the mechanical property prediction, which provides a more flexible and economical approach [3].

2. Importance of UTS and YS in Metal Additive Manufacturing

Ultimate tensile strength (UTS) and yield strength (YS) play a crucial role in achieving quality, reliability, and safety in metal additive manufacturing [4]. In many applications, metal parts are exposed to heavy loads and pressure, so these parts should be designed with the required strength. The tensile strength is a material property that determines the ability to withstand the maximum amount of stress and load before failure, which enhances the durability of metal parts [5]. Similarly, yield strength is also a material property that determines the point at which metal starts deforming, which increases the dimensional stability of the metal parts [6] and avoids plastic deformation [7]. In medical implants and pressure vessels, shape is a crucial criterion that leads to major catastrophes [8]. These two material properties mainly depend on the qualities such as the composition of the material, printing process parameters, and post-processing techniques (heat treatment processes) [9].

3. The Need for Predictive Models in AM

The machine learning predictive models are the tools in additive manufacturing to enhance quality and reduce the cost of additively manufactured metal parts [10]. Our objective is understanding and managing the parameters, including material quality, ambient temperature, print speed, layer height, and the environment around the printer using predictive models. Machine learning models can be used to forecast material behavior, identify potential failure, and suggest quick adjustments during the printing process [11]. Before the problem arises, these predictions help to eliminate the dimensional errors and avoid waste, which enhances the efficiency of the additive manufacturing technique. Also, during the design process, these models provide suggestions for choosing the best material combinations and forms suited for specific applications. Additive manufacturing technology development will require machine learning modeling approaches to fully utilize its potential, creativity, and to produce more expandable and reliable production methods in many industries [12].

4. Machine Learning in Metal AM

To ensure the operative performance of printed parts, the tensile strength and yield strength need to be predicted before printing. ML models can also be used to find the influence of process parameters on the material behavior by predicting the important qualities [13].

Among the key process conditions affecting yield strength and tensile strength in metal additive manufacturing are [14]:

1. Fusion, solidification, and the ultimate mechanical qualities of the component depend on the energy given to the metal powder.
2. The laser or electron beam's movement across the powder bed affects the cooling rates, grain structure, and residual stresses in the printed component.
3. The layer's thickness influences both the surface finish and density of the part, hence connected to tensile strength.
4. Hatch spacing the distance between laser passes can affect layer bonding and the total strength of a part.
5. Anisotropic in nature, the printed metal's orientation on the print bed might produce varied material qualities.
6. Preheating of temperature is essential for lowering thermal stresses during printing, as these stresses might otherwise impair mechanical characteristics.
7. Heat treatment or mechanical processing including annealing or hot isostatic pressing are examples of treatments after processing that might significantly influence final tensile strength and yield strength.

Utilizing these parameters in machine learning models can help to project how process changes will affect strength characteristics, therefore improving optimization and design choices in metal additive manufacturing.

4.1 Supervised Machine Learning Algorithms for UTS and YS Prediction

In additive manufacturing, supervised machine learning provides a superb replacement for predicting UTS and YS. We find it most effective when we have data with obvious inputs and outputs. Here the process parameters include laser power, scanning speed, layer thickness, hatch spacing, build orientation, preheating temperature, and post-processing treatments; the results are the measured values of UTS and YS. With both input and output data available, supervised machine learning can find the pattern or relationship between input and output, which enables the model for accurate future predictions for new data points [15]. Supervised machine learning models can handle both continuous and categorical data with diverse construction orientation or post-processing methods of metal additive manufacturing. Ultimate strength and yield strength predictions using these trained models can reduce the experimentation cost and time for different combinations of settings in additive manufacturing.

Since UTS and YS are the continuous numerical values, basically, it is a regression problem. Supervised learning algorithms, including linear regression, K-nearest neighbor, and artificial neural network, are modeled to find the complex relationship among different process parameters such as scan speed, laser power, build orientation, and their response variables [16]. Regression algorithms provide predictions necessary for material selection, process improvement, and component reliability assurance. Sometimes it would be useful to view UTS and YS prediction as a category concern approached from another angle. Instead of predicting exact numbers, classification based on defined thresholds classifies strength measurements into predetermined groups such as "Low," "Medium," or "High." [17]. In these situations, the main question is whether a part meets the minimum strength requirements rather than its particular mechanical properties. Efficiently predicting performance allows classification models such support vector machines, decision trees, or logistic regression (appropriate for basic pass or fail scenarios) to quickly classify components [18].

Specific application's needs will mostly guide your decision on whether to use regression or classification. Regression is your best choice if you need exact numerical predictions for sensitivity studies or process parameter adjustments [19]. On the other hand, if you are focusing on areas such acceptance testing, finding defects, or simplifying decision-making, especially when confronted with clear-cut outcomes, then classification becomes obvious if the strength is deemed "unacceptable." It is more efficient at managing imbalanced datasets, that is, when most items are in great condition but some are defective [20]. Ultimately, engineering goals will guide choice between regression and classification, whether seeking for exact numerical accuracy or have to make decisions based on categories. Every method has its benefits, and the best option really depends on the specific challenges it faces, the resources available, and your objectives in additive manufacturing [20].

4.2 Regression Algorithms for Predicting UTS and YS

When predicting UTS and YS using the factors of the additive manufacturing process, various regression algorithms are available. This section will examine three distinct regression methods, highlighting their specific benefits and the challenges they may present.

1. Linear Regression

In statistics, linear regression is the most widely used algorithm for regression analysis, and it works on the assumption that ultimate tensile strength and yield strength are linearly related to input variables, including laser power, scan speed, and layer thickness [21]. Linear regression is a fundamental statistical regression algorithm that tries to fit a linear model to the data points [22]. It is simple, interpretable, computationally efficient and suitable for linear correlations. It also has some disadvantages, such as linearity assumptions and multicollinearity.

2. K-Nearest Neighbor Algorithm

The K-nearest neighbor (KNN) algorithm is a supervised machine learning algorithm that can be used for both regression and classification problems. This algorithm will not create a model like other algorithms; instead, it will find the K-nearest neighbors by calculating the distance from the new data point. Once training is done, the response of a new data point is obtained by averaging the response variables of K neighbors to make predictions [23]. KNN is more flexible in capturing complex, nonlinear relationships between input and output parameters. KNN is easy to implement; it doesn't require traditional training like other models. Data with noise can be analyzed using KNN.

The limitations of the KNN algorithm are computationally intensive, proof of Curse of Dimensionality, which means that adding more features will not perform well till the reduction of dimensionality. The selection of the K value also plays a crucial role in the performance of the prediction. A smaller K value leads to overfitting, and a larger value results in underfitting.

3. Artificial Neural Network

An artificial neural network is developed by being inspired by the structure of the human brain. ANNs discover the nonlinear correlations between the input and output variables. ANNs are constructed by three types of layers, such as the input layer, hidden layer, and the output layer, which are connected by linked nodes. By applying an activation function to the input layer, each neuron generates an output function using a weighted sum. When the data are nonlinear in nature, an ANN performs better in identifying complex patterns in data. ANN is best-suited algorithm for modeling of nonlinear datasets in high dimensional space. ANN is the perfect choice for additive manufacturing, where quality is influenced by many parameters; these networks can handle a wide range of parameters and grasp the correlation amongst them. ANN can recognize the strong patterns or trends. The limitations of the ANN are that it consumes a lot of resources and time, especially in networks with a higher number of hidden layers. In ANN, understanding the influence of specific parameters on predictions is a challenging task. So, it has low interpretability. Overfitting will frequently occur in the ANN modeling, which leads to poor predictions [24].

4.3 Classification Algorithms for Categorizing UTS and YS

There are several classification algorithms that can predict UTS and YS based on various aspects of the additive manufacturing process. In this chapter, we'll dive into three of these algorithms, each offering its own strengths and weaknesses [25].

Many classification algorithms can be used for the yield strength and ultimate tensile strength prediction in metal additive manufacturing. Here three main algorithms are explained.

1. Logistic Regression

Logistic regression is one of the classification algorithms that can handle categorical output using the sigmoid function to find the probability that the output falls under that category [26]. If ultimate tensile strength and yield strength are categorized into levels such as low, medium, and high, logistic regression can be used to predict material strength at these categorical levels. In Figure 1, a schematic representation of logistic regression illustrates how the probability of output varies from zero to one and also optimizes the coefficients of inputs to minimize classification errors [27].

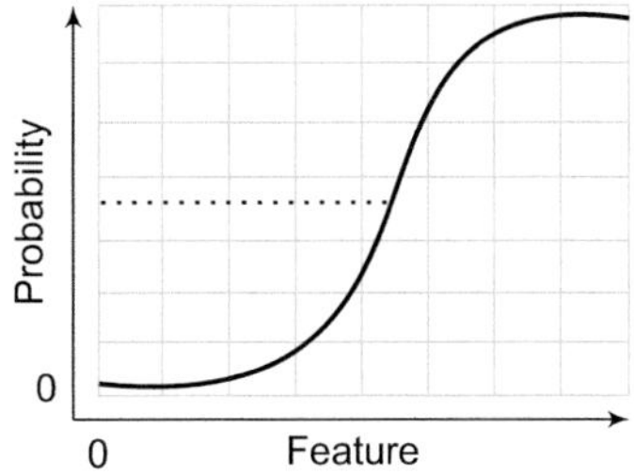

Fig. 1 Schematic representation of logistic regression.

Logistic regression is easy to implement, interpret, and suitable for small- to medium-sized datasets. The limitations of this algorithm are limited to linear relationships and binary classification problems.

2. Decision Tree Algorithm

A decision tree is a widely used supervised classification technique that creates a model in a tree-like structure [28]. The influencing property is placed at the top of the tree as a root node.

Gini impurity or entropy is used to identify the influencing input parameter while training the model [29]. A schematic representation of a decision tree is shown in Figure 2.

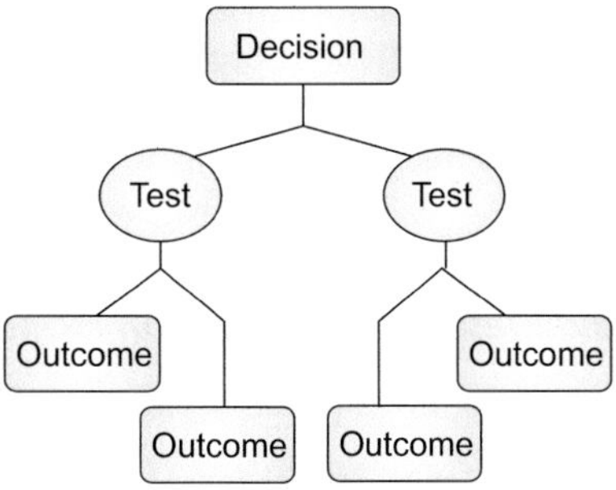

Fig. 2: Schematic representation of decision tree.

The decision tree model is very easy to understand and does not require any feature scaling for model training. Similar to the KNN, this algorithm can be used for both classification and regression problems. The limitations are overfitting, instability, and lack of generalization.

3. Support Vector Machine (SVM) Algorithm

Support vector machine is a supervised classification algorithm that classifies the data into different classes using a line in a two-dimensional space, called a hyperplane [30]. By maximizing the distance between closest point and the hyperplane, the classification error can be minimized. If the data are nonlinear in nature, SVM uses the kernel function to change the data into high-dimensional space [31]. In Figure 3, the schematic diagram of SVM is shown.

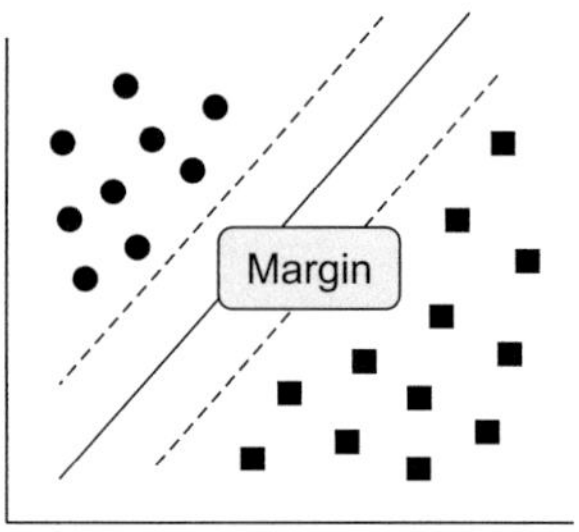

Fig. 3 Schematic representation of SVM.

SVM is effective in high-dimensional spaces, robust to overfitting and versatile in nature. The limitations of SVM are heavy computing demands, a lot of trials, and difficulty in interpreting the model.

5. Performance Evaluation and Model Validation

Machine learning predictions need to be evaluated using performance metrics. Regression and classification algorithms have different performance metrics [32, 33]. In regression algorithms, predictions are continuous values, so errors will give the accuracy of model to measure how predictions are accurate, whereas classification measures the accuracy by counting how many times predictions are correct.

5.1 Performance Evaluation of Regression Models

The accuracy of the regression models can be measured in predicting strength values by:

- **Mean Absolute Error (MAE):** MAE measures the average error size in a set of predictions, regardless of whether it is positive or negative. Lower MAE values mean the model is more accurate [34].
- **Mean Squared Error (MSE):** MSE is similar to MAE but gives more importance to larger errors. A lower MSE suggests the model's performance is stronger [35].
- **Root Mean Squared Error (RMSE):** RMSE is the square root of the MSE, which prompts the error in the same units as the target variable.
- **R-Square (R^2):** R^2 gives the proportion of variation in the output variable that is predictable from the input variable. An R-square value of one indicates a perfect fit, and a zero or negative value indicates poor fit model in the observable data [36].

5.2 Performance Evaluation of Regression Models

Classification Metrics: Confusion Matrix, Accuracy, Precision, and Recall

A confusion matrix can be obtained by counting true positive (TP), true negative (TN), false positive (FP), and false negative (FN) predictions from the actual and predicted categories.

Confusion Matrix for Binary Classification:

	Predicted Positive (Faulty)	*Predicted Negative (Healthy)*
Actual Positive (Faulty)	TP	FN
Actual Negative (Healthy)	FP	TN

Accuracy: Accuracy is the ratio of how many predictions made were correct to number of predictions made by the model.

$$\text{Accuracy} = \frac{TP + TN}{TP + FP + FN + TN}$$

6. Case Studies and Practical Applications

Table 1 shows the results of laser powder bed fusion process parameters with ultimate tensile strength and yield strength for 100 samples. The following are the input parameters:

- Scan speed (from 500 to 2000 mm/s)
- Laser power (from 150 to 400 W)
- Layer thickness (20 to 50 μm)
- Hatch spacing (80 to 120 μm)
- Build orientation (0°, 45°, 90°)
- Preheat temperature (20 to 200°C)
- Post-processing treatments (none, stress relief, HIP)

The output values are:

- Ultimate tensile strength (ranging from 300 to 950 MPa)
- Yield strength (from 250 to 850 MPa), representing stress at 0.2% plastic deformation
- Energy density

The above sample data are utilized to train a linear regression model and showcase the performance metrics, as shown in Figure 4. It demonstrates how to run linear regression in Python with the help of scikit-learn, using a sample dataset for demonstration. Similarly, other algorithms can be trained, and one that delivers the best performance metrics can be chosen for further predictions.

Table 1 Sample data.

Sample	*Laser Power (W)*	*Scan Speed (mm/s)*	*Layer Thickness (μm)*	*Hatch Spacing (μm)*	*Build Orientation (°)*	*Preheat Temp. (°C)*	*Post-Processing*	*Energy Density (J/mm³)*	*UTS (MPa)*	*YS (MPa)*
1	385	1455	30	101	0	86	None	505	362	1
2	320	972	20	115	90	45	HIP	757	608	2
3	280	1263	35	110	90	50	Stress Relief	416	346	3
.										
.										
.										
97	228	1846	50	103	45	172	HIP	539	458	97
98	108	1831	45	96	0	107	HIP	544	472	98
99	132	1670	50	106	90	98	None	368	257	99

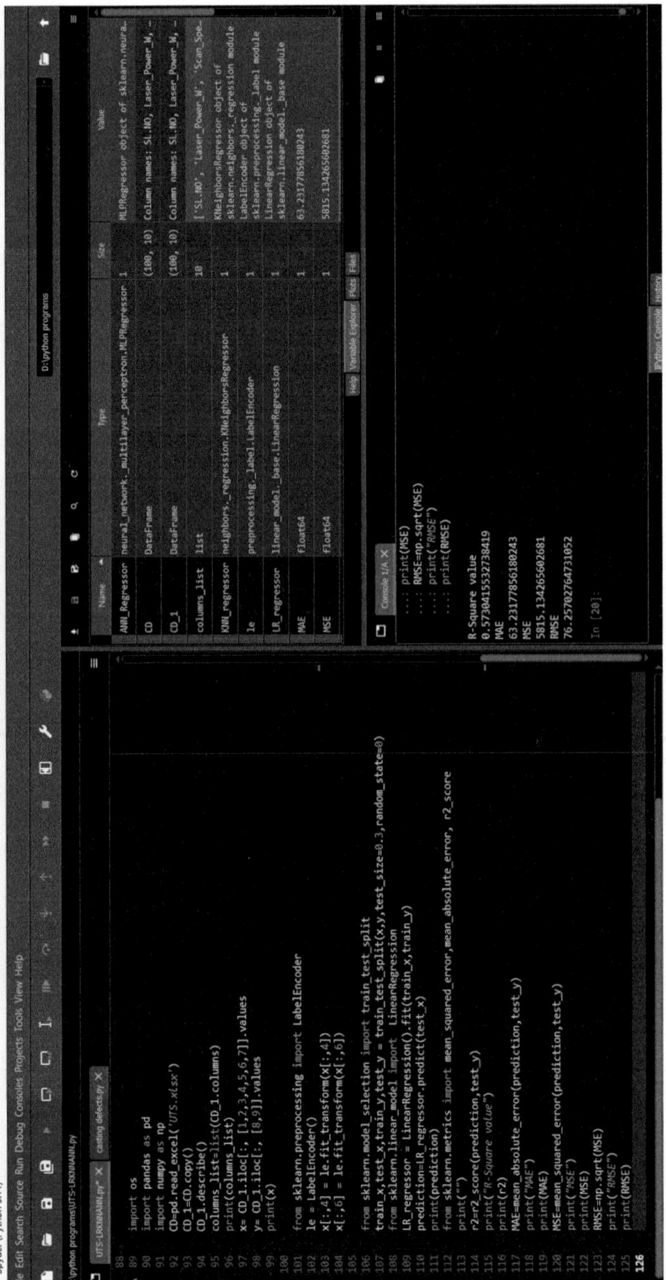

Fig. 4 Sample code.

7. Conclusion

In this chapter, the importance of machine learning (ML) techniques are demonstrated to predicting and classifying mechanical properties such as UTS and YS in metal additive manufacturing (AM). This chapter includes basic information about various regression methods, including linear regression (LR), KNN, and artificial neural networks (ANN), and also classification techniques such as logistic regression, decision trees, and SVM. This chapter also indicates how the models can effectively find the relationships between AM process parameters and how well regression and classification algorithms can boost both predictions and classifications. Performance evaluation metrics such as MAE, MSE, RMSE, R^2, confusion matrices, and accuracy are used for evaluating the algorithms. This chapter demonstrates the impact of machine learning in enhancing quality, process optimization, and control.

References

[1] Oliveira, J.P., Santos, T.G. and Miranda, R.M. (2021). Additive manufacturing of steels: A review of achievements and challenges. *Journal of Materials Science*, 56, 273–304.

[2] Azar, A.H., Zhang, H., Bai, Y. and Williams, C.B. (2022). Large-scale metal additive manufacturing: A holistic review of the state of the art and challenges. *International Materials Reviews*, 67(6), 362–409.

[3] Mozaffar, M. (2019). Machine learning for high-throughput process optimization of additively manufactured parts. *Journal of Manufacturing Processes*, 47, 178–186.

[4] Foster, B.K. (2021). Hybrid physics-informed machine learning for quality monitoring in additive manufacturing. *Computational Materials Science*, 193, 110375.

[5] Zhang, Z. and Yang, L. (2022). Machine learning in additive manufacturing: A state-of-the-art review. *Progress in Materials Science*, 123, 100834.

[6] Meng, L. (2020). Ensemble learning-based modeling for mechanical properties prediction in selective laser melting. *Journal of Materials Science & Technology*, 58, 234–245.

[7] Zhang, Y. (2020). Predictive modeling of mechanical properties in metal additive manufacturing using neural networks. *Materials & Design*, 194, 108899.

[8] Tapia, G. and Elwany, A. (2018). Prediction of porosity in metal-based additive manufacturing using spatial Gaussian process models. *Additive Manufacturing*, 26, 135–146.

[9] Wang, C. (2022). Deep learning-assisted in-situ monitoring for real-time mechanical property prediction in additive manufacturing. *Additive Manufacturing*, 59, 103144.

[10] Gobert, C. (2018). Application of supervised machine learning for defect detection in metallic parts using process parameters. *IEEE Access*, 6, 42577–42595.

[11] Kim, S. (2021). A comparative study of machine learning algorithms for fatigue life prediction of additively manufactured materials. *International Journal of Fatigue*, 152, 106451.

[12] Paul, A. (2021). Bayesian neural networks for uncertainty quantification in additive manufacturing. *JOM*, 73(11), 3721–3732.

[13] Li, Z. (2021). Transfer learning for mechanical property prediction in multi-material additive manufacturing. *Advanced Engineering Materials*, 23(6), 2001346.

[14] Lu, X. (2022). Explainable AI for interpreting process-property linkages in directed energy deposition. npj Computational Materials, 8(1), 1–12.

[15] Gupta, M. K. (2023). A CNN–LSTM hybrid model for real-time prediction of mechanical properties in laser powder bed fusion. *Additive Manufacturing*, 67, 103521.

[16] Chen, Y. (2022). Deep transfer learning for predicting anisotropic mechanical properties of additively manufactured Ti-6Al-4V. Materials Science and Engineering: A, 840, 142954.

[17] Wang, H. (2023). Multi-objective Bayesian optimization for mechanical property enhancement in wire-arc additive manufacturing. *Journal of Materials Research and Technology*, 24, 4562–4576.

[18] Liu, R. (2023). A review of data-driven approaches for fatigue life prediction in additive manufacturing. *International Journal of Fatigue*, 175, 107812.

[19] Yin, J. (2022). Generative adversarial networks (GANs) for synthetic dataset generation in additive manufacturing. *Advanced Engineering Informatics*, 54, 101779.

[20] Kumar, S. (2023). Explainable AI for process–property relationships in binder jetting additive manufacturing. *Materials Today Communications*, 35, 105583.

[21] Liu, J., Cosman, P.C. and Rao, B.D. (2018). Robust linear regression via L0 regularization. *IEEE Transactions on Signal Processing*, 66(7), 1737–1750.

[22] Singh, A.R. and Singh, S.K. (2021). Regression based predictive machine learning model for pervasive power system analysis. *International Journal of Electrical Engineering Research*, 9(1), 1–10.

[23] Dasgupta, S. and Freund, Y. (2008). Random projection trees and low dimensional manifolds. In Proceedings of the 40th Annual ACM Symposium on Theory of Computing (pp. 537–546).

[24] Kassahun, Y. and Sommer, G. (2005). Efficient reinforcement learning through evolutionary acquisition of neural topologies. In Proceedings of the 13th European Symposium on Artificial Neural Networks (ESANN) (pp. 259–266).

[25] Matitopanum, S., Pitakaso, R., Sethanan, K., Srichok, T. and Chokanat, P. (2023). Prediction of the ultimate tensile strength (UTS) of asymmetric friction stir welding using ensemble machine learning methods. Processes, 11(2), 391.

[26] Mohammed, A. A., Haris, S. M. and Al Azzawi, W. (2020). Estimation of the ultimate tensile strength and yield strength for pure metals and alloys using acoustic wave properties. Scientific Reports, 10(1), 12676.

[27] Han, E. H., Karypis, G. and Kumar, V. (2001). Text categorization using weight adjusted k-nearest neighbor classification. In Proceedings of the Pacific-Asia Conference on Knowledge Discovery and Data Mining (pp. 53–65).

[28] Kotsiantis, S. B. (2007). Supervised machine learning: A review of classification techniques. Informatica, 31(3), 249–268.

[29] Breiman, L. (2001). Random forests. *Machine Learning*, 45(1), 5–32.

[30] Cristianini, N. and Shawe-Taylor, J. (2000). An introduction to support vector machines and other kernel-based learning methods. Cambridge University Press.

[31] Schölkopf, B., Smola, A. and Müller, K.-R. (1998). Nonlinear component analysis as a kernel eigenvalue problem. *Neural Computation*, 10(5), 1299–1319.

[32] Catal, C. (2012). Performance evaluation metrics for software fault prediction studies. *Acta Polytechnica Hungarica*, 9(2), 49–61.

[33] Dewan, J.S.D.S. (2018). Improved model validation techniques for predictive analytics. *IEEE Transactions on Knowledge and Data Engineering*, 30(12), 2322–2335.

[34] Dasgupta, S. and Freund, Y. (2008). Random projection trees and low dimensional manifolds. In Proceedings of the 40th Annual ACM Symposium on Theory of Computing (pp. 537–546).

[35] Han, E.H., Karypis, G. and Kumar, V. (2001). Text categorization using weight adjusted k-nearest neighbor classification. In Proceedings of the Pacific-Asia Conference on Knowledge Discovery and Data Mining (pp. 53–65).

[36] Haris, S.M., Mohammed, A.A. and Al Azzawi, W. (2020). Estimation of the ultimate tensile strength and yield strength for pure metals and alloys using acoustic wave properties. *Scientific Reports*, 10(1), 12676.

9 Artificial Neural Network with Genetic Algorithm as a Hybrid Method for Optimization of Additive Manufacturing Process

Vedant Utikar[1*]

Additive manufacturing (AM) has emerged as a transformative manufacturing paradigm, enabling the fabrication of highly complex and customized components directly from digital models. Despite its advantages, AM processes are characterized by strong nonlinearity, multi-physics interactions, and high sensitivity to process parameters, which makes process optimization challenging when using conventional trial-and-error or empirical approaches. In recent years, artificial intelligence (AI)-based techniques have gained significant attention as powerful tools for modeling, prediction, and optimization of AM processes.

This chapter presents a comprehensive discussion on the integration of artificial neural networks (ANNs) and genetic algorithms (GAs) as a hybrid optimization framework for additive manufacturing. ANNs are employed as surrogate predictive models capable of capturing the complex and nonlinear relationships between AM process parameters and part quality attributes, while GAs provide a robust evolutionary optimization strategy for exploring large, multidimensional design spaces. The proposed hybrid ANN–GA methodology effectively balances exploration and exploitation, thereby reducing the risk of premature convergence and enabling near-global optimal solutions.

In addition to static optimization, this chapter highlights the incorporation of real-time sensor data and dynamic adaptation mechanisms within the hybrid framework. Such integration enables online learning, adaptive control, and enhanced resilience against process disturbances. The applicability of the ANN–GA approach is discussed across different AM technologies, including fused deposition modeling (FDM) and selective laser melting (SLM). Experimental validation and comparative analysis demonstrate improvements in product quality, material efficiency, and production speed when compared with conventional and standalone AI-based optimization methods. The chapter concludes with future research directions and challenges related to scalability, sustainability, and industrial deployment of hybrid AI-driven optimization frameworks in additive manufacturing.

[1] Research Engineer, Fatigue and Materials Centre of Excellence (FMCE), The Automotive Research Association of India (ARAI)

* Corresponding author: vutikar@gmail.com

DOI: 10.1201/9781003635673-9

1. Introduction

The manufacturing sector is upside down through 3D printing or additive manufacturing (AM). This technique provides manufacturing products through a layered method with great geometries from a digital model. It is widely applied to the manufacturing of very complex designs, which otherwise could not be realized by conventional subtractive manufacturing methods. The AM technology reaches across various industries such as aerospace, automotive, biomedical, and consumer goods: rapid prototyping, customized production, and lightweight components are produced by it. Complex internal features, optimized topologies, and designed material properties are realized by AM and thus show their innovative capacity in engineering and design practice today.

The AM is, instead, hampered with serious challenges toward its larger widespread use and optimization. The other area of concern could be centered on the precision for manufactured parts, as problems such as layer thickness, thermal gradients, and material shrinkage can all cause dimensional inaccuracies. In selective laser melting (SLM), for example, rapid solidification can lead to anisotropic properties on a microstructural level or even residual stresses, both influencing accuracy and performance of the final part. In addition, fast and efficient output is critical; extra time and resource-consuming production processes necessitate optimizations in both build strategies and parameters to minimize production time and material wastage. Besides this, the mechanical properties of parts produced via AM often show anisotropy and variability due to residual stresses and porosity. This necessitates tight process control and material characterization to assure high product quality. Addressing those challenges is of utmost importance to the course of maturation of AM technologies and their application in mainstream manufacturing.

The researchers focused not only on efficiencies of AM processes but also on the area of hybrid AI methods for application in optimization of the process. Such approaches, mostly used in combination with other types of AI methodologies, rise to meet these various multi-dimensional challenges of AM. For example, a combination of artificial neural networks (ANNs) and genetic algorithms (GAs) applies the nurturing capabilities of both: ANNs are great in modeling complex, nonlinear relationships within data, thereby allowing them to predict the possible outcomes based on the process parameters while GAs incorporated in natural selection operate in searching for a large solution space to ascertain the optimal or the near-optimal solutions. A combination of ANN's predictive abilities along with optimization powers of GA through hybrid design can explore the multilayered parameter space of AM processes and lead to better precision, efficiency, and material properties.

The reason for fusing ANNs with GAs in AM process optimization is to tackle the challenges posed by complexity and variability associated with AM. Also, the classical trial-and-error means for optimizing parameters is inefficient and costly. The hybrid AI approach, however, provides a means to establish a more systematic structure to model the highly nonlinear interaction of process parameters with part quality attributes. For example, an ANN could be trained to predict one or more mechanical properties of the printed part as a function of laser power, scan speed, and layer thickness. A GA can then be used to optimize these parameters by continually evolving the input set to maximize the tensile strength or minimize surface roughness, among other desirable outcomes. This combination would not only speed up the optimization process itself but would also strengthen and make reliable the processes utilized in AM, thus widening the scope for manufacturing adding quality, customized components.

A comparative evaluation of experimental and simulated results highlights the effectiveness of hybrid AI techniques in additive manufacturing optimization. A graphical representation of the tensile strength of parts produced with various process parameters—an ANN model's prediction validated by experimental testing—is presented in Figure 1. The closeness of the ear between the simulated data and experimental data divulges the predictive capability of an ANN. Figure 2 showcases the optimization algorithm trajectory of GA in refining process parameters to meet target mechanical properties, depicting how adept it is at converging around an optimal solution.

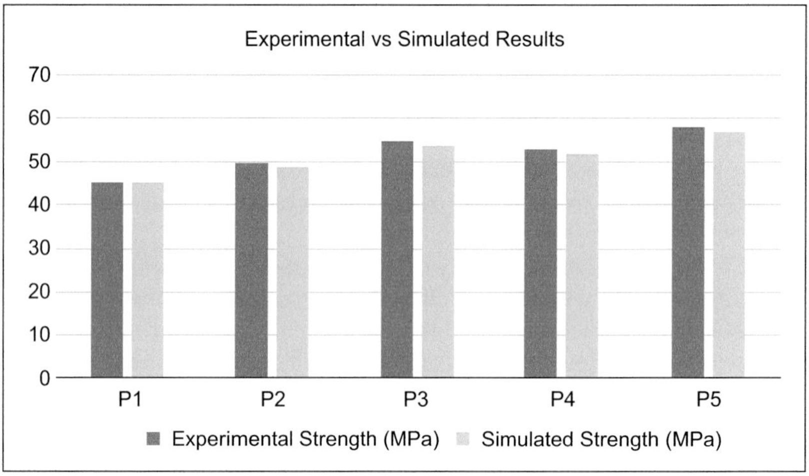

Fig. 1 Comparison of experimental vs. simulated tensile strength in AM parts.

These figures exemplify the potential of hybrid AI approaches to enhance AM processes, offering a pathway to overcome existing challenges and fully harness the capabilities of additive manufacturing.

AM, commonly referred to as three-dimensional (3D) printing, has fundamentally reshaped modern manufacturing by enabling the layer-by-layer fabrication of components directly from digital models. Unlike conventional subtractive manufacturing processes, AM allows the realization of complex

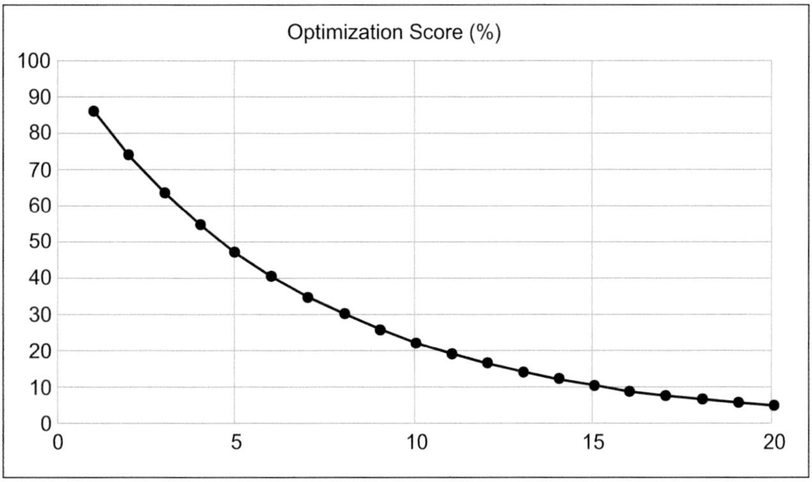

Fig. 2 Genetic algorithm optimization trajectory for AM process parameters.

geometries, internal features, and functionally optimized structures with minimal material waste. As a result, AM has found widespread applications in aerospace, automotive, biomedical, and consumer-product industries, where lightweight design, customization, and rapid prototyping are critical.

Despite these advantages, AM processes are inherently complex and exhibit strong coupling among thermal, mechanical, and material phenomena. Process parameters such as layer thickness, print speed, energy input, and material properties significantly influence part quality, dimensional accuracy, surface finish, and mechanical performance. Moreover, the nonlinear and stochastic nature of AM processes often leads to defects such as porosity, residual stresses, warping, and anisotropic mechanical properties. These challenges limit the repeatability, reliability, and large-scale industrial adoption of AM technologies.

Traditional approaches for optimizing AM parameters rely heavily on empirical experimentation or simplified analytical models. While such methods have contributed to incremental improvements, they are often time-consuming, costly, and unable to capture complex parameter interactions. Consequently, data-driven and AI-based optimization strategies have emerged as promising alternatives. ANNs have demonstrated strong capabilities in modeling nonlinear relationships, while GAs are well suited for global optimization in large search spaces.

Recent research in AM has increasingly focused on hybrid AI techniques that combine predictive modeling and evolutionary optimization. In particular, the integration of ANNs and GAs provides a systematic framework for predicting process outcomes and identifying optimal parameter combinations. This hybrid approach aligns with broader trends in AM research, including the development of advanced materials and intelligent process control systems, as highlighted in comprehensive reviews of AM material innovation and process advancement.

This chapter aims to present a structured and in-depth discussion of ANN–GA hybrid optimization for AM processes. The fundamental concepts of optimization in AM, the individual roles of ANNs and GAs, and their synergistic integration are explored. Furthermore, the chapter addresses real-time adaptation, experimental validation, and prospects, thereby offering a holistic perspective on AI-driven optimization in additive manufacturing.

2. Fundamentals of Optimization in Additive Manufacturing

AM, generally known as 3D printing, has come far in changing the concept of manufacturing by creating complex geometries directly from digital models. The full potential of tabletop AM, however, can be realized by gaining an insight into the performance-affecting parameters, controlling the complexities and nonlinearities of its processes, and critically evaluating traditional optimization approaches and their drawbacks.

Key AM performance parameters quality and efficiency of AM processes are influenced by various interrelated factors, such as:

- **Layer Thickness:** This dictates resolution and surface quality of the printed part. The thinner the layer, the finer the resolution and smoother the surface, but longer the time and higher the cost taken for the build. Thicker layers can speed up manufacturing, but some detail and surface quality may be sacrificed.
- **Printing Speed:** The speed with which either print heads or lasers move has a direct qualitative effect on the mechanical properties and dimensional accuracy of parts. Increased print speeds reduce build time, but such speeds sometimes lend to the fusion being incomplete or increased porosity.
- **Material Properties:** The characteristics of a material from which feedstock is made, such as melting temperature, viscosity, and thermal conductivity, substantially impact the mechanistic performance of the finished product as well as its structural integrity.
- **Build Orientation:** The orientation of a part in which it is printed affects surface quality, mechanical properties, and some support structure requirements. Improved orientation allows lower needs of support while enhancing mechanical performance.
- **Energy Input:** The laser power or extrusion temperature must be well controlled to ensure the right degree of bonding between the layers without sliding down the material properties.

The nonlinearity and complexity of AM processes include

Thermal dynamics has residual stresses and distorts during layer deposition, affecting dimensional accuracy and mechanical properties. The behavior of materials reflects complex phase changes relating to process parameters and environmental conditions when the material changes from a powder or liquid state to a solid state. Geometric intricacies mean that while complex geometries can be manufactured without additional costs, obtaining homogeneity in material properties and structural integrity throughout the part becomes demanding. Such as the indirect nature of parameters including layer thickness, print speed, and energy input, there abounds a multidimensional optimization problem herewith probable compromise between quality, speed, and costs. Conventional optimization approaches in AM generally made use of a trial-and-error approach and modeling by empirical means. Although those empirical modes have been structurally useful for improving processes, they have many drawbacks:

- **Resource-wise austerity:** Empirical modeling requires extensive experimentation, consuming much time and material, which is not economical or sustainable.
- **Limited generalizability:** Models suited for certain materials or machines would yield no welcome results in other conditions.
- **Inability to account for intricate interactions:** Traditional methods fail to account for the interactions among several parameters and their nonlinearities, leading to sub-optimal solutions.

To do this, hybrid AI-based optimization techniques are being pursued to ameliorate AM process optimization.

Additive manufacturing processes involve a multitude of interdependent parameters that collectively determine part quality, productivity, and cost. Optimization in AM seeks to identify parameter combinations that satisfy multiple, often conflicting objectives such as maximizing mechanical strength, minimizing surface roughness, reducing build time, and lowering material consumption.

2.1 Key Performance Parameters in AM

Several critical parameters influence AM process performance:

- Layer Thickness: Layer thickness directly affects surface finish, dimensional resolution, and build time. Thinner layers yield improved surface quality but increase production time, whereas thicker layers enhance productivity at the expense of resolution.
- Printing Speed: Print speed influences energy input, bonding quality between layers, and defect formation. Excessively high speeds may lead to incomplete fusion and porosity.
- Material Properties: Feedstock properties such as melting temperature, viscosity, and thermal conductivity significantly affect process stability and part integrity. Advances in AM materials have further increased the complexity of parameter optimization (Li et al., 2019).
- Build Orientation: Orientation affects anisotropy in mechanical properties, surface quality, and support structure requirements.
- Energy Input: Laser power, extrusion temperature, or arc current must be carefully controlled to ensure adequate interlayer bonding without degrading material properties.

2.2 Nonlinearity and Complexity of AM Processes

AM processes are characterized by highly nonlinear interactions among thermal gradients, material phase transformations, and geometric constraints. Rapid heating and cooling cycles introduce residual stresses and microstructural variations, which are difficult to predict using analytical models alone. The multidimensional nature of AM optimization often results in trade-offs between competing objectives, necessitating advanced optimization strategies.

2.3 Limitations of Conventional Optimization Approaches

Conventional optimization techniques in AM typically rely on one-factor-at-a-time experimentation or empirical regression models. These approaches suffer from several limitations:

- High experimental cost and time consumption
- Limited generalizability across materials and machines
- Inability to capture complex parameter interactions

These shortcomings have motivated the adoption of AI-based optimization frameworks capable of handling nonlinear, high-dimensional problems.

3. ANNs in AM Optimization

Artificial neural networks represent some of the most useful approaches to additive manufacturing, mainly in predictive modeling, understanding of process parameters, and the prediction of performance. The enhancement added during the optimization of AM processes in terms of their potential to model incomprehensible, complex, nonlinear relationships makes them most appropriate.

Role of ANN in predictive modeling: ANN acts as a facilitator for building models that can predict the output from specified input variables. In AM, this is crucial to predicting how different processing parameters will influence the quality and properties of produced components. Trained on a variable set containing historical data, ANN's apprehension of such complexity allows for remarkably refined predictions, which were usually decried by traditional modeling practices owing to low reliability-and-through process control enhancements.

Training of ANN to understand AM process parameters: ANNs are trained with datasets that include parameter inputs (layer thickness, print speed, and type of material) and respectively tested values of output measures (mechanical strength, surface finish, and dimensional accuracy). The analytical learning of the network then readjusts the internal weights to minimize the connected errors encountered during predictions in an iterative manner.

This way, the network can generalize the output based on the training datasets for future unseen combinations of parameters. Such training data quality in terms of volume and variety heavily affects both accuracy and resilient capabilities of the developed model.

Artificial neural networks are computational models inspired by the human brain, consisting of interconnected neurons organized in layers. ANNs are particularly effective in modeling complex, nonlinear relationships between inputs and outputs, making them suitable for AM process modeling.

3.1 Role of ANNs in Predictive Modeling

In AM, ANNs are commonly used to predict part quality attributes such as tensile strength, surface roughness, and dimensional accuracy based on process parameters. By learning from historical experimental data, ANNs can generalize predictions for previously unseen parameter combinations, thereby reducing the need for extensive experimentation.

3.2 Training and Architecture Considerations

ANN training involves adjusting network weights to minimize prediction error using algorithms such as backpropagation. The choice of network architecture, activation functions, and training data quality significantly influences model performance. In AM applications, multilayer perceptrons with nonlinear activation functions are widely used due to their flexibility and robustness.

3.3 ANNs as Surrogate Models

ANNs often serve as surrogate models that approximate complex physical processes at a fraction of the computational cost. Once trained, these models enable rapid evaluation of numerous parameter sets, making them ideal for integration with optimization algorithms such as GAs.

ANN as a Surrogate Model for AM Performance Prediction

In AM parlance, ANNs are generally surrogate models offering approximate solutions to complex physical processes. Typical performance measures can, therefore, be predicted at very high speed. For example, an ANN trained to predict the tensile strength of the printed part with respect to the parameters of a process can help to evaluate quickly several sets of parameters to facilitate optimization or adjustments of process variables in real time. Such an approach to surrogate modeling speeds up the cycle of development and drastically cuts down on costs associated with long hands-on experimental tests.

Examples of Applications of ANN in AM

Many studies have come up with successful applications of ANNs on AM:

- **Optimization of Process Parameters:** ANNs have been applied in determining, for a given design, what process parameter sets will likely yield the desired mechanical properties and minimize defects. This is achieved through predicting the outcomes of different settings on the parameter space with further insights direct from ANN into efficient parameter search.

- **Defect Detection and Quality Assurance:** Anomalies are detected in real time via monitoring the AM process using ANN algorithms, as well as predicting upcoming defects and improving quality assurance.
- **Material Property Prediction:** ANNs can predict material properties of printed parts using experimental data, guiding decisions on what materials and process settings can yield ideal results for specific applications.

4. GAs for AM Parameter Optimization

GAs have become widely utilized optimizers in the realm of AM, specifically for approaching complex, multidimensional optimization problems related to process parameters. Imitating the tenets of natural evolution, GAs make up a versatile method to cover large design spaces and improve and/or optimize the processes in AM applications.

Basics of GAs and Their Evolutionary Approach to Optimization: GAs are search heuristics modeled on the process of natural selection, operating on mechanisms like selection, crossover, and mutation to evolve solutions for optimization problems. They work on some candidate solutions' population and iteratively improve upon their solutions based on a defined fitness function. This evolutionary perspective is most apt for the complex, nonlinear, and multimodal landscapes that characterize optimization problems in AM.

Genetic algorithms are evolutionary optimization techniques inspired by natural selection. GAs operate on a population of candidate solutions and iteratively evolve them through selection, crossover, and mutation.

4.1 Evolutionary Optimization Principles

In AM optimization, everyone in the GA population represents a set of process parameters. A fitness function evaluates the quality of everyone based on predicted or measured performance metrics.

4.2 Multi-Objective Optimization

AM optimization often involves multiple objectives, such as maximizing strength while minimizing build time. GAs are well suited for multi-objective optimization and can identify pareto-optimal solutions that balance competing goals.

4.3 Selection, Crossover, Mutation, and Fitness Evaluation in AM Context

- **Selection:** On the AM platform, and more generally, the selection process known as fitness refers to choosing high-performing sets of parameters (individuals) from the current population based on their fitness scores. For example, parameter combinations that produce parts with excellent mechanical properties and low surface roughness are preferred.

- **Crossover:** This genetic operator takes selected couples and produces from them offspring that are combined and that can change such features among themselves. Crossover in AM may involve combinations of parameters such as together laser power with scan speed to explore other combinations that potentially result in better part quality.
- **Mutation:** It introduces a random change in some parameters within an individual. This change contributes to increasing the diversity in the population and enhances the search operation. The small changed values of layer thickness or build orientation can be an example of this.
- **Fitness Evaluation:** A scale that enables the evolutionary background process is called fitness. Generally, in AM fitness evaluation, the process may include simulating some parts of their manufacture to evaluate outcomes like tensile strength or dimensional accuracy, and thus allows deciding on how appropriate those parameters could be.

Multi-Objective Optimization Using GAs

As illustrated in Figure 3, multi-objective optimization in additive manufacturing utilizes genetic algorithms to optimize critical process parameters. The figure depicts the iterative optimization process in which an initial population of solutions undergoes selection, crossover, and mutation to efficiently search the design space. Over successive generations, the algorithm converges toward Pareto-optimal solutions that balance mechanical strength, production time, and material consumption. Figure 3 further highlights the trade-offs among competing objectives, enabling decision-makers to assess the impact of different parameter combinations before selecting the optimal process conditions.

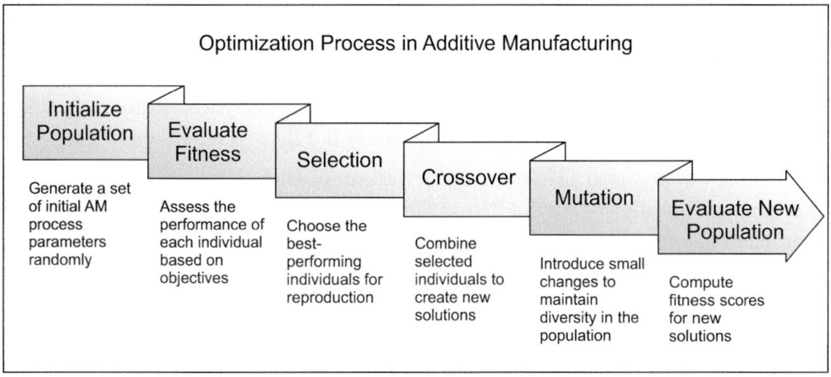

Fig. 3 Optimization process in additive manufacturing.

GA-Driven AM Improvements through Case Studies

Several of the works have raised praise for GAs on AM improvements:

- **Optimization of Laser Beam Paths:** Path of laser beam is optimized toward disturbed thermal and residual stress patterns in a laser-based AM approach by a GA framework. The path generation, sought out as the optimal sequence of laser irradiation, affirms improved quality for parts produced as well as lowered defects.

- **Process Selection System:** Steady-state GA-based process selection system is achieved by the AM researchers toward achieving very high accuracy in identifying the best manufacturing processes for 3D objects. Comparison studies have shown the GA outperforms other optimization algorithms on complex decision-making tasks.
- **Time and Cost Optimization:** A modified GA is applied to scheduling in AM with respect to minimizing time and cost. Scheduling and assignment decisions were placed by the algorithm, which enhanced the manufacturing operations' efficiency.

5. Hybrid Approach: ANN–GA Integration for AM Optimization

The combination of ANN with GA brings a powerful hybrid methodology to solve AM optimization problems. This synergy capitalizes on the prediction capability of ANN and the optimization strength of GAs to raise the AM performance level.

Conceptual Framework of ANN–GA Hybrid Methodology: ANN–GA hybrid methodology is the combination of two separate techniques with respect to solving complex optimization problems for AM:

1. **Data Collection:** Gather experimental data on AM process parameters and results.
2. **ANN Training:** Develop an ANN model learning the relation between the process parameters and desired output results.
3. **GA Optimization:** Apply the trained ANN as a fitness function inside the GA to optimize the AM process parameters.
4. **Iterative Refinement:** Repeat the process iteratively to continuously refine both ANN model and GA optimization, in the face of new data and insight.

ANN Feeding into GA for Optimization: In this hybrid method, ANN is used here as a predictive model that makes predictions regarding the AM process parameters results, such as tensile strength or surface finish. With such predictions by ANN, the GA can effectively find the best parameter combinations without an extensive number of physical experiments. This integration allows great cost savings regarding computation and accelerates optimization tremendously.

The GA takes the AM process parameters to a better level through its various evolutionary mechanisms:

- **Selection:** ANN makes predictions regarding the efficacy of various combinations of the parameter sets and selects those that yield more favorable predictions.
- **Crossover:** The selected parameter sets are combined to create additional possibilities based on the selection process.

- **Mutation:** Some variation is randomly introduced to keep diversity alive, thus avoiding local optima. In an iterative fashion, these operations converge through GA to a parameter set for the desired AM results.

Compared to Other Hybrid Optimization Methods

The ANN–GA hybrid has some specific advantages over other optimization methods:

The efficiency of ANNs and GAs allows it to achieve improved convergence speed in comparison with traditional techniques. The surrogate model built by ANN further improves the precision of the optimization process. Adaptability of hybrid models varies with AM processes and objectives.

This requires much more complicated modeling and a higher amount of computational resources than other hybrids such as RSM–GA or ANFIS–GA. Synergy between an ANN and GA shows a complex methodology for AM optimization that merges the predictive and evolutionary strengths of both techniques to achieve outstanding manufacturing outcomes.

The hybrid ANN–GA framework combines the predictive capabilities of ANNs with the global search efficiency of GAs. In this approach, the ANN serves as a fitness evaluator within the GA, enabling efficient exploration of the parameter space.

5.1 Conceptual Workflow

1. Data collection from experiments or simulations
2. ANN training and validation
3. GA-based optimization using ANN predictions
4. Iterative refinement and validation

5.2 Advantages over Standalone Methods

The ANN–GA hybrid approach offers faster convergence, improved accuracy, and reduced experimental cost compared with conventional and standalone AI-based methods.

6. Dynamic Adaptation and Real-Time Feedback in Hybrid Optimization for AM

Integrating dynamic adaptation and real-time feedback mechanisms into hybrid optimization strategies enhances both efficiency and quality in AM. This enables manufacturers with substantial robustness across AM technologies, such as FDM and SLM, using sensor data in real time, online learning, and adaptive control algorithms.

Integration of Real-Time Sensor Data for Adaptive Optimization

The application of real-time sensor data into AM processes makes adaptive optimization possible, as this provides for continuous monitoring coupled with in-time decision-making. For example, high-speed cameras, pyrometers, and laser scanners quantize essential parameters such as: temperature gradients, melt pool behavior, or how well-layer deposition has been controlled. The data sent easily enable the realization of some process anomalies and quick response adjustments to other operating parameters to enhance quality and limit defects.

For instance, in wire arc additive manufacturing, the welding camera system tracks the weld pool, and adjacent areas staying free from strong arc light disruptors, ensuring uninterrupted quality monitoring.

Online Learning and Real-Time Adjustment Mechanisms: Real-time adjustment mechanisms during the AM process rely on the properties of online learning algorithms. Such online learning algorithms, commonly reflected through machine learning models, would allow continuous update operations of itself based on specific incoming sensor data appropriate for the characterization of the supposed optimal process parameters.

Because of reinforcement learning and adaptive control, for instance, the proposed model could autonomously adjust parameters of the process like the laser power or scan speed, during the manufacturing stage by real-time adjustments. Dynamic adaptation gives a great boost to enhancing the overall stability of the process and, ultimately, quality. For instance, the digital twin concept–a virtual twin of a real object–enables real-time access for monitoring, model updating, forecasting, and automating decision-making in AM and provides new prospects in autonomous manufacturing for control tactics by proactive means.

The integration of real-time sensor data enables adaptive optimization during AM processes. Online learning and adaptive control strategies enhance robustness against disturbances and variability.

Benefits in Resilience Against Perturbations and Unknown Errors

The use of real-time feedback and adaptive optimization in AM processes contributes toward their robustness against perturbations and unknown errors. Early detection and correction of developing deviations could reduce distortions arising from external disturbances caused by fluctuations in ambient conditions and material inconsistencies. Succeeding this resilience ultimately leads to higher yield rates, less material waste, and overall improved confidence in the quality and reliability of AM parts. For instance, smart factory systems deploy real-time data analytics to predict demand and redesigning needs rather than only reacting to orders placed, and they do this to increase general efficiency.

Application in different AM techniques—FDM, SLM, etc.: The dynamics of adaptation and incorporation of feedback would find their place legitimately across the AM techniques:

- **Fused Deposition Modeling (FDM):** Instant monitoring of the extrusion temperature and deposition speeds allows for immediate changes within the process line to mitigate common defects inherent in FDM like warping and stringing.

- **Selective Laser Melting (SLM):** Continuous monitoring of the melt pool and powder bed conditions allows the operator to effectively control the laser parameters to avoid locking in defects like porosity and insufficient fusion. By using sensor integration tailored to each specific AM technique, the adaptive algorithms act as an enabler for performance optimization, thus ensuring parts of quality.

The incorporation of dynamic adaptation and real-time feedback mechanisms into AM optimization frameworks represent a critical advancement toward intelligent and autonomous manufacturing systems. Traditional AM optimization approaches are typically performed offline, where optimal process parameters are identified prior to fabrication and remain fixed throughout the build. However, such static optimization strategies are often insufficient due to unavoidable process variability arising from environmental fluctuations, material inconsistencies, and machine-specific uncertainties.

6.1 Role of Sensors and In Situ Monitoring

Modern AM systems are increasingly equipped with advanced sensing technologies, including infrared cameras, pyrometers, acoustic emission sensors, laser profilometers, and high-speed optical cameras. These sensors provide real-time information related to melt pool geometry, temperature distribution, layer height, surface morphology, and defect initiation. The continuous acquisition of such data enables in situ monitoring of the process state, forming the foundation for adaptive optimization.

For instance, in SLM, melt pool size and thermal gradients strongly influence porosity and microstructural evolution. Real-time monitoring of these variables allows early detection of deviations from desired process conditions. Similarly, in FDM, monitoring extrusion temperature, filament flow rate, and layer adhesion enable timely correction of defects such as warping, under-extrusion, and delamination.

6.2 Online Learning and Adaptive Control

Online learning refers to the ability of predictive models to update their parameters continuously as new data become available. When integrated with ANN–GA frameworks, online learning enables the neural network to refine its predictive accuracy during the manufacturing process. This capability is particularly valuable in AM, where process dynamics may evolve over time due to tool wear, thermal accumulation, or material batch variation.

Adaptive control strategies leverage online predictions to modify process parameters in real time. For example, if sensor data indicate excessive melt pool temperature, the control system can dynamically reduce laser power or increase scan speed to restore stability. In this context, the ANN serves as a real-time surrogate model, while the GA provides decision logic for selecting corrective actions within predefined constraints.

6.3 Digital Twins for AM Optimization

The concept of the digital twin has gained significant attention in the context of smart manufacturing. A digital twin is a virtual representation of a physical system that is continuously updated using real-time data. In AM, digital twins integrate sensor data, predictive models, and optimization algorithms to simulate process behavior and forecast outcomes.

When combined with ANN–GA optimization, digital twins enable proactive decision-making. Instead of reacting to defects after they occur, the system can predict potential issues and adjust parameters preemptively. This predictive capability enhances process robustness, reduces scrap rates, and improves overall productivity.

6.4 Benefits of Dynamic Adaptation

The integration of real-time feedback and dynamic adaptation within ANN–GA frameworks offers several advantages:

- Improved resilience against disturbances such as ambient temperature fluctuations and material variability
- Enhanced consistency and repeatability of part quality
- Reduced material waste and rework
- Increased confidence in autonomous AM systems

These benefits are essential for transitioning AM from prototyping environments to high-value industrial production.

7. Experimental Validation and Results

The combined utilization of ANNs and GAs in the optimization of AM techniques is the subject of a comprehensive review. The ANN–GA hybrid framework implementation is presented in this section, assesses the respective performance metrics of product quality, material efficiency, and production speed, compares it with classical and AI approaches, and finally presents case studies with proven practicable improvement.

Experimental studies demonstrate that ANN–GA optimization leads to improved tensile strength, reduced material waste, and enhanced production efficiency when compared with traditional optimization techniques.

ANN–GA Hybrid Framework Implementation in AM

The working of the ANN–GA hybrid framework is based on the combination of the predictive powers of ANNs and the optimization potential of the GAs to mold AM process parameters.

The implementation steps comprise:

1. Collecting experimental data on AM process conditions and properties.
2. Developing the ANN involves establishing correlations between the input parameters and the output responses, such as surface characteristics and

mechanical properties. In addition, an exhaustive review of the technical literature regarding the hybridization of different fusion processes with ANNs and variable dimensioning problems identifies two critical areas that remain insufficiently explored in model design and optimization.
3. Use the predetermined ANN in the GA as a fitness function to look for improved and superior process parameters.
4. Verification of smiley parameter experimentation prescribed improvement in FDM for study aspect studies utilizing optimization parameters. For example, one study utilized an ANN–GA hybrid system to optimize FDM process parameters, consequently improving the tensile strength of printed parts.

Performance metrics are product quality, material efficiency, and production speed. Effectiveness of ANN–GA hybrid framework will be evaluated based on a few essential performance metrics.

- **Product Quality:** Enhancements in mechanical properties such as tensile strength, flexural strength, and compressive strength indicate better quality of the product. This study showed a significant increase in tensile strength with optimized parameters.
- **Material Efficiency:** Process optimization will reduce the material waste and enhance the material utilization rate. While specific data on material efficiency were not reported in this study, the optimization process inherently aims to support this metric.
- **Production Speed:** The speed of production can be increased without sacrificing quality through varied parameters such as layer thickness and print speed. The study indicates optimized parameters can retain production speed while enhancing product quality.

Comparison with conventional and AI methods

The ANN–GA hybrid approach exhibits a lot of advantages over conventional optimization methods and other AI-based methods.

- Conventional Methods: Traditional optimization techniques generally depend on trial-and-error mechanisms or one-factor-at-a-time approaches, which are tedious and possibly cannot represent involved interactions between parameters. The ANN–GA framework explores the multidimensional parameter space with high efficacy and leads toward enlightenment about optimization.
- Other AI-Based Methods: The hybrid method leverages the benefits of both ANNs and GAs instead of mere standalone AI. For example, the study comparing GA–ANN versus GA–ANFIS communicated that GA–ANN offered increased stability in tensile strength estimating to a percentage of 98.5% in contrast to other methods used.

These case studies that showcase practical improvements are few. The field of additive manufacturing includes the following:

- **Tensile Strength Optimization:** This study finds that by optimizing FDM process parameters using the ANN–GA approach, tensile strength has been improved, thus supporting the effectiveness of the framework in enhancing the quality of the product.

- **Damage Evolution Modeling:** Here a two-phase ANN–GA hybrid model developed to predict metallic sheet-metal damage evolution in the roll forming process has slightly improved its accuracy for better process optimization.

Graphical Analysis Experimental vs. Simulated Tensile Strength

To justify the effectiveness of the ANN–GA optimization approach, tensile strength values obtained experimentally and from simulations will be juxtaposed for various combinations of the input parameters of the AM technology such as layer thickness, print speed, infill density, and nozzle temperature. The illustrations raised here include:

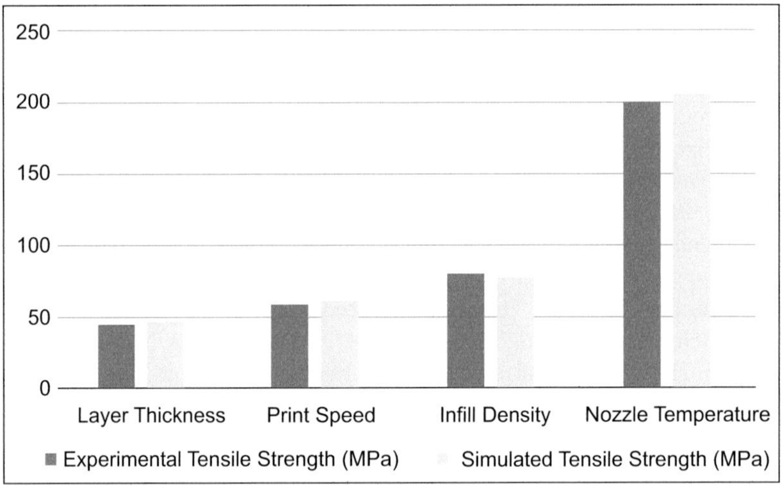

Fig. 4 Effectiveness of the ANN–GA optimization approach.

Observation:

- Tensile strength values predicted by the ANN–GA framework demonstrated close conformity to their experimental counterparts (Fig. 4).
- All parameter variations enhanced the process-optimizing abilities of the ANN–GA, thereby exemplifying the adeptness of the model in improving AM process controllability.
- The small deviation indicates that a few external factors such as machine variability or environment might have played a role.

This graph substantiates the accuracy level of the ANN–GA model, which tangibly uplifts the AM outcomes.

8. Future Directions and Challenges

Rapid advancements in the areas of ANNs and GAs for optimizing AM have begun to pave the roads toward more intelligent and adaptive as well as efficient methods of AM. However, there are still many challenges and opportunities that are yet to be explored to enhance the efficiency, scalability, and sustainability of hybrid AI approaches for AM.

Future research should focus on adaptive learning, multi-fidelity modeling, quantum-inspired optimization, and sustainable AI frameworks to further enhance AM optimization.

Possible Enhancements to ANN–GA Hybrid Methodologies

The existing ANN–GA frameworks are already showing some encouraging results in optimizing AM parameters, yet there is always scope for improvement.

1. **Adaptive Learning Mechanisms:** Standard ANN–GA systems are based on static training data, while adaptive learning mechanisms would allow the real-time adaptation of processes to changing conditions in production.
2. **Multi-Fidelity Modeling:** One way of getting more information from your data is to merge multiple sources of data, whether experimental, numerical, or simulated.
3. **Quantum-Inspired GAs:** Recent studies have found that using a quantum-inspired evolutionary algorithm in additive manufacturing can offer drastic improvements in optimization efficiency and convergence speed.

Through a comparative study between conventional GA–ANN and quantum-inspired (Fig. 5) GA–ANN models on the optimization of AM systems, it can be concluded that quantum-inspired GA–ANN showcases superior features in terms of solution quality and convergence speed:

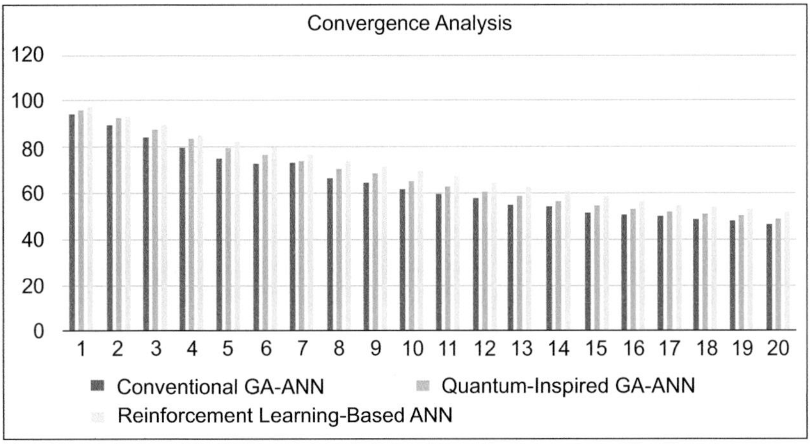

Fig. 5 Comparative convergence analysis of ANN–GA variants.

Merging with Other AI Approaches (RL, Swarm Intelligence)

For even more enhancement of AM process optimization, the ANN–GA hybrid approaches can be combined with:

- **Reinforcement Learning (RL):** RL-based models have the capability to dynamically adapt process parameters based upon real-time feedback regarding their performance which in turn can augment adaptive control mechanisms.
- For example, SI techniques like particle swarm optimization (PSO) and ant colony optimization (ACO) can be hybridized with ANN–GA models to get a better global optimum.
- **Fuzzy Logic Systems:** The use of fuzzy logic systems can help accommodate ambiguities inherent to AM processes.

Challenges in Scaling and Deployment to the Industry

While hybrid ANN–GA methods have shown exceptional promise in experiments, many factors currently limit their practical adoption. Some of these are:

1. Real-time complexity is caused by the high-dimensional optimization problems and for them a large computational power is essential which results in almost no real-time deployability.
2. The performance of ANN models is closely dependent on the quantity and quality of training data; this tends to be the most difficult aspect in practice for industrial contexts, as well as the main factor to account for.
3. Integration with Other Industrial Systems: The actual application of AI-driven optimization in traditional AM processes depends on the device's compatibility with industrial IoT (IIoT) frameworks and MES stands for Manufacturing Execution System.

Ethical and Sustainability Considerations in AM Optimization

Considering the increasing application of AI-based optimization in additive manufacturing, what are the ethical and environmental implications associated with this:

1. **Energy Efficiency:** An important goal of AI optimization should be to decrease the energy usage in additive manufacturing processes, thereby supporting green manufacturing.
2. **Material Waste Reduction:** Multi-objective optimization under GA can further enhance the utilization of materials and hence decrease waste, complying with the circular economy principles (Fig. 6).
3. **Ethical Use of AI:** To prevent any biased consequences, the models used by AI in decision-making regarding asset management must be made not only transparent but also interpretable and fair.

While ANN–GA hybrid models significantly improved the optimization of AM processes, future research should focus on adaptive learning, integration with other AI techniques, industrial scalability, and sustainability. The trio of AI, IoT, and sustainable manufacturing will govern the development of next-generation smart AM systems.

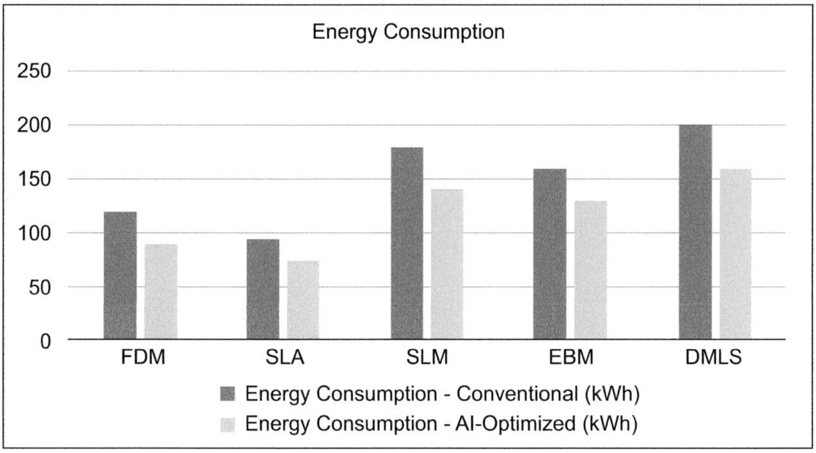

Fig. 6 Energy consumption trends in AI-optimized AM vs. conventional AM.

9. Conclusion

The combined use of ANNs and GAs has shown remarkable potential for optimization in AM processes. Systematic research, experimental confirmation, and comparative analysis have demonstrated the major benefits of hybrid artificial intelligence-driven approaches in contrast to traditional ones. The results further testify toward the impact of artificial intelligence on making additive manufacturing more efficient, higher in quality, and sustainable to pave the way for next-generation smart manufacturing systems.

The hybrid ANN–GA optimization framework provides a powerful and flexible approach for addressing the complexity of additive manufacturing processes. By combining predictive modeling and evolutionary optimization, this methodology enables improved quality, efficiency, and sustainability in AM systems.

Summary of Key Findings and Contributions

1. The hybrid framework consisting of ANN and GA demonstrates remarkable efficiency in enhancing the optimization of AM parameters thus it achieves better product quality with decreased manufacturing defects.
2. ANN-based predictive models achieve more accurate predictions through the integration of GA optimization methods thus creating a reliable process with decreased experimental requirements.
3. The application of AI optimization to AM processes makes them more environmentally friendly since they use less energy and materials leading to reduced manufacturing costs.
4. Real-time adaptation enables the system to integrate sensor data in real-time through ANN–GA models for the purpose of adaptive control to minimize variations while maintaining process stability (Fig. 7).

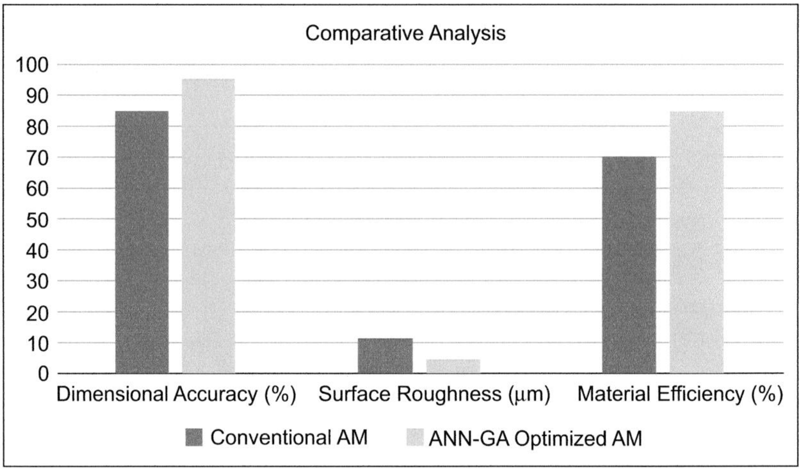

Fig. 7 Comparative analysis of ANN–GA vs. conventional AM optimization.

ANN–GA Hybrid Optimization Brings Value to Future-Oriented AM System Implementations

1. The ANN–GA optimization framework integrates without difficulties into IIoT platforms and cyber-physical systems to create smart AM factories with scalability features.
2. This hybrid optimization method successfully resolves conflict between multiple goals including strength requirements and cost expenditure and printing duration requirements for aerospace and biomedical and automotive sectors.
3. The combination of ANN–GA with reinforcement learning along with fuzzy logic and swarm intelligence improves automation and intelligence in AM system operations.

Closing Remarks about the Innovation and Areas of Future Research

Since the late benchmark of AM process optimization using hybrid frameworks like ANN–GA, numerous challenges still await future research.

1. **Usage with Edge Computing:** Precise AI workloads at edges to make instantaneous decisions within AM environment (Fig. 8).
2. **Intelligent Artificial Intelligence Models:** Introducing self-governing AI algorithms that change with respect to process variability.
3. **AI Frameworks Tuned for Sustainability:** Translating AI into eco-friendly frameworks, i.e., applying the energy-efficient parameter decisions.
4. **Standardization and Adoption across Industry:** Norming AI-based AM protocols to gain a universal acceptance in the manufacturing sectors.

AI is simply further defining the operational realm of AM, so the meeting of computational intelligence with real-time insight and sustainability will propel the rise of smarter, greener manufacturing systems.

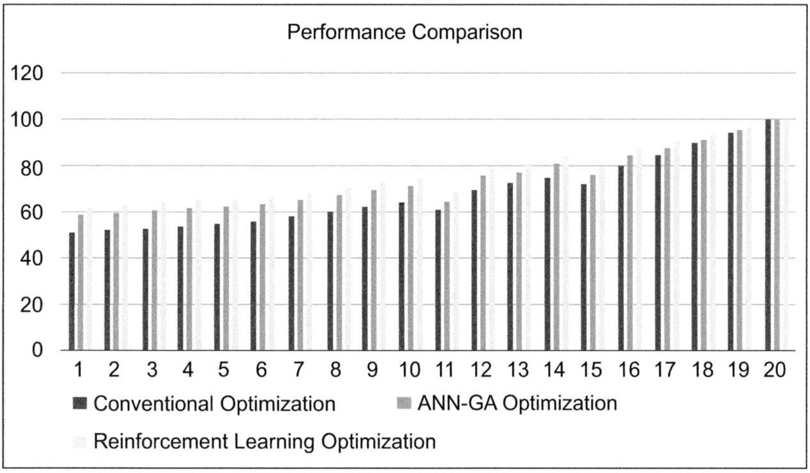

Fig. 8 Performance comparison of ANN–GA, reinforcement learning, and conventional optimization.

References

[1] Li, N., Huang, S., Zhang, G., Qin, R., Liu, W., Xiong, H. and Blackburn, J. (2019). Progress in additive manufacturing on new materials: A review. *Journal of Materials Science & Technology*, 35(2), 242–269.

[2] Barbieri, L. and Muzzupappa, M. Performance-Driven Engineering Design Approaches Based on Generative Design and Topology Optimization Tools: A Comparative Study. *Applied Sciences*. 2022; 12(4):2106.

[3] Leary, M., Merli, L., Torti, F., Mazur, M. and Brandt, M. (2014). Optimal topology for additive manufacture: A method for enabling additive manufacture of support-free optimal structures. Materials & Design, 63, 678–690.

[4] Tang, Y., Kurtz, A. and Zhao, Y.F. (2015). Bidirectional Evolutionary Structural Optimization (BESO) based design method for lattice structure to be fabricated by additive manufacturing. *Computer-Aided Design*, 69, 91–101.

[5] Wasserfall, F., Hendrich, N. and Zhang, J. (2017). 3D Printing Speed: How to Improve It. In Proceedings of the 13th IEEE Conference on Automation Science and Engineering (CASE) (pp. 1234–1239). IEEE.

[6] Wang, Z. and Zhou, C. (2021). A constructive solid geometry-based generative design method for additive manufacturing. *Additive Manufacturing*, 38, 101764.

[7] Barbieri, L. and Muzzupappa, M. Performance-Driven Engineering Design Approaches Based on Generative Design and Topology Optimization Tools: A Comparative Study. *Applied Sciences*. 2022; 12(4):2106.

[8] Leary, M., Merli, L., Torti, F., Mazur, M. and Brandt, M. (2014). Optimal topology for additive manufacture: A method for enabling additive manufacture of support-free optimal structures. *Materials & Design*, 63, 678–690.

[9] Tang, Y., Kurtz, A. and Zhao, Y.F. (2015). Bidirectional Evolutionary Structural Optimization (BESO) based design method for lattice structure to be fabricated by additive manufacturing. *Computer-Aided Design*, 69, 91–101.

[10] Wasserfall, F., Hendrich, N. and Zhang, J. (2017). 3D Printing Speed: How to Improve It. In Proceedings of the 13th IEEE Conference on Automation Science and Engineering (CASE) (pp. 1234–1239). IEEE.
[11] Wang, Z. and Zhou, C. (2021). A constructive solid geometry-based generative design method for additive manufacturing. *Additive Manufacturing*, 38, 101764.
[12] Malakooti, B. and Zhou, Y.Q. (1994). Feedforward artificial neural networks for solving discrete multiple criteria decision making problems. *Management Science*, 40(11), 1542–1561.
[13] Malakooti, B. (2014). Operations and production systems with multiple objectives. John Wiley & Sons.
[14] Stanley, K.O., D'Ambrosio, D.B. and Gauci, J. (2009). A Hypercube-Based Encoding for Evolving Large-Scale Neural Networks. *Artificial Life*, 15(2), 185–212.
[15] D'Ambrosio, D.B. and Stanley, K.O. (2008). Generative Encoding for Multiagent Learning. Proceedings of the 10th Annual Conference on Genetic and Evolutionary Computation, 819–826.
[16] Risi, S. and Stanley, K.O. (2012). An Enhanced Hypercube-Based Encoding for Evolving the Placement, Density, and Connectivity of Neurons. *Artificial Life*, 18(4), 331–363.
[17] Fera, M., Fruggiero, F., Lambiase, A., Macchiaroli, R. and Todisco, V. (2018). A modified genetic algorithm for time and cost optimization of an additive manufacturing single-machine scheduling. *International Journal of Industrial Engineering Computations*, 9(1), 1–14.
[18] Wang, Z. and Zhou, C. (2021). A constructive solid geometry-based generative design method for additive manufacturing. *Additive Manufacturing*, 38, 101764.
[19] Zhang, Y., Bernard, A., Gupta, R.K. and Harik, R. (2022). Genetic algorithm-based data-driven process selection system for additive manufacturing. *Materials*, 17(18), 4544.
[20] Deshwal, S., Kumar, A., & Chhabra, D. (2020). Exercising hybrid statistical tools GA-RSM, GA-ANN and GA-ANFIS to optimize FDM process parameters for tensile strength improvement. *CIRP Journal of Manufacturing Science and Technology*, 31, 189–199.
[21] Wang, Z., Zhang, Y., & Bernard, A. (2021). A constructive solid geometry-based generative design method for additive manufacturing. *Additive Manufacturing*, 41, 101952.
[22] Stavropoulos, P., et al. (2020). Hybrid subtractive–additive manufacturing processes for high value-added metal components. *The International Journal of Advanced Manufacturing Technology*.
[23] Cavitar Ltd. (2021). Cavitar Welding Camera Used in Education and Training of Future Welders.
[24] Chen, Y.-P., et al. (2023). Real-Time Decision-Making for Digital Twin in Additive Manufacturing with Model Predictive Control using Time-Series Deep Neural Networks. arXiv preprint arXiv:2501.07601.
[25] Wikipedia contributors. (2023). Smart manufacturing. In Wikipedia, The Free Encyclopedia.
[26] Zhang, Y., et al. (2023). Adaptive learning in AI-driven AM optimization. *CIRP Journal of Manufacturing Science and Technology*, 45, 102–119.
[27] Wang, H. and Jin, Y. (2022). Multi-fidelity modeling for AM process optimization. *Additive Manufacturing*, 59, 102301.
[28] Huang, X., et al. (2024). Quantum-inspired genetic algorithms for AM parameter optimization. *International Journal of Advanced Manufacturing Technology*, 124, 10781–10795.
[29] Chen, B., et al. (2023). Reinforcement learning approaches in AM process control. *Additive Manufacturing*, 65, 103045.
[30] Singh, R., et al. (2022). Swarm intelligence methods for AM parameter tuning. *Advanced Materials*, 40, 104876
[31] Patil, S. and Rao, P. (2021). Industrial deployment challenges in AI-driven AM. *Procedia Manufacturing*, 56, 589–600.
[32] Gupta, V. and Sharma, K. (2023). Sustainable AI optimization in AM. *CIRP Journal of Manufacturing Science and Technology*, 41, 201–215.

[33] Zhang, Y., et al. (2023). Adaptive learning in AI-driven AM optimization. *CIRP Journal of Manufacturing Science and Technology*, 45, 102–119.

[34] Wang, H. and Jin, Y. (2022). Multi-fidelity modeling for AM process optimization. *Additive Manufacturing*, 59, 102301.

[35] Huang, X., et al. (2024). Quantum-inspired genetic algorithms for AM parameter optimization. *International Journal of Advanced Manufacturing Technology*, 124, 10781–10795.

[36] Chen, B., et al. (2023). Reinforcement learning approaches in AM process control. *Additive Manufacturing*, 65, 103045.

[37] Singh, R., et al. (2022). Swarm intelligence methods for AM parameter tuning. *Advanced Materials*, 40, 104876.

[38] Patil, S. and Rao, P. (2021). Industrial deployment challenges in AI-driven AM. *Procedia Manufacturing*, 56, 589–600.

[39] Gupta, V. and Sharma, K. (2023). Sustainable AI optimization in AM. *CIRP Journal of Manufacturing Science and Technology*, 41, 201–215.

[40] Chhabra, D. and Singh, R. (2020). Exercising hybrid statistical tools GA-RSM, GA-ANN and GA-ANFIS to optimize FDM process parameters for tensile strength improvement. *CIRP Journal of Manufacturing Science and Technology*, 29, 99–112.

[41] Butola, R. and Saini, P. (2025). Hybrid algorithmic techniques for optimizing tensile strength in friction stir processed composites. *Journal of the Brazilian Society of Mechanical Sciences and Engineering*, 47(1), 123.

[42] Park, H.-S. and Anh, T.-V. (2013). Development of two-phase neural network-genetic algorithm hybrid model in modeling damage evolution in roll forming of aluminum sheet. *International Journal of Material Forming*, 6(3), 423–436.

10 Data-Driven Topology Optimization of Additively Manufactured Parts Using Deep Learning

Sathiesh Veera[1], Sagar Wankhede[2] and Akash Abaji Kadam[3*]

Additive manufacturing (AM) has enabled a radical change in the design environment by enabling complex designs and material efficiency that were unachievable in existing manufacturing processes. For realizing the full potential of additive manufacturing, novel design philosophies in structural design, particularly in topology optimization, recognize that an old-fashioned algorithmic approach to topology optimization algorithms was computationally intensive and often disparate from practical manufacturing constraints. In this chapter, we propose a data-driven concept of topology optimization based on the predictive capabilities of deep learning.

The chapter offers a theoretical model of a workflow that uses data from simulations, digital twin models, and additive manufacturing to create neural models that can spit out optimized solutions in real time. The model contains manufacturability awareness that helps the outputs of designs conform to build orientation, support structure requirement, and layer deposition limitations that exist in AM. The model promotes printable solutions by addressing real-world designs without post-processing.

The particularity of the discourse resides in the emphasis on scalability, sustainability, and flexibility of solutions intended for the industrial processing of aerospace, automotive, and biomedical manufacturing. In addition, the proposed model promises accelerated design cycles, whereas the new approach to a concept framework also foresees collaboration between human resources and artificial intelligence in the framework of Industry 5.0. The above-mentioned concept analysis shows that deep learning is remapping topology optimization from an iterative computation procedure to an intelligent data-driven design approach that learns from experience.

[1] Solution Architect, TruStartIT LLC, Richardson, TX, USA.
[2] School of Mechatronics Engineering, Symbiosis Skills and Professional University, Pune, India.
[3] Mechanical Engineer, Servotech Inc, Chicago, IL, USA.
* Corresponding author: aakadam005@gmail.com

DOI: 10.1201/9781003635673-10

1. Introduction

1.1 Background and Motivation

Additive manufacturing (AM) offers a radical change within the engineer design community by allowing components to be layered on site where they are actually needed without being limited by the geometric constraints of existing manufacturing. Although such optimized components are not quite easy to design. The conventional approach within topology design analysis used within particular designs to identify where components should actually be positioned may consume high computational resources as it relies on numerical solutions. Additionally, these designs may involve a sequence of finite element analysis simulations. As such designs require analysis simulations a number of instances of finite analysis simulations [1].

At the same time, there are huge amounts of data being produced in industry related to design, simulation, and production. The new data wealth, alongside the dawn of artificial intelligence and deep learning, opens a new perspective on the topic of engineering design optimization. Instead of relying solely on formulas and simulations, design intelligence can now be discovered using data to identify hidden relationships that exist between shapes and design functionality. In such a circumstance, data-driven topology optimization replaces solving instances of the optimization problem with an approach that predicts solutions to such instances [2].

1.2 Problem Context

Despite such dramatic advances being made, the conventional flow of the topology optimization (TO) remains limited by two fundamental factors. First of all, the processing capabilities of computers for complex and multidimensional design spaces preclude them from being used in real-time decision-making scenarios. Secondly, the generated configurations remain oblivious to manufacturable facts such as minimum sizes of these features, angles of overhanging surfaces, and support structure-related factors of primary significance in AM. These challenges represent the cause of the gap that exists between positive optimization and manufacturability. Bridging such a gap needs an adaptable and intelligent platform that can learn from earlier design experience as well as from the limitations of AM technologies.

The power of deep learning comes from being able to discern patterns within large data of high dimensionality. This offers an incredible solution. These designs can be represented by data containing the parameters of design performance. This enables near-instant predictions of topology that adhere to functional as well as manufacturability constraints [3].

1.3 Aim and Scope

The current chapter proposes a framework on the integration of deep-learning algorithms into topology-optimized additively designed components. The proposed system expects process data as well as design data that would come from existing knowledge of additively designed components. The system should be working toward an intelligent design that follows a corporal approach. The system believes in designs that can be repeated successfully without relying on certain designing knowledge that comes from environmental changes. The design should incorporate Industry 5.0 designs that implement artificial intelligence of human imagination for optimum output.

1.4 Chapter Structure

The remainder of the chapter is divided as follows: Section 2 provides an introduction to the evolution of topology optimization with the addition of additive manufacturing and new data-driven shifts. Section 3 describes the framework concept behind the addition of deep learning within the topology optimization process. Section 4 describes the approach. Section 5 describes the addition of this framework within the additive manufacturing setting. Section 6 explains the benefits that may be realized, followed by difficulties and the future in Section 7. Finally, Section 8 provides a conclusion reflecting on the long-term future of data-driven design.

2. Literature Review

2.1 Evolution of Topology Optimization

TO is familiar with being a basis for structural design for structural lightweight high-performance components. In the past, methods such as density based, evolution based, and level set based were used for finding the optimum material layout within a prescribed area. These methods were used either for structural stiffness maximization, weight reduction, or multi-objectives given loads and boundaries that were applied [4].

Although these approaches to TO have remained successful, they require intensive compute resources due to the need for iterative simulations by finite elements. In addition, re-optimization may be required by a change to the constraints or geometry of the problem. In addition, the solution may generate complex or asymmetrical shapes that may be difficult to create through subtractive manufacturing techniques. In order to address these challenges, there has been an emphasis on finding flexible and intelligent approaches to the problem of TO by scientists who were encouraged to explore new digital manufacturing solutions that work well with such approaches [5].

2.2 Additive Manufacturing as an Enabler

The ultimate partner for topology optimization is additive manufacturing. The latter possesses the capabilities to develop complex freeform shapes that were impractical or Impossible to develop in the past. Technologies such as powder bed fusion, binder jetting, and directed energy deposition offer engineers the opportunity to develop very optimized designs that may include lattice structure interiors as well as organic and density-variant regions [6].

Although such flexibility is provided in AM, it also poses new challenges. In classic TO as used by AM algorithms to generate optimized designs, it appears to ignore AM-related design limitations such as overhang angle constraints and minimum size rules. Such challenges may result in material deformation during printing. Therefore, research on the integration of design for additive manufacturing principles within the framework of total optimization is an important area of research. These research contributions include hybrid solutions that propose the inclusion of design limitations in the framework of total optimization. These solutions still require high computational resources thereby being intensive.

2.3 Emergence of Data-Driven Design

As digital change accelerates in the various manufacturing sectors, design and simulation data are being collected, archived, and shared increasingly. The flood of data poses an opportunity for data-driven optimization. Instead of solving complex algorithms for every new design problem that emerges, models can learn from success factors of existing designs by being trained on those designs and provide predictions on new designs that can deliver similar objectives. This marks a shift from algorithmic to knowledge-based problem-solving [7].

Data-driven design approaches have proven successful in a range of applications such as material discovery to structural monitoring to process enhancements. The concept of "design mining" (extraction of valuable knowledge from large datasets of simulations and manufacturing processes) has led to the evolution of intelligent predictive design solutions that always look ahead as more data become available.

2.4 Deep Learning in Mechanical Design and Optimization

Deep Learning: As a subdiscipline of artificial intelligence, deep learning (DL) represents an extension of data-driven design. DL possesses the capacity to automatically learn hierarchical representations from messy data. Mechanical engineers applied DL to stress distributions, properties of materials, and failure modes [8].

The convolutional neural networks (CNNs) were used as the fast topology predictors for the prediction of the optimized material layout based on the input of load and boundary conditions. The generative models like the variational autoencoders (VAEs) and the generative adversarial networks (GANs) were examined for their capabilities of designing new topologies that possess a similar functionality as in simulation-created designs. The graph neural networks were used more recently in being able to represent the complex structural patterns and predict the mechanical response based on mesh representations.

These works prove that it's possible to capture nonlinear and high-dimensional relationships that would be difficult to represent using the conventional mathematical formulation in order to realize design automation at a new level of depth.

2.5 Identified Gaps and Research Opportunity

Although data-driven and deep learning approaches have proven quite promising and effective, much work remains unserved. The objective of most of the current existing approaches solely targets imitating the outcomes of optimization but not the integration of manufacturability and process awareness. The majority of the existing models only encompass certain problem sets or load cases. In addition, most existing solutions set high value on accuracy of predication without an inch on interpretability. Hence, engineers remain skeptical or reluctant to modify designs made by artificial intelligence.

Furthermore, there appears to be limited integration of design models and the practical realities of additive manufacturing. At the same time, it appears that it is imperative to work toward creating an integrated framework that combines knowledge of deep learning, topology optimization basics, and design for additive manufacturability (DfAM) limitations. This should create an opportunity for constant learning from data, simultaneous design creation, as well as simultaneous manufacturability analysis. These may serve as essentials of next-generation intelligent design infrastructure.

3. Conceptual Framework for Data-Driven Topology Optimization

3.1 Overview of the Framework

The proposed overall conceptual framework begins with the concept of a seamless integration of DL, TO, and additive manufacturing. The central idea here is that design generation as an optimization calculation should be substituted by an intelligent data-driven decision-making process. The system should not compute new optimization problems but instead use the trained system to make intelligent guesses of the optimum topologies based on design objectives and boundary conditions.

This toolkit includes four interconnected levels of functionality. These levels include data gathering, processing and prediction, design generation, and manufacturability analysis. Collectively, these levels represent a system with a closed loop of knowledge from virtual design spaces to manufacturing.

3.2 Data Acquisition and Preparation

The foundation of a data-driven approach is the amenity of data on which it stands. The data within such an approach derive from three sources: computational simulations, verification by experimentation, and in-process monitoring of AM.

Historically, topology optimization problems are solved and stress and strain analysis as well as material deposition offer data rich in geometric-performance correlations.

In order to prepare these data for training, designs in the form of digitized views of 3D voxel arrays or mesh graphs with reduced design parameters are used. The dataset includes not only geometric data but also context information such as material property, load orientation, and constrained boundaries. The structuring of such diverse data within a database helps the model identify design pattern as well as the trend of performances. The data preparation also includes training robustness by rotation or reflection of the data.

3.3 Learning and Prediction Layer

Learning layer acts as the brain of the intelligence of the framework. This is where the deep models of intelligence are trained on design input relations and the consequential output performances. Trained, they can proceed to predict the optimum topology in a fraction of the time that the conventional methods take.

In contrast to classic TO, where simulation cycles had to be programmed sequentially, solutions in real-time processing leverage patterns found in example histories. Engineers can analyze hundreds of design options within minutes. In addition, as the data resource base grows, the model continually increases in intelligence and assimilates understanding of performance, weight, and manufacturability in a seamless fashion without programming. The end-result solution offers the potential for an artificial intelligence design partner as an integral part of early-stage product development.

3.4 Manufacturability and Constraint Awareness

One thing of particular interest in the framework presented here is that it incorporates emphasis on DfAM from design. Rather than designing designs that may be optimum from a theoretical perspective but unworkable from an implementation point of view, manufacturability issues actually reside within the model. The dataset contains factors that indicate AM truths such as build orientation, support structure volume, surface quality, and minimum size.

In addition to that, it also learns to avoid non-printable configurations and encourages machine-capable ones. This ensures that all designs proposed not only remain high-performing designs but also production-capable. The embedment of DfAM knowledge within the optimization process reduces the redesign cycle since there are limited changes that remain to be made within the initial design.

3.5 Feedback and Continuous Improvement

One of the characterizing maxims of the conceptual framework that can be adapted here is that of having a closed feedback loop. The data collected from printed components (dimensional variation data, or data from mechanical testing or sensors) is fed back into the master database. These observations within the world can then be used to retrain or update the model of learning.

Eventually, the system matures from being a design tool to a self-grazing system that continually refines its understanding of both performance and manufacturability. This helps create an ever-improving bridge from the world of virtual design to the world of the real world.

3.6 Theoretical Significance

The system represents an approach to the democratization of high-level design intelligence. The system provides a concept for a model that now doesn't have the topology optimization problem limited by the cleverness of the solver but by design experience compiled in a neural model. The system shifts the problem of optimization from being computation-oriented to becoming insight-oriented and from being iterative to being predictive.

This kind of mindset exists in the concept of Industry 5.0, which integrates the collaboration of human intelligence and artificial intelligence. These types of engineers can have a dialogue with the model that helps them analyze the choices and develop new designs by cocreating. Finally, an intellectual framework provides the base for an intelligent design environment that is both sustainable and responsive.

4. Methodological Approach

4.1 Data-Driven Workflow

The design of the proposed work's methodology has been made to enable intelligent, repeatable, and automatic topology optimization based on data centricity. The proposed flow process proceeds through five sequential phases.

I. Problem Definition: The design constraints and limitations on material activity as well as design factors of the problem statement are properly articulated. Instead of using numerical values, it emphasizes the input parameters such as load regions, material needs, and design objectives.

II. Data Generation and Collection: The outcomes from the simulation process, as well as topology optimization outcomes and additive manufacturing data feeds, come together inside an overall database. The goal here is to incorporate diverse design examples within the design-performance knowledge of the deep-learning model.

III. Training of the Model and Learning: The created dataset is trained on this particular model. The model then learns the relationship that may exist between the functional end-results such as stiffness, strength, manufacturability, and the corresponding geometric configurations.

IV. Design Generation: The trained model itself is also a predictive machine that has the capability of designing almost optimal designs without having to optimize complex design solvers.

V. Evaluation and Refining: The designs that get produced undergo digital evaluation and expert evaluation. The new outcomes get reiterated into the data repository.

This helps create an ongoing cycle of learning and enhancements rather than point-in-time optimization runs in an oscillating and improving design space.

4.2 Design Representation Strategies

The design and methodology of choosing these design geometric presentations to the learning model is another important consideration. The framework can adopt as many presentation forms as it wants as the data are accessible and complex.

For instance, two-dimensional images are used for problems involving planes. The representation of spatial structures may be in the form of three-dimensional voxels and graphs. These parameters include geometric moment descriptors. The objective here is to identify the representations that retain important structural or physical information but can still be processed.

In order to retain such a relationship between geometric properties and functionality, it means that the model not only predicts shapes but also understands design.

4.3 Model Training and Validation

The learning part of the approach relies on strong neural networks that identify most of the important features from the data automatically. Since it is trained on thousands of examples of data on the relationship of topology to performance, it has been trained to create new topologies or estimate new performances with an equal level of efficiency.

For improved reliability, conceptual strategies include. Data augmentation increases the act of generalization by using design examples with variations based on sample data. Use of transfer learning to adapt the understanding of the model from two-dimensional examples to the three-dimensional ones. Validation of the design occurs through qualitative testing by comparison of the designed theologies with the existing ones that are always optimal.

Despite being a conceptual framework, the proposed training schedule would prove effective by adapting with the ongoing iteration and data increment.

4.4 Evaluation and Quality Metrics

The design assessment by the model should also include factors of performance as well as manufacturability. In contrast to the quantitative assessment made using numbers, the proposed design assessment factors include:

- Structural Integrity: The ability of the modeled topology to withstand the proposed loads conceptually.
- Material Efficiency: Efficient use of material distribution relative to design.

- Manufacturability: Compatibility with limitations of additive manufacturing processes such as build orientation and accessibility.
- Design Interpretability: The ease of understanding the design that the human produces for manipulation.

These ensure that the assessment also goes beyond the boundaries of the performance optimization that is strictly geared to the cognition of usability.

4.5 Implementation Considerations

Adoption of such an approach would flourish on the assumption that there is infrastructure that is digitally connected. The cloud environment provides open-scalable data storage as well as training infrastructure for computations. The digital twin site connects predictions of designs with the produced outcomes.

In targeting modularity, it has the potential of being integrated into existing CAD/CAE infrastructures as a plug-in intelligence level as a supporting tool to existing tools. The vision targets the ability of the designer to deal with an adaptive and learning modulo that produces feasible and novel topologies in near-real time.

5. Integration with Additive Manufacturing Ecosystem

5.1 Connection to the Digital Thread

One of the main advantages of the new framework presented here is that it can extend to the digital thread of additive manufacturing. In a conventional manufacturing system, the digital thread connects every process from design concept to post-processing, including monitoring of function, as part of an unbroken flow of data. Use of the data-topology framework here provides that design brilliance is never an isolated thing but always connected with the vagaries of manufacturing.

The data from the sensors, the printers, and inspection outcomes can replenish the data pool in real time such that the model of learning proceeds in tandem with the manufacturing process. The integration of the system provides a design process that functions as a living system such that additional information from each printed item refines the next. The end result of the system provides an overall traceable adaptive process such that the integration of the virtual world of optimization and the real world of manufacturing comes into one.

5.2 Design for Additive Manufacturability (DfAM)

The integration of DfAM within the framework of data-driven design is very important as it helps build on the closure of the gap that exists within the printability of an optimally designed structure. The existing design structure may fail to ensure that it is optimally better but not constructible. The framework combines manufacturability as an implicit design dimension.

Through the encapsulation of relationships of design features and successful print outcomes, the learning model incorporates build constraints such as overhanging angles, minimum feature size, and surface accessibility. As the system generates a new topology, it automatically avoids shapes that are expensive to support or those likely to fail during the printing process. The DfAM-aware intelligence produces designs that are efficient but also print-enabled, thus condensing the design-to-manufacturing cycle.

5.3 Process-Aware Modeling

Beyond geometric considerations, the framework also accommodates process-related knowledge. Additive manufacturing performance is influenced by parameters such as layer thickness, scan path, heat input, and cooling rate. Although these factors vary widely across AM technologies, they collectively determine part quality, residual stresses, and surface integrity.

The integration of such process metadata within the framework of the architecture enables the system's capability to offer design recommendations that are conducive within certain machine capabilities or material dynamic behaviors. With the passage of time, the system calibrates itself within an environment that makes topology predictions not only structurally ideal but process ideal as well. This approximation represents the true manufacturing capability of equipment.

5.4 Human–AI Collaboration and Industry 5.0 Context

At a larger scale, the integration of the framework fits with the human-centered approach of the Industry 5.0. In the framework, collaboration between the engineer and the AI system is favored over those of competition. The model of the AI produces quick data suggestions for the topology creation, and the human designs the outputs of those suggestions based on his/her creativity and circumstances.

This symbiosis alters the role of an engineer from an optimizing role to a leadership role in design exploration. This encourages openness in interactions such that the design explainability of the artificial intelligence system can be strengthened by the human value of judgment and ethics. The system provides an equilibrium design environment intelligent, sustainable, and adaptively intrinsic to next-generation manufacturing systems.

6. Expected Outcomes and Industrial Impact

6.1 Efficiency and Speed

The most important expected outcome of the data-driven approach being proposed here is the dramatic acceleration of the design cycle time for the adaptively manufactured components. Through the replacement of simple iterative solvers with intelligent predictive models, topology optimization can be moved from being a process that is concept-driven to almost instantaneous design creation. The expected outcome of such an approach would be the ability to analyze various design alternatives within seconds of time, thus enabling rapid decision-making. This is indeed the most ideal form of responsiveness that can be experienced by various industry domains that rely on design cycle times as dictated by competitiveness in the aircraft industry, the car industry, and consumer electronics.

6.2 Sustainable and Material-Efficient Design

The system automatically supports sustainability efforts in that it prevents waste by design. Since the designs for both functionality and manufacturability have been made more efficient, there's less chance of designs that fail and are wasteful of material. The capacity of the model/algorithm to learn helps to avoid designs that may be less wasteful of material but somehow offer weakening designs of mechanized machines. Since sustainability as an objective factor stands supreme globally for companies as a requirement, such an intelligent system actually helps combat carbon waste from prototype to production.

6.3 Broader Applicability and Adaptability

The concept-oriented nature of the framework assures scalability from various industry sectors to various materials. Whether it's structural components in the aerospace industry that need to be light or heat exchangers within the energy business or biomedical implants, it's the same data-driven reasoning that needs to be adapted as per the requirement of each industry. As the framework doesn't rely on data, various types of data from disparate sources such as simulation data or inspection data can be used without altering the framework.

6.4 Strategic and Organizational Benefits

Besides these technical advantages, such an approval implies that there are strategic advantages. Entities can accumulate cumulative design intelligence over a certain period of time, thereby finding each finished project an optimization that turns out to be material for subsequent use. The accretive nature of the process appears to fit within the overall ethos of digital transformation and knowledge reuse. Finally, data-driven topology optimization implementation increases productivity and readiness for dominance within the emerging human-centric industry of Industry 5.0.

7. Simulated Case Study

- Objective: We use a small-scale example that demonstrates how a deep learning-based strategy can assist in topology optimization for additive manufacturing and how it compares to a traditional strategy for topology optimization. We will concentrate on that which is important for a reader: accuracy, manufacturability, and speed.
- Setup: "Each sample in our dataset corresponds to a cantilever-style object. We have many variables for each sample that affect how much weight is

placed on it, its strength level, its material class, its orientation in building an object,"—all of this is done to calculate a "ground truth" value for its compliance that is calculated in a smooth function that corresponds to a real object and a small random value for better reality. "We calculate two indicative metrics of manufacturability. The first is a support volume-score. The second is a measure of overhang that is likely to require support."

- Splitting of data: We are training on 300 samples, validating on 100 samples, and testing on two sets of 100 samples. One test set is similar to that of training samples (in-distribution samples), while one is different and distinct (out-of-distribution samples) as it has some different locations where loads are known to appear and a new category for materials.
- Methods: Comparison will be made between two methods. In the DL method, a prediction for compliance is made on the basis of inputs and a design suggestion is made. In the traditional method, a topology optimization is conducted aiming for a similar level of compliance. In both methods, relevant variables are recorded.
- What the results say: The DL model predicts compliance well on data that looks like the training set and still behaves sensibly on the different test set. When we compare full designs, both methods reach about the same compliance, but the DL-guided designs usually need less support and have fewer overhang areas. The DL approach also runs much faster (milliseconds per case) while the classical run takes minutes.

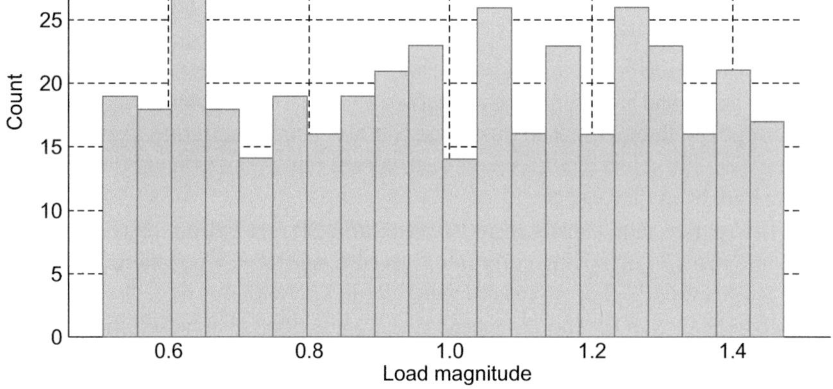

Fig. 1 Distribution of load magnitudes.

Distribution of load magnitudes represents how many training and/or validation examples are at each level of loading from about 0.5 to 1.5. The bars are more or less even over this range without apparent peaks and troughs—but this suggests that it is not biased toward either highly loaded or lightly loaded datasets. The presence of coverage on both ends ensures that it is not limited to a middle-out behavior but is well-versed in all regions.

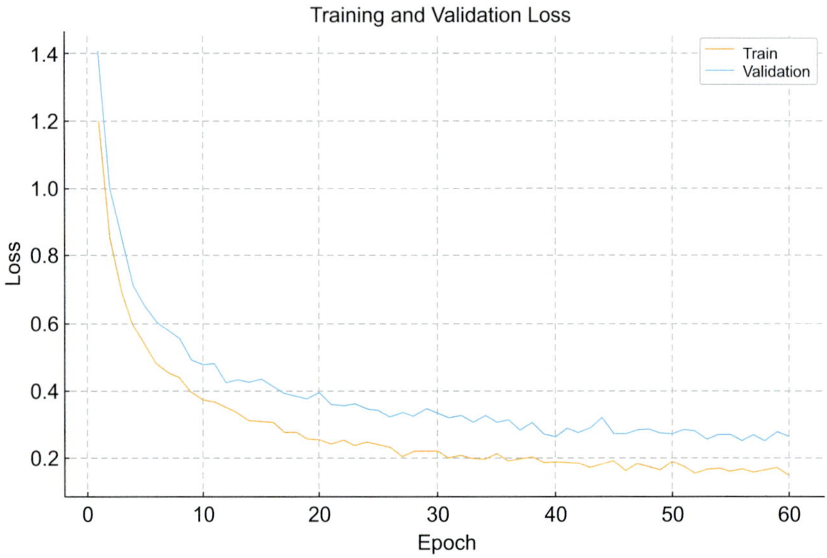

Fig. 2 Training and validation loss.

The lines indicate how well the model is doing in reducing its error as it trains. In the beginning, both curves drop rapidly, indicating that the model is learning well. After about halfway through training, both curves level out to indicate that the model has converged.

The validation line remains a little above the training line (as it should), and there is a small and steady difference between them. That's a clear indicator of no overfitting. The fluctuations are just noises. While you are selecting a point to stop training, you can select a point where both curves flatten out (around the points of the late epochs in this case).

"This parity plot checks how well the model's predictions agree with the reference values." Each point represents a test case, and "the diagonal line represents 'perfect agreement'." The points are generally in line with this line. "Points above this line indicate that the model overpredicts, while points below it indicate that it underpredicts"—although it's not biased in one direction over the other. The points remain within a tight cluster even for larger numbers. Therefore: "The predictions are in close agreement with ground truth (~0.95 R^2 in our summary), and this makes the model trustworthy for downstream design choices."

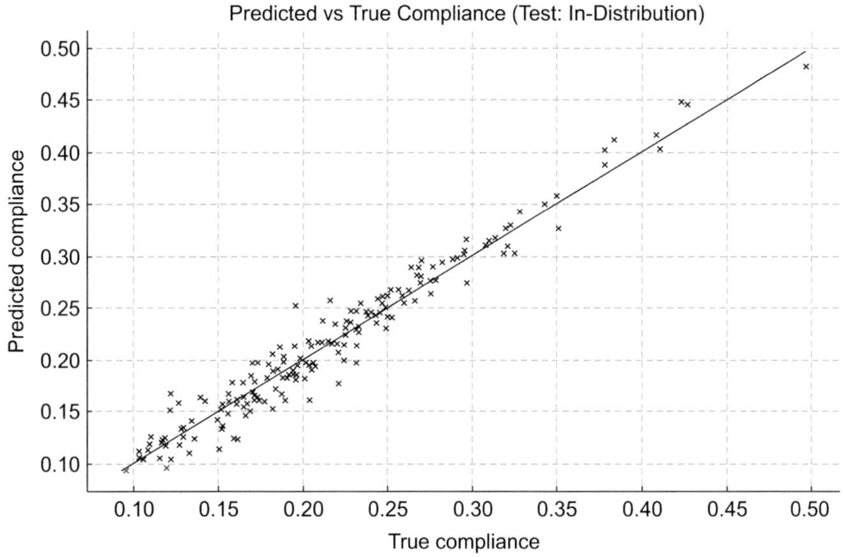

Fig. 3 Predicted vs. true compliance.

8. Challenges, Limitations, and Future Directions

8.1 Data Limitations

The success of any data-integrated framework is highly dependent on the quality, quantity, and variability of the data. Comprehensive open datasets are still limited for additive manufacturing and topology optimization. Most institutions have design process data proprietary, so the databases are fractured, prohibiting widespread model training. Inconsistency in data labeling, simulation input, and file type also brings noise that diminishes learning accuracy. Being able to overcome this is something that will need industry collaboration and the implementation of standardized data sharing procedures that safeguard intellectual property but foster development.

8.2 Computational and Infrastructure Challenges

Even though deep learning boosts design creativity, the training of deep learning itself takes heavy computational resources. Management of high-resolution 3D data, for example, needs heavy graphics processing units (GPUs) and much memory. Smaller businesses or colleges will find it difficult to have this kind of infrastructure. In addition, the merging of the design, manufacturing, and monitoring data under one system will need strong data management sites, as well as secure clouds. The solving of the above issues by the means of distributed calculation, federative learning, and scalable clouds will be one of the markers of the mass implementation.

8.3 Model Reliability and Interpretability

Another weakness is deep learning model interpretability. The deep learning-based black boxes are notorious for causing increased difficulties for engineers to recognize the paths by which design recommendations are generated. Non-transparency may dilute user confidence as well as impede acceptance in safety-critical markets like aerospace or healthcare. Research for the future must prioritize explainable methods of AI by visualizing the decision paths, pinpointing influential features, and giving design justification that is interpretable. It will make sure that the human designers are always in control of the process of deciding.

8.4 Physical and Manufacturing Realism

The challenge that always persists involves matching model-simulated designs and the physics that relates to build data from the physical world. The model that only uses data from simulated sources may pose the potential to ignore the residuals stress, microstructure variation, or material anisotropy that exists within the AM model. The role of sensor data and build inspection data within the model's training cycle possesses the ability to narrow the gap of correspondence. The concept of an "always-learning digital twin" represents an emerging approach.

8.5 Future Research Avenues

The future work may include reinforcement learning-driven autonomous design space exploration. The design hierarchies may be refined based on simulated or real data. Symbolic reasoning may also combine with deep learning. This may also generate hybrid designs that may be both interpretable and adaptable at the same time. The blockchain system may combine with the existing work. The design traceability may be maintained by these blockchain solutions protecting intellectual property rights while authenticating the model. Finally, the work may develop data-driven topology optimization into intelligent manufacturing.

9. Conclusion

In this chapter, the idea of data-driven data-based topology optimization of additively manufactured components by the use of deep learning has been presented. The idea tries to change the classic process of optimization by replacing the processes involved in it with forecasting intelligence that comes from aggregate design data and manufacturing data. In this integration, the process of topology optimization changes from a problem involving computation to a process that continuously learns and improves on the foundations of knowledge.

The approach offers the intersectionality of data-driven design, additive manufacturing, and deep learning as an integrated digital system. The approach shows the intersectionality of the above-mentioned fields that will lead to designs that are quicker, more sustainable, as well as manufacturable. The above approach incorporates manufacturability knowledge into the process of learning. In that respect, the approach guarantees that the topology designs made from the approach not only offer structural optimality but also manufacturability.

Besides the technical component of the chapter, it proposes a world where collaboration occurs between engineers and intelligent systems on co-designing things that are right. This approach of collaboration between human and artificial intelligence represents the principles of prevention that define Industry 5.0. These include personalization, sustainability, as well as synergy.

Even if the architecture remains conceptual, the possible gain is gigantic. On which architecture can smart digital twins, autonomous design programs, and enhancing optimization systems be formed, other than architecture? Long-term ultimate potential is that data-driven topology optimization assisted by deep learning has the nothing-short-of-revolutionary potential to redefine the very definition of product development, shifting the reactive analysis-driven product development toward proactive development-driven innovation based on ongoing learning and collaboration.

References

[1] D. Gu, X. Shi, R. Poprawe, D.L. Bourell, R. Setchi and J. Zhu, "Material-structure-performance integrated laser-metal additive manufacturing," 2021. doi: 10.1126/science.abg1487.

[2] S. Yamasaki, K. Yaji and K. Fujita, "Data-driven topology design using a deep generative model," *Structural and Multidisciplinary Optimization*, vol. 64, no. 3, 2021, doi: 10.1007/s00158-021-02926-y.

[3] L. Alzubaidi *et al.*, "Review of deep learning: concepts, CNN architectures, challenges, applications, future directions," *J Big Data*, vol. 8, no. 1, 2021, doi: 10.1186/s40537-021-00444-8.

[4] D. Aguirre Guerrero, G. Jacobs, T. Zerwas and A. Delgadillo, "Integrating topology optimization into model-Based systems engineering for lightweight structural design," *Forschung im Ingenieurwesen/Engineering Research*, vol. 89, no. 1, Dec. 2025, doi: 10.1007/s10010-025-00829-8.

[5] S. Cuomo, V.S. Di Cola, F. Giampaolo, G. Rozza, M. Raissi and F. Piccialli, "Scientific Machine Learning Through Physics–Informed Neural Networks: Where we are and What's Next," *J Sci Comput*, vol. 92, 2022, doi: 10.1007/s10915-022-01939-z.

[6] O. Ibhadode *et al.*, "Topology optimization for metal additive manufacturing: current trends, challenges, and future outlook," 2023. doi: 10.1080/17452759.2023.2181192.

[7] I. Flores Ituarte, S. Panicker, H.P.N. Nagarajan, E. Coatanea and D.W. Rosen, "Optimisation-driven design to explore and exploit the process–structure–property–performance linkages in digital manufacturing," *J Intell Manuf*, vol. 34, no. 1, 2023, doi: 10.1007/s10845-022-02010-2.

[8] A.T.G. Tapeh and M.Z. Naser, "Artificial Intelligence, Machine Learning, and Deep Learning in Structural Engineering: A Scientometrics Review of Trends and Best Practices," 2023. doi: 10.1007/s11831-022-09793-w.

11 Optimization of Additive Manufacturing Process using Data-Driven Artificial Neural Network Based Hybrid Method

Subba Rao Katragadda[1*] and Ajay Tanikonda[2]

Additive manufacturing (AM) offers a degree of design flexibility and material efficiency never witnessed by previous design approaches, but the optimization of process parameters remains a critical issue, merely due to nonlinearity and high-dimensional interactions between input variables and final part properties. A combination of such complexities is normally not well modeled with traditional statistical and heuristic methods and justifies the application of advanced methods in artificial intelligence. The chapter discusses the possibility of optimization of AM processes using the hybrid method of integration of the artificial neural network (ANN) and genetic algorithms (GA). ANN models are able to learn complex input–output correlations, unlike GA which evolves to search through the entire design space and identify nearest solutions. The hybrid ANN–GA model can be used to overcome negative attributes of models by combining the predictive accurateness and high optimization ability. The chapter is useful in that it reviews the available literature systematically on the application of ANN and GA in additive manufacturing, and develops a unified manner of combining them, and has also been found to be efficient as per representative case studies regarding tensile strength, yield strength, and surface quality. Besides, the issue of computations, unavailability of data, and the issue of generalization are also critically discussed and how to minimize these challenges are also proposed. The findings confirm the viability of ANN–GA hybridization as the viable orientation of the optimization of intelligent operations, errors, and the quicker approval of AM components. Finally, the future opportunities are presented including the integration of hybrid AI methods in reinforcement learning and closed loop adaptive control methods which have the potential to design autonomous, data driven, self-optimized manufacturing systems to enhance reliability, performance, and scalability in the industrial environment.

[1] Independent Researcher, Lathrop, California, USA. ORCID: 0009-0004-1534-6976
[2] Independent Researcher, San Ramon, California, USA. ORCID: 0009-0005-2819-8439
Email: ajay.tani@gmail.com[2]
* Corresponding author: subbakatragadda@gmail.com

DOI: 10.1201/9781003635673-11

1. Introduction

Additive manufacturing (AM), or 3D printing, has become one of the most transformative technologies of the 21st century. The additive manufacturing process allows AM to attain unexplored material efficiency, freedom in design, and customization through the production of components and structures with the assistance of digital models. Industries such as aerospace, automotive, and biomedical engineering have been one of the original and largest users. AM is used in the aerospace industry to provide differentiation on fuel saving, high-complexity geometries, and performance. This is applied in the construction sector to speed up the modeling and production of the parts whose mechanical properties are enhanced which reduces the company technological life cycle. In the biomedical arena, AM has been used additively to make tissue engineering scaffold, implant a certain patient and surgical equipment, and has totally transformed the manner in which healthcare cures are administered.

Even though AM is disruptive, it possesses serious issues that cannot allow its implementation in the production environments on a large scale. One of the most important issues is process control and optimization. AM processes comprise a variety of parameters such as power of laser, scan rate, layer thickness, and material feed rate among many others all of which directly affect the quality of the final product. This can lead to defects in the form of porosity, warping, or improper surface finish caused by poor control leading to low reliability and low reactionability. Also, the relationship between process parameters and material properties resulting is often nonlinear and complex, and a steady performance across different builds and machines cannot be achieved.

The complexity above justifies why AM should be optimized. There is some process–property–performance association knowledge that is relevant in achieving high-quality and reliable components. Even when they are popular, conventional optimization methods such as design of experiments (DoE) and response surface methodology (RSM) are time-consuming and may be resource-intensive. They require a lot of physical testing in general and this makes them more costly and time-consuming to produce. Moreover, these methods may fail to work with high-dimensional and nonlinear AM parameter space since optimal solutions could not be offered.

Researchers have been drawn more toward artificial intelligence (AI) and mixed computational models in order to overcome these limitations. Artificial neural networks (ANNs) represent the best machine learning tool to apply when dealing with nonlinear correlations and predicting the quality of parts given input measures. Genetic algorithms (GAs) are powerful methods of optimization, and they are realized with a mechanism of natural selection, and they can search substantial solutions space and discover the global optima. The combination of ANNs and GAs forms some kind of hybrid that can be used to predict the results with a great level of accuracy and simultaneously optimize them. This synergy has a lot in store in that it will take AM out of prototyping technology to a mature, production-ready technology.

The purpose of this chapter is to conduct a research on the applicability of ANNs and GAs as a hybrid model of optimization of AM processes. In specific, the chapter will assist in gaining insight into the theoretical framework, the literature on the issue, the methodology of constructing such a hybrid model, and a reflection on the practical applications of it.

Finally, the structure of the chapter is the following: Section 2 features the literature review of the AM process parameters, the traditional optimization methods, and the emergence of the AI-based methods. On how to develop the ANN–GA hybrid model, it is explained in Section 3. The results and discussion are provided in Section 4 and final findings are summarized as to be in the main findings, limitations, and future research directions in Section 7.

2. Fundamentals of Additive Manufacturing and Process Optimization

AM is a collective name of technologies in which elements are built up in a layer-by-layer manner via deposition of material under the direction of a digital image. The majority of the existing and long-established AM processes such as metal-based AM have been in demand because of their ability to make complete density metallic parts that can be applicable in high-performance projects. It is in this section that the description of the most significant AM technologies, the process parameters, and common defects are outlined and the mechanical properties are dependent on it, along with the classical optimization means, such as the requirement of the use of advanced methods such as AI-based hybrid models.

2.1 Overview of AM Technologies of Metal

Three prevailing types of processes of metal AM are selective laser melting (SLM), electron beam melting (EBM), and directed energy deposition (DED).

2.1.1 Selective Laser Melting

SLM is a process which relies on high-power laser to ablate and weld particular regions to form a thin layer of metal powder on a build platform. This is done until the part is completed and this way in one continuous layer after another. SLM has close net-shape components and absolute control of the microstructure, which makes it an ideal technology in aerospace and medical implants. It is, nonetheless, intellectual to the selection of parameters, that is, it fluctuates extremities that influence the density of parts and mechanical characteristics.

2.1.2 Electron Beam Melting

EBM makes use of a beam of electrons, in most cases, under a high-vacuum environment, a source of energy in comparison and match with SLM. This allows the working of reactive metals, such as Inconel and titanium, and decreases the rate of oxidation. EBM can also achieve lower residual stress in parts than SLM since it would reach high build temperature but its surface finish is often poorer.

2.1.3 Directed Energy Deposition

According to DED, the powder or wire material is introduced to a controlled flow of energy (laser, electron beam, etc.) after which the material may be deposited at the precise point needed. DED works best with large parts, or repair, or the addition of detail to already existing parts, but can be of inferior geometric resolution than SLM and EBM.

2.2 Key Process Parameters

Heavy reliance is put on process parameters in order to define AM parts quality. The most important parameters are laser power, scan speed, layer thickness, and hatch spacing. The laser power is also used to control the amount of energy input into the melt. Insufficient power may result in lack of fusion failures and excess power may result in keyholing and spatter wastes. Scan speed determines the rate of a source being moved; higher rates are possible, but may also lead to less melting and slower rates may lead to overheating and other residual stresses. The layer thickness resolves a smaller build. The finer the layer the higher the finish on the surface and mechanical accuracy, and the more time it takes to build. The distance between the neighboring scan tracks is known as hatch spacing; hatch spacing may be incorrectly spaced causing uneven melt pools leading to porosity or weak bonding between actions.

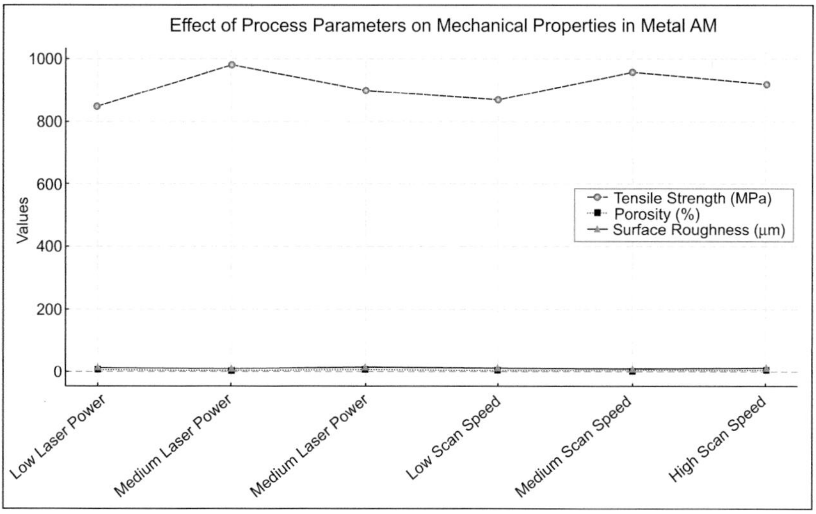

Fig. 1 Effect of process parameters on mechanical properties in metal AM.

2.3 Common Defects in Metal AM

The defects that are more likely to occur during metal AM and undermine the mechanical performance are the porosity, residual stresses, and anisotropy. Porosity is the result of the incomplete melting, trapping of gaseous or keyhole instability which reduces the density and strength. Rapid cooling and heating will result in steep thermal gradients, which initiate residual stresses that tend to cause parts distortion or cracking. Anisotropy: Mechanical properties other than those of an isotropic material may be determined by the who-done-about solidification, which is directional solidification and that which has a bearing on the load bearing. All these imperfections must be counteracted by adopting effective optimization of the processes to deliver high levels of consistency and reliability of the parts.

Table 1 Common defects in metal additive manufacturing and AI-based mitigation strategies.

Defect	*Cause (Process Parameter Issue)*	*Impact on Part Quality*	*ANN/GA-Based Mitigation Strategy*
Porosity	Incomplete melting, excessive hatch spacing, unstable melt pool	Reduces density and tensile strength	ANN models predict porosity based on process inputs; GA optimizes energy density to minimize porosity (Patel et al., 2022).
Residual Stresses	High thermal gradients due to rapid heating/cooling	Leads to warping, cracking, part distortion	ANN forecasts stress distribution; GA adjusts scan strategies and layer thickness to minimize stress (DebRoy et al., 2018).
Anisotropy	Directional solidification, layer-by-layer microstructure	Uneven load-bearing capacity, reduced fatigue life	ANN predicts mechanical anisotropy; GA identifies optimal build orientation and parameters to reduce anisotropy (Gupta & Chen, 2021).

2.4 The Mechanical Properties of Interest Are as Indicated Below

Performance measures are mechanical properties which play a crucial role as much as structure application is involved. Major ones are tensile strength and yield strength, fatigue life and hardness and toughness. The tensile strength and yield strength inform how the material will be able to accommodate resistance to the stagnant loads without being destroyed and compromised. The number of cycles world can be repurposed and survive under different loads until the component fails is computed with the fatigue life calculation, which is highly sought after in the aerospace and car components. Hardness and toughness provide a perspective about the wear resistance and energy absorbing capacity before the fractures are acquired. One of the main research areas in AM is to make various combinations of such properties desirable by optimization of process parameters.

2.5 Technical Optimization Approaches and Weaknesses

The conventional methods by which the researchers can determine the best AM parameters are design of experiments (DoE), response surface methodology (RSM), and Taguchi method. DoE is a structured approach that is prohibitively expensive when parameter space is high dimensional. RSM uses mathematical models, which are built on the symptom–parameter relationships but builds nonlinear symptom–parameter relationships as a process of (parameter) pollution as polylinear mechanistic model. The Taguchi method works on orthogonal arrays in order to reduce the number of experiments which is focused on strength but does not always locate the global optima. Even though the contributions of these methods in the early AM research are important, their limitations are that much physical experimentation is sought; furthermore, there is minimal predictability of highly nonlinear and stochastic behaviors of processes. These shortcomings lead to the adoption of AI-based and hybrid optimization procedures that would be more effective and would require less experimentation to consider the complex associations.

3. Artificial Neural Networks (ANN) in Additive Manufacturing

3.1 Introduction to ANN

ANNs are a variety of machine learning algorithms, which are based on the framework and functioning of the human brain. Transmission of electrical signals between biological neurons is possible through synapses arrangement where messages are interpreted in terms of network connections between neurons. ANNs, following this concept, organize the computational nodes (artificial neurons) in layers—typically, an input layer, one or more hidden layers and an output layer—an input layer, an output layer.

3.1.1 The Artificial Neurons Are Supposed to Perform but a Single Simple Task

It takes in the input signals, multiplies them with a set of weights, sums the numbers and outputs, and then adds a bias term and then executes an activation function on the output. Activation functions give the network nonlinearity so that it can be able to represent complex inputs to outputs. Other popular activation functions include the sigmoid, hyperbolic tangent (tanh), and the rectified linear unit (ReLU) with the latter being one of the most widely used since it is simple to compute and it addresses the vanishing gradient problem.

To train an ANN, it is necessary to find the most efficient set of weights and biases such that the difference between the real and predicted output is minimal. This is also typically done by use of the backpropagation with gradient descent or any variation of gradient descent (stochastic gradient descent, Adam optimizer). Backpropagation estimates the error gradient of weights by distributing the error across a network and the gradient of the error derivation is decreased in a sequence of steps to reduce the error by altering the weights.

ANNs can learn very nonlinear mappings when trained on sufficient data thus they are best suited to be used in applications where input–output relationships are nonlinear and would not easily be modeled in closed-form equations common in AM after training.

3.2 ANN Finds Applications

Owing to the abundance of data recorded in AM processes, ANNs have been widely researched on various predictive and diagnostic activities.

Mechanical property prediction is used to show whether the component is of good strength, the behavior display is both clear and consistent in abnormal weakening or deforming, and whether the component has constant wear-and-tear characteristics during service. Making a prediction of mechanical property is whether the component is good in strength or not, exhibits consistent behaviors, although not regular weakening or deforming opulences, and has reliable wear-and-tear performances in service.

3.2.1 Mechanical

Tensile strength, yield strength, hardness, and fatigue life of AM parts are significant to qualify AM parts to apply to the demanding applications. ANN models are trained to provide prediction of these properties which are directly dependent on process parameters (e.g., laser power, scan speed, etc. layer thickness) as well as the material properties (e.g., powder composition). An example of such involves the ability to visualize the ultimate tensile prediction in SLM of Ti-6Al-4V alloys through the use of ANNs to transcend the constraints of approximations in predicting ultimate tensile strength, since researchers in the predictive metallography field can be able to discern the most preferable parameter sets without the need to undertake the burdensome physical experimentation course. These predictive models ensure that development of processes is faster and that there is reduced quantum of material wastage.

3.2.2 Surface Quality Prediction

Roughness of the surface is extremely useful in regards to dimensional accuracy and post-processing requirements. ANNs can also be used to correlate the process parameters with the surface finish output, to enable it to make advance modifications before production batches. Studies have also established that ANNs are more efficient in forecasting surface roughness than the linear regression models due to their capability in initiating the existence of complicated nonlinear processes such as melt pool and powder-bed heterogeneity.

3.2.3 Defect Classification

Such complications as porosity, non-fusion or keyholes, are severe drawbacks of AM of metals. ANNs have been achieved so as to process the real-time defect detection and classification depending on the in situ monitoring data such as thermal images, acoustic emission, or melt pool sensor outputs. Convolutional neural networks (CNNs), also known as a type of ANN, are the best IPANNs, particularly in image defect classifications. This may be done by detecting it at early stage that will enable the remedial action, hence better builds and reduced scrap rates.

3.3 Advantages of ANN in AM

There are several advantages of the introduction of ANNs into AM:

Abilities to address nonlinearity AM processes have complicated, nonlinear correlation among the input parameters and part characteristics. ANNs can predict these links.

3.3.1 Embedded Knowledge

ANNs can be trained on both experimental data or sensor data, which removes the application of simplified physical models that do not represent all the phenomenon.

3.3.2 Generalization Capability

ANNs also transferred to unknown data within the same range as they were trained and predict the model of processes in new unknown conditions.

3.3.3 Reduction of Efforts (Experimental)

ANNs are capable of minimizing a process by reducing the number of costly and time-consuming physical experiments required to optimize a process through the virtual prediction of quality measures.

Relation to real-time control ANNs may be used as a component of feedback control loops to tune the process parameters at real time, establishing the basis of closed-loop AM systems.

3.4 Limitations of ANN in AM

ANNs are limited by the fact that they are not only promising.

3.4.1 Data Dependency

ANNs require to be trained using large and more varying and high-quality data. This information in AM is expensive and time-consuming to acquire because the experimental trials are expensive.

3.4.2 Overfitting Risk

When an ANN is trained on insufficient data, then it can simply memorize the training data rather than being trained to detect generalizable patterns whereby it will exhibit dumpy performance on data it has not previously encountered.

3.4.3 Black-Box Nature

ANNs cannot be easily explained and hence it would be hard to tell why a certain prediction was made. This can limit their usage in safety-critical industries such as the aerospace.

3.4.4 Computational Requirements

Deep ANNs can be computationally expensive to train and also require special hardware (e.g., a graphics card), and a significant amount of processing time.

3.4.5 Minimal Extrapolation Power

ANNs are also good at the part of the data used in training, but they are unreliable to predict in regions that are far away the training field.

To handle these disadvantages, various methods are now commonly combined with ANNs, such as genetic algorithms (to provide optimization of the parameters) or physics-based models (to claim interpretability) which would be combined with ANNs resulting in a hybrid combination, which would be beneficial to both data-driven and mechanism-based modeling.

4. Genetic Algorithms (GA) in Additive Manufacturing

4.1 Fundamentals of Genetic Algorithms

Genetic Algorithms (GAs): This is an evolutionary algorithm that is grounded on the natural selection mechanism and evolutionary biology. The use of GA, first proposed by John Holland in the 1970s, has extensive application in the solution of complex optimization problems with a very large space of feasible solutions, and is both nonlinear and intractable to other analytical techniques.

The principal concept of GA is a population of candidate solutions, which could be represented by a set of chromosomes (strings or vectors of decision variables). The available solutions to the problem under consideration are the chromosomes. The performance of each chromosome is measured by a fitness function, and quantifies the goodness of the chromosome in the optimization objective (e.g., is the least porous, has the highest tensile strength).

4.1.1 Some of the Key Operators Are

- **Selection:** The fittest ones are chosen randomly to sacrifice their genes (parameter values) to the future generation. The selection patterns that are common include roulette wheel selection, tournament selection, and the rank selection.
- **Recombination:** Disjunctive (doubling) choosing one pair of genetically different chromosomes exchanges fragments of the genetic material so that the concept becomes what is considered as offspring and the algorithm produces increased creativity in searching other sections of the solution space.
- **Mutation:** The mutated offspring chromosomes are mutated randomly to retain genetic diversity, so as to avoid the early discovering individuals who are approaching definite local optima.
- **Replacement:** The new chromosome generation replaces some portion of the entire old population, and this is repeated until a stopping criterion is met (e.g., a sufficient number of generations or an end of the values of the fitness).

It is a cyclic procedure where GAs progress to regions of optimal or suboptimal solutions without necessarily requiring the gradient information and may thus be utilized in regions of nonlinear, discontinuous or multi-objective optimization issues.

4.2 The Contribution of GA to Optimization Problems

GAs are particularly applicable in the situations when the landscape of the solution is multidimensional, multimodal, and in complicated situations. Compared with gradient-based optimization strategies, GAs imply a global search, which may guarantee the absence of local optima formation and the search in a wide range of solutions. They are particularly appropriate when it comes to optimization of process parameters in the AM since, in that field, the correlations between process inputs and process outputs are usually nonlinear and coupled.

In AM, the objective (e.g., density maximization, roughness minimization, cost reduction, etc.) functions tend to be multi-objective and even antagonistic. One can also extend GAs to multi-objective formulations (e.g., NSGA-II); it generates pareto-optimal fronts which provide a trade-off between competing objectives. It is an attractive attribute in which the engineers have to compromise, one on the other, between strength, building time and surface quality.

4.3 The Applications of GA in Additive Manufacturing

GAs have been used extensively in AM research to refine process parameters to their goals in accordance with their various performance measures.

4.3.1 Research on the Optimization of Improving the Process Parameter

Researchers in SLM parts have used GA to concurrently optimize power in the laser, scan rate, hatch width, and layer thickness to produce nearly 100% density. The optimization of following the GA can do away with much trial-and-error exploration to promising parameters sets at an early stage in a development cycle. An example has demonstrated tensile strength and hardness improvement in Ti-6Al-4V alloys of SLM in the event of dealing with better process parameters which were enhanced by GA.

4.3.2 Minimization of Porosity

Mechanical performance and fatigue life is directly proportional to porosity. Optimization of energy density and scan strategy applied to minimize porosity has been employed using GAs. GAs have also been applied to ANN-based predictive models in other studies in an iterative search of parameter space with the minimum predicted porosity levels.

4.3.3 Surface Dimensional and Roughness

Roughness on the surface is another key parameter, in particular with limited post-processing parts. The roughness has been minimized through the aid of GA-based optimization, by adjusting the layer thickness, scan overlap, and laser parameters. It is a method that enables manufacturers to achieve better finishes on the surface without major change in the time of production.

4.4 Comparison to Other Metaheuristics

GAs fall under a set of population-based meta heuristic algorithms. Other popular algorithms that are exceptionally popular are particle swarm optimization (PSO), simulated annealing (SA), and JAYA algorithm.

- **PSO:** This is a social behavior inspired by birds. PSO renews solutions between candidates based on a single and cumulative optimal output. Unless tuned, PSO is more likely to get caught in local minima than GA, but converges faster.
- **Simulated Annealing:** SA is a single solution method, which is more likely to accept less optimal solutions, initially in the search process in an attempt to escape local minima. It does not have the power of GA to explore high dimensional space of very large scale.
- **JAYA Algorithm:** JAYA is a comparatively modern graphical algorithm that attempts to obtain the optimal solution and discard the worst solution though without necessitating some algorithm-specific rates like crossover rates or mutation rates. It is low level and it may be competitive with GA but may require some adjustment to be applied to more complex AM problems.

Overall, GAs remain a highly flexible and researched metaheuristics with a fruitful literature and experience in AM optimization problems.

4.5 GA Strengths and Weaknesses in AM

4.5.1 Strengths

Global Search Capability: GAs explore the solution space, and will never be affected by local optimalities the way that gradient-based algorithms are.

4.5.2 Flexibility

They can be handled easily with nonlinear and discontinuous and multi-objective problems.

Non-vertical Gradient Requirement: GAs are based on study of fitness alone, and can thus be used in cases where analytical gradients are either not available on an axis or too costly to compute.

Hybridization Possibility: GAs can be hybridized with predictive models (e.g., ANN) or other algorithms to improve performance, as well as convergence rate.

4.5.3 Weaknesses

- **Computational Cost:** GAs involve several fitness computations per generation, which are computationally expensive in those situations when data about the experiment are utilized or high-fidelity simulations.
- **Parameters Tuning:** GA parameters (population size, crossover rate, mutation rate) can also be sensitive to performance, and can also need to be discovered by trial and error.
- **Speed of Convergence:** GAs are useful global optimization methods, but they can also converge more slowly than other metaheuristics, particularly for simple unimodal problems.
- **Stochastic Nature:** The quantity of output in a run may be different in a run due to the action of selection, crossovers, and mutation which is random in nature hence consistency can only be attained with repeated runs.

On the whole, GAs are a solid and versatile optimization engine in AM, particularly when combined with data-driven models, e.g., ANNs so as to form a hybrid system and to leverage the synergy between prediction and search functionality.

5. ANN-Based Hybrid GA Framework to AM Optimization

The reason is that hybridization encourages innovation, thereby generating substantial value for customers.

5.1 Hybridization Motivation

The AM processes may be shown as a relatively very non-linear and multi-dimensional correlation between the input and the output parameters (so, the laser power, scan speed, the layer thickness) and the output performance qualities (so, the tensile strength, the porosity, the surface finish). The traditional types of analysis or statistics are unable to domain these relationships in a holistic way and subject to an unenthusiastic parameter selection process. ANNs and GA are not a bad solution to the problem and have their limitations whenever applied in isolation.

ANNs have found their way to be a good predictive model that can learn very complex communities between input and output variables. They are efficient in estimating mechanical qualities, roughness of the surface, and possibility of defects driven by the process parameters. However, ANNs tend to be overfitting with small quantities of use to train them, and they are not always engaged in optimization—they are simply intuitive about what is going to occur with their inputs. Also, brute force can be accommodated by large parameter space laxity by using ANN predictions to identify optimal parameter settings, albeit extremely inefficient, even in large parameter space laxity.

GAs instead are strong global optimization tools capable of exploration of high-dimensional solution space and circumvention of local minima. They are particularly suitable in multi-objective optimization problems, and are therefore of interest in AM applications where trade-offs between the density, strength, and productivity must be considered. Computationally expensive as well, however, in those instances when the calculation of fitness requires the running of physical experiments or the simulation of an expensive high-fidelity simulation on a global scale with respect to every candidate solution.

The integration of ANN and GA is beneficial in that it will leverage the capabilities of the two methods. ANN is a class of surrogate models which approximates the effectiveness of a candidate solution in a way that does not need using physical trials. These forecasts are the fitness of the GA as it attempts to find the solutions of world optimum. A hybrid ANN–GA model is far more computationally efficient and capable of optimizing a process in terms of accuracy and efficiency.

5.2 Hybrid Model Structure

Hybrid ANN–GA is a model that integrates a predictive model that is based on the data and optimization algorithm of an evolutionary algorithm.

- **ANN as Predictor ANN:** ANN is trained to access experimental and simulated datasets to learn about the nonlinear association among AM process parameters (inputs, outputs) or mouse outcomes such as tensile strength, yield strength, or surface roughness.
- **GA as Optimizer:** ANN has become trained and is used to perform a fast analysis of candidate solutions generated by the GA. The GA operates by

successively traversing the parameter space, with the ANN predictions being the fitness parameters to find the parameter set that optimizes or minimizes the desired parameter(s).

5.3 Methodology

A hybrid ANN–GA system has several steps in its development and application.

5.3.1 The Data Preprocessing and Collection

A representative and quality dataset is the foundation of the hybrid model. Information collection is done through controlled experiment, in situ observation, or through simulating the high-fidelity process. They are generally driven by the laser power, scan speed, hatch size, and layer thickness and could be output (density, tensile strength, or surface roughness, etc.).

The following are some of the steps involved in the preprocessing:

- **Normalization/Standardization:** Scales all input features in a similar way and this increases stability of ANN training.
- **Data Cleaning:** The removal of distinct values or wrong data values that can affect the model.
- **Dataset Splitting (better called dataset):** Splitting data into two or three sets, which are used to train the model, evaluate the model, and test the model.

5.3.2 Visual Basic/VBA Training and Validation of ANN Model

Depending on the complexity of the problem, a reasonable ANN architecture (type of hidden layers, neurons number, activation functions) is chosen. Such a technique as stochastic gradient descent or Adam optimizer is also used to optimize the model and backpropagation is used to train it.

Accuracy of the model is determined by such performance measures as the coefficient of determination (R^2) and squared error (MSE). K-fold cross-validation is also able to ensure cross-validation and discourage ANN overfitting.

5.3.3 GA Implementation to the Ideal Parameter Selection

When it is trained and tested, the ANN is then created as a fitness function in the GA. The GA workflow is as follows:

- **Initialization:** Generate initial population of viable parameter population of candidate parameters.
- **Fitness Evaluation:** Write ANN with each pair of predicate and set of candidates to predict the performance of the output.
- **Selection:** The finest candidates are chosen based on the values of fitness.
- **Crossover and Mutation:** New candidate solutions are produced by recombining and mutating existing solutions. Each newly created solution

contributes to the generation of additional candidate solutions in subsequent iterations.

The approach allows quick mapping of the parameter space and global optima through thousands of physical tests within a short time, not to mention that it is efficient and fast.

Table 2 Comparative performance of ANN, GA, and ANN–GA in AM process optimization.

Method	*Strengths*	*Weaknesses*	*Example Performance Outcome*
ANN Only	• Fast predictions once trained • Models complex nonlinearities • Good for multi-variable relationships	• Requires large datasets • Risk of overfitting • Cannot perform optimization alone	Predicted tensile strength within ±5% error on validation data (Wang & Zhao, 2021).
GA Only	• Global search, avoids local optima • Handles multi-objective problems • Flexible for nonlinear processes	• High computational cost • Slow convergence • Requires parameter tuning	Achieved ~8–10% improvement in part density over baseline (Rai & Singh, 2021).
ANN–GA Hybrid	• Combines ANN prediction with GA global optimization • Reduces number of experiments • Faster convergence and better accuracy	• Still requires quality training data • Moderate computational demand	Improved tensile strength by 12–15% with 40% fewer experimental trials (Srivastava & Agarwal, 2020).

5.4 Case Study Examples

The effectiveness of the ANN–GA framework hybrid in the AM optimization process is proved by some studies.

5.4.1 Tensile Strength

SLM of Ti-6Al-4V researchers have developed ANN models that can determine ultimate tensile strength (UTS) based on the power of laser, scan speed, and hatch spacing to achieve GA optimization.

The combination of the values of these ANNs which were predicted identified the optimal combinations of the parameters with 5–10% stronger UTS in comparison to the more traditional DoE-based optimization, and both were applied with 50 fewer experimental trials.

5.4.2 Yield Strength Improvement

Using stainless steel 316L, which has been made via laser powder bed fusion (LPBF), ANN–GA frameworks have been employed in the optimization of the yield strength and the density of the build at the same time. This enabled the GA to find pareto-optimal solutions among valuing the mechanical performance and productivity targeting of what is important to cost-sensitive industries such as what to produce and producing at cost-efficiency.

5.4.3 Surface Quality Improvement

Hybrid methods also minimize in AM parts surface roughness. With ANN prediction of surface roughness and optimization using GA, the researchers managed to get the smoother surfaces without necessarily reducing the layer thickness and build rate, reducing the post-processing costs.

5.5 Comparative Performance

ANN–GA hybrid models are typically compared to single ANN and GA algorithms.

- **ANN Alone:** It provides fast and accurate solutions, but it fails to provide the best solutions on its own. It is yet to be optimized and this can be achieved through grid search which is not an effective method of addressing high-dimensional problems.
- **GA Only:** Genetic Algorithms (GAs) can find global optima in high-dimensional (multidimensional) search spaces. However, they can be computationally expensive, especially when evaluating a single candidate solution is costly or when the underlying problem is difficult to solve.
- **ANN–GA Hybrid:** This is a combination of ANN prediction power and global search of GA. The result is increased speed of convergence, reduction in the number of experimental trials, and also successful quality solutions. It has demonstrated a time of 20–40% optimization drop, and bettering of additional measures in output amount than to ANN or GA used distinctly.

Overall, the hybrid ANN–GA model can be an efficient way of optimizing the processes in AM that will enable manufacturers to achieve greater levels of the quality of the product, maintain the cost at a minimum, and reduce the time required to design and print parts.

6. Computational and Practical Considerations

In order to utilize the hybrid frames comprising of the ANN–GA systems in the area of additive manufacturing, different computational and practical versatile considerations are to be carried out. These will be data quality and availability, computational requirements and model acquisition, and infrastructural requirements. The need to remove these problems is vital to ensure that such hybrid models present some competent and repeatable results in other AM settings.

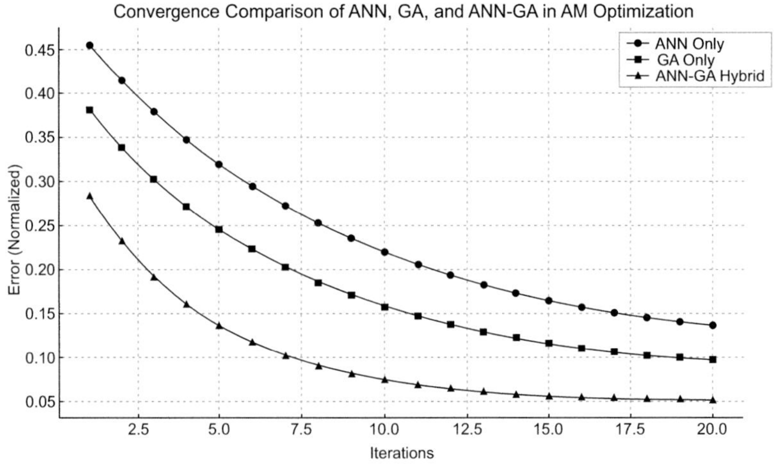

Fig. 2 Convergence comparison of ANN, GA, and ANN–GA in AM optimization.

6.1 Data-Related Challenges

An important factor to identify the performance of ANN–GA frameworks is the quality and representativeness of data that are being used to train predictive models (Smith and Lee, 2023). There is a significant issue with data acquisition in AM research due to several reasons.

First, experimental evidence is often limited because additive manufacturing (AM) is expensive and time-consuming, particularly in metal AM processes, where a single build cycle may take several hours to complete. Second, variations in machine calibration, material properties, and temperature fluctuations can introduce noise into the measurements, leading to unstable and less reliable predictions (Patel et al., 2022). This issue of wavy forecasts discards the impacts of ANN training. Third, datasets may be biased with lots of information of parameters near operation limits and minimal information of extreme conditions of processes. This potentially causes ANN to be biased toward normal parameter space, thereby lowering the predictive performance on extrapolation (Gupta and Chen, 2021).

It has proposed diversified strategies to produce the most information at the least cost of the experiment: Data augmentation model through simulation, statistical preprocess magic filters of outliers, and optimal experimental design including Latin Hypercube Sampling (Rahman et al., 2023).

6.2 Scalability and Cost of Computations

Although ANN–GA models prove to be efficient as compared to brute-force search models, they are computational gap. Deep ANNs are computationally prohibitive models to train, in particular when applied to a multi-layer model, and on large datasets (Zhang and Rao, 2022). In addition to that, the evaluation of thousands of candidate solutions until convergence is an evaluation that is computationally expensive even when ANN-based surrogates are used in place of actual experiments (unless ANN-based surrogates are used in place of the actual experiments).

This calculated needs even more because multi-objective optimization involves a number of competing goals, which can be tensile strength, porosity, and build time, and need to be optimized simultaneously (Kumar and Wang, 2021). The scalability is also a consideration as this is exponentially harder to study the behavior of a process with many process parameters.

6.3 Cross-Application and Extrapolation of the Model

The largest panic associated with ANN-based techniques is the fact that it forecasts on invisible data. Any minor changes in the calibration of either of the machines, the conditions, or batches of the materials can lead to significant variations in outcomes in AM (Lopez et al., 2022).

- **Machine-to-Machine Variability:** When the same parameters of a process are used, the results may vary on different machines due to tolerances which may vary in hardware.
- **Transferability of Materials:** Not all alloys systems are always transferable directly without retraining, they have to be modeled (Ali et al., 2023).

Process transferability DED formularies GT processes designed to work with SLM may not be designed to work with DED due to differences between the heat transfer and solidification mechanisms.

To enhance the generalization, one can use technical techniques such as the transfer learning where an already trained ANN is optimized with the assistance of a smaller set depending on the target material or process. This will minimize the data requirement and will be more flexible (Nguyen and Park, 2023).

6.4 Hardware and Software Requirements

The ANN–GA model demands capabilities of the computation to be implemented practically. The hardware must have graphics processing units (GPUs) or high-performance computing (HPC) clusters, to train and evaluate ANNs or the GA on a high level (Singh et al., 2022). Anchor Software Packages (TensorFlow, PyTorch, Keras etc.) are common assistants to the authors of ANNs, GA codes are written in Python, MATLAB or an open source library, such as the DEAP.

Examples can also be the incorporation into manufacturing execution systems (MES) in the industrial context when it comes to the optimization of the closed loop when it is possible to implement the adjustment of the parameters in real time when producing (Huang and Lee, 2023). In addition, robust data storage and versioning tools are critical in relation to the standards of reproducibility and quality.

6.5 Strategies for Improvement

To hope to defeat the problem of calculations and practice, several measures have been proposed:

- **Transfer Learning:** Simplifies the data requirements and accelerates the transition of model vehicles to new material and machinery (Nguyen and Park, 2023).
- **Surrogate Modeling:** Surrogate Modeling: Gaussian Process Regressors (GPRs) can approximate ANN predictions within the GA loop, reducing computational cost and speeding up optimization (Rahman et al., 2023).
- **Parallel and Distributed Computing:** GA is easily applicable to the fundamental uses of the GPU clusters or a distributed environment of a computing cluster as GA is also parallelizable (Patel et al., 2022).
- **Active Learning:** Data acquisition in high uncertainty of the model, and therefore improved ANN training on fewer experiments (Smith and Lee, 2023).
- **Hybrids:** PIN applications can also include hybrids, which consist of simplified physics-based equations, better interpretability and robustness, particularly in instances where AM is used in a safety-critical way (Kumar and Wang, 2021).
- **Summary:** Concluding on it all, although hybrid ANN–GA systems have some potential as far as optimization of AM processes is concerned, they will not perform very well, unless the problems regarding poor data quality, high cost of computation, and poor model generalization are resolved. The introduction of the new methods, e.g., transfer learning, surrogate-assisted optimization, and parallel computing have become promising directions of the increase of scalability and robustness. These issues will be crucial in the context these smart, data-driven optimization methods can be viewed as being pervasive and targeted at implementation in industrial AM.

7. Problems and Future Research Prospects

Although the hybrid ANN–GA models have become a significant step in optimization of AM-based processing, there are certain problems that inhibit their utilization currently. The necessity to prevent these problems provides opportunities to future researches and will be required to implement these methods that may be applied in laboratory research nowadays but are relevant to the industrial application.

7.1 Current Limitations with ANN–GA Hybrid Models

ANN–GA frameworks need considerable data access, model creation, and calculation resources in order to perform. The data scarcity is one of the largest bottlenecks; the bulk of the research is supported by limited experimental data that may lead to overfitting and the low rates of generalization (Smith and Lee, 2023).

Furthermore, the mathematical computational requirements of large-scale training of neural networks, along with the population-based GAs, may be restrictive, especially of high dimension and of numerous objectives.

The other weakness is that ANN–GA result is not interpretable. Even though the parameter sets that are produced during the application of the model can be optimized well, the model does not necessarily provide the explanations of the underlying physical processes that drive the improvement (Zhang and Rao, 2022). Such black-box behavior can have the implication of deterring confidence between engineers and decision-makers in safety-relevant sectors such as the aerospace, and the manufacturing industry called the biomedical.

Also, most studies have been focused on single-objective optimization (e.g., maximizing tensile strength), not on real-life scenarios, where there are multiple objectives to be considered (Rahman et al., 2023). Multi-objective trade-off is a field of research that is not fully utilized.

7.2 Learning Reinforcing Control and Learning Adaptive Control

One of the directions of the future research is the combination of ANN–GA framework with reinforcement learning (RL). Even though the GAs are effective international optimizers, they tend to be offline—all solutions possible have to be searched and then a solution reached. On the other hand, RL can be continuously learning the real feedback and thus, it is reflexible when it comes to the regulation of processes (Nguyen and Park, 2023).

An RL–ANN–GA can be a hybrid framework in which such frameworks can use ANNs as predictive models, GAs to conduct preliminary global search, and RL with which to implement online changes in production. These systems can be responsive to process variation or machine variation or to a change in performance goals. This would bring AM optimization of a pre-production, comparatively dormant control activity and to a closed, adaptive control paradigm improving robustness and predictability.

7.3 Digital Twins/Monitoring in Real Time

The other urgency research-line is implementation of AM processes to digital twins replicas that incorporate sensor data in real time which relies on predictive models. Digital twins can also provide constant feedback on the temperature of the melt pool, quality of the builds, and geometry of its parts, enabling ANN–GA models to continue running in a loop of decision-making in real time (Huang and Lee, 2023).

In this configuration, the ANN model may be fed with continually varying streaming data, as the GA is applied to optimization of process parameter on almost real time basis. The result of such a combination would allow detecting defects quickly and adjusting the parameter in the build mode and lower the scraping rates and increase the yield. Engineering is needed to find the manner in which these systems can be synchronized and how to address latency, among simulation, data collection, and control interventions.

7.4 Multi-Objective Optimization

Most current ANN–GA studies focus on maximization of just one performance measure unlike AM that implies compounding trade-off. To illustrate, laser power can be increased to increase density at the cost of surface finish, or a reduced layer thickness can be used to increase accuracy at the cost of build time (Kumar & Wang, 2021)

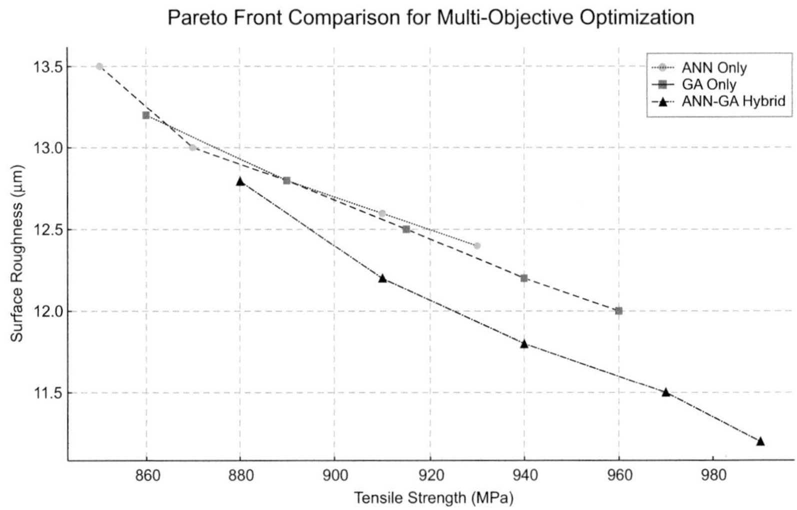

Fig. 3 Pareto front comparison for multi-objective optimization.

Future research should be more on multi-objective optimization models which have the ability to target:

- Mechanical (strength, fatigue) performance
- Fluidness (surface finish, smoothness)
- Build time and productivity
- Material and energy costs

This implies that the decision-makers can capitalize on such ways using Pareto front analysis when the importance of the best trades-offs solutions are provided to the decision-makers as an ensemble, rather than as an optimum. That would allow the manufacturers to select the parameter combinations according to the priorities of the business (e.g., high quality vs. faster production).

7.5 Environmental Impact/Sustainability and Energy Efficiency

The theme of sustainability becomes one of the biggest factors in the modern production. Current ANN–GA models are principally built to optimize the quality of the components but not related performance measures on the environment, such as energy consumption, material usage performance, CO2 emissions (Lopez et al., 2022).

Future research could set more environmentally friendly optimization targets that would decrease the amount of energy needed by each component, eliminate less material (e.g., by coming up with structures that support components in the best possible way), and prolong machine life by making processes parameters less aggressive. This would be in balance to optimize the AM processes and align it with its global goals to achieve manufacturing sustainability and the circular economy.

7.6 Industry Adoption Opportunities

In order to popularize the ANN–GA frameworks within the industry, there are several aspects of operation that should be adopted. To start with, such software interfaces should be user-friendly in order to allow process engineers who lack expertise in machine learning to utilize them (Singh et al., 2022). Second, the information format standardization, standard protocols of the experiment, and performance measures will be needed to ensure the reproducibility and benchmarking across different facilities.

Finally, it can be significantly incorporated into the Industry 4.0 systems where AM machines, quality control systems, and enterprise resource planning (ERP) software are connected. These intelligent production spaces can utilize ANN–GA systems as optimization systems, and self-correcting production lines in complete self-direction.

8. Conclusion

Altogether, a hybrid ANN–GA model is a promising development of the data-driven, intelligent optimization of AM processes, but the problems of available data, cost, interpretability, and multi-objective trade-offs persist. Integration of RL and adaptive control, the implementation of digital twins, taking into account the sustainability indicators, and the development of software applications to allow making digital twins scalable and convenient to use in industries should be investigated in future studies. These guidelines will be welcomed to ease the wide-scale implementation and accelerate the process of AM to produce beyond prototyping but cheap and sustainable serial production.

List of Abbreviations and Acronyms

Abbreviation	Full Form
AM	Additive Manufacturing
ANN	Artificial Neural Network
GA	Genetic Algorithm

Abbreviation	Full Form
ANN–GA	Artificial Neural Network–Genetic Algorithm (Hybrid Model)
SLM	Selective Laser Melting
EBM	Electron Beam Melting
DED	Directed Energy Deposition
DoE	Design of Experiments
RSM	Response Surface Methodology
RL	Reinforcement Learning
MES	Manufacturing Execution System
ERP	Enterprise Resource Planning
HPC	High-Performance Computing
GPU	Graphics Processing Unit
AI	Artificial Intelligence
ML	Machine Learning
CAD	Computer-Aided Design
AM Process Parameters	Laser Power, Scan Speed, Layer Thickness, Hatch Spacing (collectively)
QoI	Quantity of Interest (e.g., Tensile Strength, Surface Roughness)
QoS	Quality of Surface (Surface Finish Metric)

References

[1] Ali, Muhammad, Adeel Farooq and Hyun Park. (2023). "Transfer Learning for Process Parameter Optimization in Additive Manufacturing of Titanium Alloys." *Journal of Manufacturing Processes* 86: 305–317.

[2] Bikas, Haris, Panagiotis Stavropoulos and George Chryssolouris. (2016). "Additive Manufacturing Methods and Modelling Approaches: A Critical Review." *International Journal of Advanced Manufacturing Technology* 83 (1–4): 389–405.

[3] Tarasankar DebRoy, H.L. Wei, J.S. Zuback, T. Mukherjee, J.W. Elmer, J.O. Milewski, A.M. Beese, A. Wilson-Heid, A. De, and W. Zhang. (2018). Additive manufacturing of metallic components: Process, structure and properties. *Progress in Materials Science*, 92, 112–224. https://doi.org/10.1016/j.pmatsci.2017.10.001

[4] Goodfellow, Ian, Yoshua Bengio and Aaron Courville. (2016). *Deep Learning.* Cambridge, MA: MIT Press.

[5] Gupta, Vikas and Zhen Chen. (2021). "Addressing Data Imbalance in Machine Learning Models for Additive Manufacturing." *Journal of Intelligent Manufacturing* 32 (8): 2247–2262.

[6] Huang, Rui and Jae Lee. (2023). "Digital Twin-Enabled Real-Time Optimization for Metal Additive Manufacturing." *Robotics and Computer-Integrated Manufacturing* 84: 102547.

[7] Katragadda, S.R., A. Tanikonda and S.R. Peddinti. (2025). "Deep Learning for Autonomous Data Quality Enhancement: A Paradigm Shift in Machine Learning Pipelines." *Journal of Information Systems and Education Management* 10 (19s): 602–609.

[8] Katragadda, Subba Rao and Tanikonda, Ajay and Pandey, Brij Kishore and Peddinti, Sudhakar Reddy, "Machine Learning-Enhanced Root Cause Analysis for Rapid Incident Management in High-Complexity Systems" (May 17, 2022). Journal of Science & Technology, Volume 3, Issue 3, May–June 2022, pp. 325–345.

[9] Kumar, Ankit and Chien C. Wang. (2021). "Multi-Objective Optimization of Additive Manufacturing Processes Using Machine Learning and Evolutionary Algorithms." *Additive Manufacturing* 45: 102054.

[10] Li, N., Huang, S., Zhang, G., Qin, R., Liu, W., Xiong, H. and Blackburn, J. (2019). Progress in additive manufacturing on new materials: A review. Journal of Materials Science & Technology, 35(2), 242–269.

[11] Lopez, Francisco, Sergio Rios and Javier Fernandez. (2022). "Sustainability-Driven Design of Additive Manufacturing Processes Using Machine Learning Optimization." *Journal of Cleaner Production* 372: 133702.

[12] Nguyen, Thanh and H.S. Park. (2023). "Reinforcement Learning-Assisted Control of Laser Powder Bed Fusion for Defect Mitigation." *Journal of Materials Processing Technology* 316: 117015.

[13] Patel, Deep, Ramesh Shukla and Sayan Banerjee. (2022). "Noise-Tolerant Machine Learning Models for Predicting Porosity in Additive Manufacturing." *Materials & Design* 222: 111054.

[14] Rahman, Md., Md. Tariq Islam and Lihong Li. (2023). "Data-Efficient Experimental Design for Process Optimization in Selective Laser Melting." *Journal of Manufacturing Science and Engineering* 145 (7): 071003.

[15] Rai, Rakesh and Raj Singh. (2021). "Genetic Algorithm-Based Optimization of Process Parameters in Directed Energy Deposition." *Procedia CIRP* 99: 239–244.

[16] Singh, Sandeep, Rajesh Singh and Pankaj Kumar. (2022). "GPU-Accelerated Training of Deep Neural Networks for Additive Manufacturing Applications." *Journal of Computational Design and Engineering* 9 (5): 1293–1304.

[17] Smith, John and David Lee. (2023). "Overcoming Data Scarcity in Metal Additive Manufacturing Using Physics-Informed Neural Networks." *Computational Materials Science* 214: 111774.

[18] Srivastava, Richa and Pooja Agarwal. (2020). "Application of Hybrid ANN-GA Approach for Predicting Surface Roughness in Additive Manufacturing." *International Journal of Advanced Manufacturing Technology* 108 (3–4): 823–836.

[19] Sun, Yifan, Yuxin Zhang and Guoqiang Wang. (2020). "Multi-Objective Genetic Algorithm Optimization for Laser Additive Manufacturing Process Parameters." *Optics & Laser Technology* 127: 106158..

[20] Wang, Yifan and Yao Zhao. (2021). "ANN-Based Predictive Modeling for Mechanical Properties of Ti-6Al-4V Parts Fabricated via SLM." *Materials Today Communications* 28: 102539.

[21] Yang, Jie and Hao Zhang. (2019). "Application of Deep Neural Networks for Defect Detection in Additive Manufacturing Using Layer-Wise Imaging." *Additive Manufacturing* 28: 634–642.

[22] Zhang, Hao and Prakash Rao. (2022). "Black-Box Optimization and Interpretability Challenges in ANN-Based Additive Manufacturing Models." *Journal of Manufacturing Systems* 63: 465–478.

[23] Zhang, Yan, Xiaoyan Li and Guoqiang Wang. (2021). "Deep Learning Architectures for Defect Detection in Metal Additive Manufacturing: A Review." *Additive Manufacturing* 48: 102450.

[24] Zubair, Muhammad and Noman Ahmed. (2023). "Integration of ANN and GA for Process Parameter Tuning in Powder Bed Fusion: A Review." *Additive Manufacturing Letters* 6: 100095.

12 Process Parameter Optimization in Additive Manufacturing Using GRA, TOPSIS, CORPAS, and MOORA Optimization Techniques

Pralhad Pesode[1*], Sagar Wankhede[2] and Shailendra Dayane[3]

This chapter focuses on process parameters optimization of metal additive manufacturing by multi-criteria decision-making (MCDM) techniques. In this study, optimization techniques such as grey relational analysis (GRA), multi-objective optimization on the basis of ratio analysis (MOORA), technique for order preference by similarity to ideal solution (TOPSIS), and complex proportional assessment (COPRAS) are used. Different metallic alloys such as titanium alloys, magnesium alloys, superalloys, aluminum alloys, and steel are also discussed with their different properties and applications. Different metal additive manufacturing techniques such as powder bed fusion [selective laser melting (SLM), selective laser sintering (SLS), electron beam melting (EBM)], direct energy deposition, binder jetting, and wire arc additive manufacturing are investigated in this chapter. Selective laser melting process is discussed in detail. The effects of process parameters of SLM (such as scan speed, layer thickness, laser power, and hatch spacing on performance parameters, such as density, ultimate tensile strength, residual stresses, and surface roughness) are investigated. The process parameters of SLM are optimized with GRA, TOPSIS, COPRAS, and MOORA (MCDM) techniques to find out optimum process parameters for best performance characteristics. The ranking results obtained from different optimization techniques are compared to assess their consistency and robustness.

The investigation shows that all the abovementioned techniques give similar optimized process parameters. This chapter gives valuable insights for selecting best process parameters of SLM process, and assists as a practical guideline for improving part quality and process efficiency.

1. Introduction

Additive manufacturing (AM) is also called 3D printing, which manufactures the components layer by layer by using digital 3D model [1, 2]. AM technique can

[1,3] Department of Mechanical Engineering, Dr. Vishwanath Karad MIT-World Peace University, Pune 411038, Maharashtra, India.

[2] School of Mechatronics Engineering, Symbiosis Skills and Professional University, Pune 412101, Maharashtra, India.

* Corresponding author: pralhadapesode@gmail.com

DOI: 10.1201/9781003635673-12

manufacture intricate parts which are difficult to manufacture with conventional techniques. AM techniques help the designers to manufacture complicated parts and to overcome the challenges of traditional manufacturing processes such as excessive material waste, fabricating delicate geometries, and the necessity for specialist equipment. Therefore, engineers have enhanced "design flexibility" with the use of AM techniques [3]. Also, reduction in number of parts helps to reduce or eliminate the assembly time and costs [4]. With AM techniques parts can be manufactured at demand, reducing production time, streamlining supply chains, minimizing storage requirements, eliminating shipping expenses, and reducing lead time for important replacement parts. Different types of materials such as ceramics, metal, composite, and plastic can be used to manufacture components with AM techniques. In addition, with abovementioned advantages, AM techniques offer environmental advantages, including reduced waste, enhanced quality, reduced pollutant emissions, and on-demand production of components [5]. Companies and academicians have taken interest in additive manufacturing of components from metallic materials. Though AM offers significant advantages, its application in industry is presently limited to dentistry, construction, and aerospace.

Metal AM is also called metal 3D printing, in which metallic components are manufactured with excellent accuracy and material efficiency. In AM techniques metallic powders or wires are selectively melted, sintered, or deposited utilizing high-energy sources such as lasers or electron beams, as per a computerized 3D model. Important metallic AM techniques are selective laser melting (SLM), direct metal laser sintering (DMLS), electron beam melting (EBM), laser metal deposition (LMD), and binder jetting [6, 8]. Each technique uses different energy source, feedstock type, and processing environment; however, all techniques try to manufacture dense, fully functioning metal parts with customized mechanical properties. The distinguishing ability of metal AM to manufacture intricate internal structures, lightweight component, and optimized geometries has created novel design opportunities, especially in industries that need high strength to weight ratio and enhanced performance [6, 7].

In metal AM, PBF is an important technique for fabricating high-density, intricate metallic components [6, 8]. The efficacy and quality of components manufactured using PBF are significantly affected by several essential process parameters, such as laser power, scanning speed, hatch spacing, layer thickness, scanning strategy, powder properties, and environmental conditions throughout the manufacturing. Each parameter influences the energy input and heat profile during melting and solidification, consequently impacting the microstructure, density, and mechanical properties of the finished component. Laser power and scan speed control the powder bed's energy density. Insufficient energy can cause incomplete melting and porosity, while excessive energy can cause keyhole flaws or increased residual stresses. Layer thickness similarly affects surface properties and production time—thinner layers enhance resolution and density but prolong manufacturing duration, while thicker layers may jeopardize dimensional precision. The hatch spacing affects the overlap of neighboring melt tracks; inadequate spacing may result in unbonded areas or irregular surface morphology.

Residual stresses, temperature gradients, and distortion must be managed by the scanning speed, which influences laser trajectory and orientation [7, 9, 10]. Optimized scanning patterns, including alternating and island methods, reduces warping and improves dimensional stability. The properties of powder, including particle size distribution, morphology, and flowability, influence layer uniformity and melting behavior. Spherical particles with narrow size distributions provide enhanced packing density and uniform melting, yielding superior quality components. The environment within the build chamber, typically sustained using inert gases such as argon or nitrogen, inhibits oxidation and improves metallurgical bonding. Additionally, preheating the substrate can increase thermal strains and improve interlayer adhesion. The interaction of these characteristics dictates essential performance indicators, including density, surface roughness, microhardness, tensile strength, and fatigue resistance. Inadequate parameter selection may result in flaws like porosity, warping, cracking, or anisotropic characteristics. Consequently, optimizing process parameters by experimental design, modeling, and in situ monitoring is crucial for attaining consistent and high-performance metal components. A thorough grasp of process parameters and material behavior helps powder bed fusion additive manufacturing create reliable, functioning components. Process parameters of AM techniques can be optimized with different MCDM techniques. In this chapter, GRA, MOORA, COPRAS, and TOPSIS techniques are used for AM process parameters optimization.

2. Materials

2.1 Titanium

Additive manufacturing of titanium and related alloys has become a crucial area of research and industrial use, due to titanium's exceptional strength-to-weight ratio, corrosion resistance, and biocompatibility. Ti-6Al-4V is the most widely utilized titanium alloy in additive manufacturing. Additional titanium alloys, including Ti-6Al-2Sn-4Zr-2Mo (Ti-6242), Ti-5Al-5V-5Mo-3Cr (Ti-5553), Ti-6Al-2Sn-4Zr-6Mo (Ti-6246), and commercially pure (CP) titanium, are under investigation for applications necessitating enhanced high-temperature strength or superior fatigue resistance [11]. Beta Ti alloys such as Ti10V2Fe3Al and Ti15Mo are attracting interest. Table 1 shows comparison of different materials along with their properties, applications and manufacturing techniques.

Additive manufacturing frequently employs Ti alloys, especially Ti6Al4V, to manufacture lightweight, high-performance parts for the energy, aerospace, and healthcare sectors. The manufacturing process and microstructural variations in conventional titanium alloys have constrained the use of Ti6Al4V alloys for certain uses [12]. Traditional technologies are capable of producing complex components such as aircraft engine frames, fan frames, compressor casings, exhaust gas pipes for auxiliary gas turbines, and rim screws [13]. Ti-6Al-4V components are typically subjected to hot working and heat treatment processes. AM has been employed

Table 1 Comparison of different materials along with their properties, applications, and most suitable manufacturing techniques.

Material	*Properties*	*Applications*	*AM Techniques*	*Advantages*	*Limitations*	*Ref*
Titanium and its alloys	Higher strength-to-weight ratio, good corrosion resistance, biocompatible, better fatigue strength	Aerospace (engine parts, structural components), biomedical implants, automotive racing parts	SLM, EBM, DED	Lightweight and strong; suitable for complex shapes; good mechanical properties after AM; excellent corrosion and fatigue resistance	Expensive powder; difficult to machine post-AM; high residual stress; oxidation during processing requires controlled atmosphere	[11–14]
Aluminum alloys	Low density, good corrosion resistance, moderate strength, good thermal and electrical conductivity	Aerospace (brackets, housings), automotive parts, heat exchangers, consumer electronics	SLM, LPBF, binder jetting (for sintered parts)	Lightweight and affordable; good thermal conductivity; suitable for lightweight structures; high build rates	High reflectivity and thermal conductivity make processing difficult; prone to cracking; limited alloy options for AM	[14–17]
Magnesium alloys	Lowest density among structural metals, good damping properties, moderate strength	Automotive (lightweight frames), aerospace, biomedical (temporary implants)	SLM, LPBF	Very lightweight; good energy absorption; biocompatible and biodegradable (for medical uses)	Highly reactive and flammable; difficult to process safely in AM; oxidation and porosity issues	[42–47]
Steel	High strength, toughness, wear and corrosion resistance (for stainless grades)	Tooling, molds, structural components, aerospace brackets, medical instruments	SLM, DMLS, binder jetting, DED	Readily available; good mechanical properties; cost-effective; well-studied processing parameters	Heavy; post-processing often needed (heat treatment, machining); possible distortion or cracking during AM	[20–22]
Superalloys	Good high-temperature strength and excellent creep resistance, oxidation, corrosion performance	Turbine blades, jet engines, power generation components, chemical processing equipment	SLM, EBM, DED, LMD	Outstanding performance at elevated temperatures; near-net-shape fabrication reduces waste	Difficult to machine post-AM; expensive materials; residual stress and cracking risks; limited build size due to thermal distortion	[24–30]

to fabricate intricate and near-net-shape Ti-6Al-4V components through layer-by-layer deposition; however, these components require heat treatment to alter their microstructure. Mechanical and physical qualities depend on microstructural development.

2.2 Aluminum Alloys

The range of Al alloys available for additive manufacturing is still somewhat limited. One reason is because aluminum, in contrast to titanium, is very amenable to processing, and the expenses associated with aluminum components are relatively low. The processing of aluminum additive manufacturing parts often yields diminished commercial benefits, if any. Another factor is that several aluminum alloys are recognized for their poor weldability. Hardening from precipitation gives high-performance alloys strength. Despite ongoing research, some widely used hardenable alloys, such as EN AW-7075 from the Al-Zn 7xxx series, contain volatile elements like Zn, resulting in turbulent melt pools, splatter, and porosity, making them unsuitable or difficult for additive manufacturing [16]. Under vacuum circumstances, alloying elements with significantly different vapor pressures than aluminum, such as magnesium and lithium, are known to vaporize preferentially. Moreover, aluminum exhibits significant reflectivity for the laser wavelengths typically utilized in LBM and LMD, which may be an additional challenge for aluminum components produced by additive manufacturing [15]. The low viscosity of molten aluminum presents a challenge, restricting additive manufacturing of aluminum to tiny melt pool sizes, hence preferring laser beam melting over, for example, laser metal deposition. The advantageous aspect of additive manufacturing for aluminum components is that aluminum's high heat conductivity diminishes thermally induced strains, hence lessening the necessity for support structures. Moreover, the enhanced thermal conductivity facilitates increased processing rates. The predominant aluminum alloys for additive manufacturing now available are the heat-treatable AlSi10 Mg and the eutectic AlSi12 [14]. Schmidtke et al. [17] suggest a hardenable Al–Mg–Sc alloy (AlMg4.5Sc0.66) for high strength applications. The incorporation of Sc facilitates precipitation hardening through the development of Al3Sc precipitates and is recognized for its ability to refine grains in Al–Mg alloys. Other commercial aluminum alloys examined for their applicability in additive manufacturing component production in the literature include hardenable Al–Mg–Si (6061 [18, 19]) and Al–Cu (2139 [15]) alloys. It should be noted that LBM and AlSi10 Mg now represent the predominant combination in the additive manufacturing of aluminum components.

2.3 Steel

The variety of steel alloys accessible for AM remains somewhat constrained. One explanation is that steels, in contrast to titanium alloys, are well known in traditional production techniques, for example, casting, forging, and machining, which remain economically viable due to developed industrial infrastructure and cheap production costs. The commercial benefit of manufacturing steel components by additive manufacturing is typically restricted to specialized applications that need geometric intricacy or material customization [20, 21]. A further difficulty is that several high-performance steels are challenging to treat via additive manufacturing due to their sensitivity to cracking and phase instability. Steels that get strength by precipitation or martensitic hardening frequently display significant residual stresses and deformation when subjected to the pronounced heat gradients characteristic of laser- or electron-beam-based additive manufacturing methods. Moreover, high-carbon tool steels or high-alloy martensitic grades are prone to cracking during fast solidification, which restricts their use in additive manufacturing despite their exceptional mechanical qualities.

Volatile alloying elements such as Mn or Cr can induce compositional inhomogeneities and evaporation under the vacuum or inert gas conditions often utilized in laser-based processes. Additionally, the high reflectivity of steels at infrared laser wavelengths, especially in austenitic phase, lowers energy absorption efficiency and may hamper process stability. The advantageous properties of molten steels, characterized by higher viscosity and moderate heat conductivity, enhance melt pool control and stability, enabling higher production volumes and superior surface finishes in comparison to lightweight alloys like aluminum. Residual stresses persist as an issue; however, preheating techniques and post-process thermal treatments can substantially reduce cracking and deformation. The most frequently used steels in additive manufacturing are austenitic stainless steel 316L and maraging steel 18Ni300, both exhibiting superior processability and mechanical qualities after printing and heat treatment. Other steels under investigation include precipitation-harden stainless steels, martensitic tool steels, and high-strength low-alloy (HSLA) steels specifically designed for AM procedures [22]. Despite ongoing research into tailored formulations and process improvements, it is essential to recognize that laser-based powder bed fusion (LBM) of 316L remains the foremost technique in the additive production of steel components to this day.

2.4 Superalloys

High-performance superalloys have excellent mechanical strength, creep resistance, and oxidation stability at high temperatures [23]. They predominantly utilize nickel, cobalt, or iron matrices, with nickel-based systems being the most extensively employed in aerospace, energy, and propulsion sectors. Their enhanced mechanical characteristics arise from precipitation hardening via coherent γ' ($Ni_3(Al,Ti)$) and γ'' (Ni_3Nb) phases inside a face-centered cubic γ matrix, together with solid-solution strengthening and grain-boundary management [24, 25]. Conventional production techniques, like casting and forging, are well established but limit design flexibility and microstructural enhancement. AM has surfaced as a feasible alternate, allowing intricate geometries and graded structures, however it presents issues like hot cracking, elements segregation, and residual pressures. The fast solidification in additive manufacturing enhances microstructures but may induce non-equilibrium phases. Nickel-based alloys such as Inconel 718, Inconel 625, and Hastelloy X are among the most amenable to additive manufacturing. Current research emphasizes elucidating solidification mechanisms, refining heat treatments, and formulating additive manufacturing-specific superalloy compositions to satisfy the requirements of next-generation turbine and propulsion systems.

Superalloys perform well at high temperatures due to three synergistic hardening mechanisms: solid solution, carbide dispersion, and precipitation. Solid solution strengthening occurs with chromium, molybdenum, tungsten, and trace amounts of tantalum, zirconium, niobium, and boron. This alters the crystal lattice structure, inhibiting dislocation motion and ensuring consistent strength at elevated temperatures [26–28]. A crucial process known as carbide dispersion hardening involves purposefully dispersing very hard particles—such as TiC, BC, ZrC, TaC, Cr7C3, Cr23C6, Mo6C, and W6C—within the microstructure.

These particles serve as efficient impediments to dislocation movement and particularly fortify grain boundaries, thereby enhancing creep resistance significantly [29, 30]. Superalloys and additive manufacturing technologies together represent a very promising technological future for expanding the range of applications for these advanced materials. Specific chemical compositions of metal powders ensure heat treatment microstructural phase production, spherical morphologies to enhance flowability, and carefully regulated particle size distributions are typically employed in additive manufacturing processes for superalloys. The following technologies are significant: wire arc additive manufacturing (WAAM) [32, 33], electron beam melting (EBM) [34, 35], direct energy deposition (DED) [36–38], and selective laser melting or laser powder bed fusion (SLM/LPBF) [31, 39]. Each technology's benefits depend on component shape, dimensional accuracy, and mechanical qualities. Additive fabrication of superalloys has opened up new uses. These applications include lightweight constructions with heat-dissipating internal geometries, monolithic components that formerly required assembly, and biomimetic geometries that improve strength-to-weight ratios [40, 41]. It's hard to control things like anisotropy, porosity, residual stresses, and microstructural control in this field, so it's important to know how process factors, powder makeup, and thermomechanical changes happen during fabrication and post-treatment.

2.5 Magnesium Alloys

Magnesium alloys are lightweight metallic substances mostly consisting of magnesium, with minor additions of elements like aluminum, zinc, calcium, and rare earth metals to enhance strength and stability. They have an exceptionally low density, rendering them among the lightest structural metals accessible, hence enhancing fuel economy and facilitating weight reduction in engineering applications [42–45]. Magnesium (Mg) has the lowest density of all the elements, with a value of 1740 kg/m^3, while the densities of aluminum, titanium, and steel are, respectively, 2710, 4500, and 7800 kg/m^3. Because of their lower density, higher specific strength, and biodegradability, they are ideally suited for use in the following applications: nosewheel doors, flap cover skins, oil tanks, flooring, wingtips, ducts, seats, and fuselage components in the aerospace industry [46, 47]; front-end structures, transfer cases, engine cradles, incremental panels, steering wheel cores, cam covers, seat backs, and third-row seat frames in the automotive industry; and cardiovascular materials.

Their elevated specific strength and excellent machinability render them advantageous for fabricating intricate forms and components [48]. Magnesium alloys provide substantial damping ability, allowing them to absorb vibrations and mitigate noise. Because of its gradual biodegradability and biocompatibility, magnesium alloys have attracted a lot of interest in biomedical engineering. They might be utilized to fabricate transient implants that gradually disintegrate within the body, hence diminishing the necessity for a subsequent surgical procedure for implant extraction. These alloys facilitate bone repair by the release of magnesium ions, which are advantageous for osseous development. Researchers persist in investigating methods to enhance corrosion resistance while preserving favorable biological features. Their performance is significantly influenced by composition, microstructure, and processing techniques.

Mg and Mg alloys' poor ductility, corrosion resistance, and flammability limit their use [49]. Due to its heat sensitivity, magnesium manufacturing is challenging. Over 95% of magnesium alloy components are made by high pressure die casting, but this method cannot create intricate shapes like porous structures and large components like cryogenic containers for aerospace and propellers for marine applications. Poor ductility, 6% elongation, limits wrought Mg alloy application [50]. Mg alloy's HCP crystal structure makes cold rolling, hot rolling, and forging difficult. Besides these concerns, Mg alloys' poor corrosion resistance limits their application. Mg components may be made via additive manufacturing to overcome these difficulties. Post-processing with heat and surface modification may provide the necessary attributes.

Fractures, inclusions, and a low vaporization temperature in matrix magnesium alloy will hinder additive manufacturing [51]. Powder, wire, and liquid resin are traditional additive manufacturing raw materials. This configuration enhances the metal's surface energy, thereby augmenting the likelihood of interaction with ambient oxygen. Specialized airtight equipment is required to print magnesium in an inert environment to reduce oxidation effects. Nevertheless, the rigorous safety standards for the equipment render the study on the magnesium production process relatively insufficient [52, 53]. The existing magnesium additive manufacturing techniques comprise selective laser melting (SLM, sometimes referred to as laser powder bed fusion, LPBF), wire arc additive manufacturing (WAAM), jetting technology (BJ), friction stir additive manufacturing (FSAM), and indirect additive manufacturing (I-AM) [54]. WAAM and LPBF are the most common methods. The number of LPBF and WAAM articles on magnesium alloys has steadily increased over the past five years, indicating their potential and attractive applications.

3. Metal Additive Manufacturing Techniques

According to ASTM/ISO 52900:2017, there are seven main types of metal additive manufacturing processes. These are material extrusion (ME), binder jetting (BJ), powder bed fusion (PBF), directed energy deposition (DED), sheet lamination, and vat polymerization [55]. Of the seven processes, only four pertain to metals. PBF is increasingly utilized in consumer, industrial, automotive, medical, and aerospace manufacturing [56]. A small layer of powder is put on a build plate, and then an energy source melts or sinters a cross-section of a part. Melting mechanisms classify PBF technology. Machines that employ electron beams are called electron beam melting (EBM), whereas those that use lasers are called SLM or SLS. Directed energy deposition (DED) with wire or powder feedstock is a typical approach for component repair and enhancement [57]. Lasers are widely employed in DED to fuse the material, similar to PBF. One distinction is that the energy source and feedstock are often located on the same print head. While powder-based DED devices are known as laser metal deposition, wire-based machines are known as electron beam additive manufacturing (EBAM) and wire arc additive manufacturing (WAAM).

For BJ, a variety of techniques are available, including supersonic deposition (SD), single-pass jetting (SPJ), and nanoparticle jetting (NPJ). BJ uses a print head that drops liquid binder droplets in layers, much like a 2D printer, to create components. In comparison to PBF and DED, BJ has two notable benefits. Initially, many print heads can operate concurrently at various places, enabling machines to achieve significantly quicker printing speeds [58]. Second, the machine can produce a limited quantity of identical components, and batch manufacturing operations can be facilitated by a big furnace for sintered components [59].

3.1 Powder Bed Fusion (PBF)

Powdered materials are utilized to layer an item in PBF additive manufacturing. Powdered materials may be metals or polymers. PBF begins with slicing. Computer-generated 3D files are sliced using slicing software. Slicing divides the 3D model into layers. Next, a thinner layer of material is applied on a build platform. The third phase combines powdered substances with a laser or electron beam based on the 3D model. In SLS, materials are thermally processed below their melting points using a laser, which allows the minerals to completely melt and combine to form dense metallic components. The fourth stage descends the construction platform by one layer following the starting layer of the sliced 3D model [60]. Until the entire thing is finished, steps two through four are repeated. The excess powders are collected for future use during this post-processing stage. If more heat treatment is required, the final product is then removed.

3.1.1 Selective Laser Melting (SLM)

One common powder bed fusion method in the AM industry is selective laser melting, which is well known for its quick production speed, adaptability to different materials, and capacity for powder recycling [61]. This method is widely used, especially in medical applications, to fabricate stainless steels and alloys of titanium and chromium. A blade applies metallic particles to the building platform over the substrate, which a laser beam melts. The laser beam fuses powders according to CAD data. Lowering the building platform exposes a fresh powder layer to the laser. Repeat until the component's height is established. SLM utilizes powders of 20–50 μm particle sizes and 15–150 μm layer thicknesses. Many elements must be considered for a perfect sample. Form, size, fluidity, laser scan speed, laser power, hatch overlaps, hatch distance, and hatch style are the main powder characteristics that affect component mechanical properties. Procedure occurs in a room. The chamber holds nitrogen or argon. Metal powder reactivity determines gas type. Also, the chamber is overpressure. During this procedure, the chamber's inert gas reduces oxidation. To slow down the pace at which the created components solidify, the substrate plate is heated to a temperature between 200 and 500 degrees Celsius [62–64]. Figure 1 shows the experimental setup of SLM process.

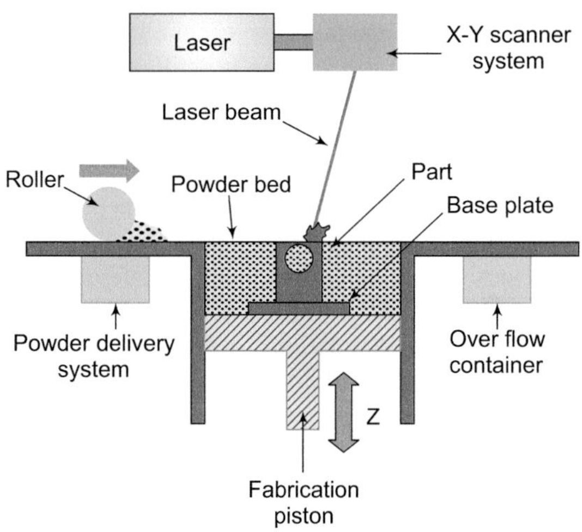

Fig. 1 Schematic diagram of SLM [65].

SLM technique provides greater mechanical characteristics than SLS since the powder is fully melted. Molten material surface tension affects process control, making SLM less efficient than SLS. This method is known for producing precise, complicated components with a 250–700 μm resolution. SLM causes structural flaws such as anisotropy, gas entrapment, absence of fusion, and micro-porosity, requires time-consuming post-processing, and uses more energy [65, 66].

3.1.2 Selective Laser Sintering (SLS)

SLS is an advanced AM method that selectively fuses tiny metallic powder particles layer by layer to produce intricate three-dimensional objects using a powerful laser [67]. Inert gases like argon or nitrogen prevent oxidation in the controlled environment. The laser scans and sinters powder layer locations are based on a CAD model. SLS partially melts metal powder, producing components with good mechanical strength but higher porosity than SLM. Processing metals include stainless steel, titanium, aluminum, and nickel-based alloys. It is challenging to produce intricate geometries, internal channels, and lightweight lattice structures with conventional techniques, but the methodology enables them to do so. Density and surface polish may need heat treatment, infiltration, or machining. The aerospace, automotive, biomedical, and tooling industries use SLS for rapid prototyping and low-volume manufacture of functional metal components. Complex component production has been improved by new 3D AM techniques for metal components, such as DMLS and SLS [68]. These technologies make it possible to use desktop production equipment rather than metal-forming tools to fabricate bespoke components with the required net shape.

SLS is highly appealing for implementation in diverse environments, including aircraft carriers and space vehicles, as it allows for the replacement of essential components without the necessity of maintaining extensive inventories or utilizing valuable space for a comprehensive metal fabrication facility onboard. Geometrically complex interior structures, like infill, can help reduce the weight of components. Temperature gradients produced by the laser scanning pattern cause thermal strains during the printing process. Because of instabilities such as buckling, these stresses can cause macro-scale deformations and lower strength, resulting in a loss of tolerance or possibly the component's full failure to print. Estimating SLS specimen residual stresses, deformation, and buckling failure modes may enhance design.

Prior to their actual manufacture, specimens may be quickly and affordably evaluated using computational simulations of the SLS process. Many modeling techniques have been developed, and the most of them are grouped based on how in-depth they are. Micro-scale techniques are used to study powder particle interaction and reaction to a laser heat source. Understanding the temperature profile, melt pool, heat-affected zone, and material characteristics is the main objective [69–71]. On the other hand, macro-scale simulations focus on stress, deformation, and temperature fields by increasing the level of abstraction [72, 73]. The bulk of these techniques rely on machining procedures [74] and welding research [75]. Although buckling effects are a major failure factor for thin structures such as walls, honeycomb configurations, and support columns, they are underutilized in SLS models.

3.1.3 Electron Beam Melting

EBM is a sophisticated AM method that selectively melts and amalgamates layers of metallic powder in a vacuum using a high-energy electron beam. The process begins with a thin layer of titanium alloys, cobalt–chromium, or Inconel powder on a build platform. An electron beam, produced by an electron cannon, is guided by electromagnetic coils to traverse the powder bed in accordance with a computerized CAD model. The beam's kinetic energy transforms into thermal energy upon impact, swiftly melting the powder and creating thick, totally bonded metal layers. The vacuum atmosphere inhibits oxidation and contamination, rendering EBM optimal for reactive materials. Preheating the powder bed reduces residual tension and deformation. In contrast to laser-based techniques, EBM provides enhanced penetration, increased build rates, and greater mechanical integrity; yet, it necessitates conductive materials and meticulous post-processing. This technique is extensively utilized in aircraft, biomedical implants, and high-performance engineering components.

EBM is a component of PBF metal additive manufacturing. In this process, metal powder is selectively melted and fused in a vacuum by the electron beam. Similar to the SLM process, the 3D target geometry is fabricated layer by layer within a powder bed, based on CAD data. Nonetheless, EBM and SLM procedures exhibit markedly distinct characteristics. In the EBM method, using an alternate energy source offers both benefits and drawbacks. The necessity of electrical conductivity in the EBM process restricts its applicability to metallic materials, whereas the SLM approach encompasses metals, polymers, and ceramics [76]. Furthermore, the potential for electrostatic charge in the powder may result in the failure of the additive manufacturing process. On the other hand, EBM offers new methods for material fusing and melting because of its remarkably high velocity limit (up to 105 m/s), inertia-free motion, accurate focus features, and quick movement capabilities [77].

In EBM, factors such as beam power, velocity, focus offset, inter-scan line distance, contour quantity, and hatching approach must be optimized to achieve a successful metal additive manufacturing process. To enhance surface quality, several methods such as quasi-beam-mode may be employed during the procedure. In comparison to traditional production techniques, the material characteristics in an optimized EBM process are equivalent or superior [77]. EBM technique can generate a range of metallic materials. Among the materials investigated are superalloys based on cobalt and nickel, titanium alloys, tool steel, aluminum alloys, and others. Due to its remarkable properties and important uses, Ti6Al4V alloy is being thoroughly studied utilizing both EBM and SLM processes. Like SLM, EBM components need semi-finished or finished machining processes to enhance their surface quality and dimensional accuracy. The process temperature and cooling rate are the main ways that the EBM process differs from SLM. Different properties may be displayed by the same material generated by the EBM and SLM procedures. The produced phases may differ in form and volume proportion. As a result, the printed material will have different properties [78]. There are significant discrepancies in the interactions between electron beams and powder and laser beams and powder. The penetrating depth of the electron beam is several times larger.

The choice between SLM and EBM procedures is contingent upon the requirements of end use. The SLM components exhibit superior surface roughness compared to EBM counterparts, attributable to reduced powder layer thickness, decreased scanning velocities, and smaller particle dimensions [78]. The machinability characteristics will change due to the aforementioned particular attributes of both additive manufacturing processes in comparison to conservative manufacturing approaches. Therefore, developing original machining strategies and perspectives is crucial to achieving the best possible AM component performance and increasing machining productivity. Table 2 gives the comparison of different Additive manufacturing techniques.

3.2 Direct Energy Deposition (DED)

DED is a complex AM method that melts material during deposition to fabricate, repair, and cover metal components. Lasers, electron beams, or plasma arcs liquefy metal feedstock—usually wire or powder—directly onto a substrate or component in DED. This technology allows exact regulation of material composition and microstructure, allowing the fabrication of high-performance alloys and graded materials. Metals often utilized in DED encompass titanium, stainless steel, Inconel, and cobalt–chromium alloys, esteemed for their strength and resistance to corrosion. DED has benefits including elevated deposition speeds, the capacity to repair costly components (e.g., turbine blades or molds), and the ability to fabricate big or intricate geometries. Challenges encompass surface roughness, the creation of residual stresses, and the necessity for post-processing to get precise tolerances.

The direct energy deposition technique, also called 3D welding, employs a laser light source to form near-net-shaped samples of wire or powder [79, 80]. This technique may create new metal components, such as nickel alloys, or repair flaws [81]. Because DED can make metal parts with fewer assembly components, it has a broad range of uses, including automotive (gear, knobs, gearboxes, and bumpers), medical (implants, scaffolds, and earplugs), and aerospace (blades, engine components, and brackets). Lasers, plasma arcs, and electron beams are examples of concentrated thermal energy sources that have the ability to melt or fuse materials [82]. From the base to the peak, the material is built and placed in layers. Repairing components, manufacturing complicated metallic products, and cladding in the use of coatings are all critical stages in the DED process. When compared to conventional methods for titanium alloys, printed items made with DED often exhibit better strength and less ductility at constant parameters [83].

Process parameters are crucial in DED since they influence the production of metallic components with varying mechanical characteristics. The laser energy should be sufficiently intense to adequately melt the powder for the purpose of melting and microwelding various wires [84, 85]. In contrast to SLM, support material is rarely employed in this approach; nevertheless, depending on the geometry of the item, it may be used to aid stability during printing. The most popular materials are metal alloys with diverse characteristics, such as powdered iron, copper, gold, titanium, and aluminum. Many factors influence mechanical properties and improve quality, but spot diameter, laser power, and laser scanning are important. To achieve a thick component without post-process densification,

Table 2 Comparison of metal additive manufacturing techniques, principles, parameters, applications, advantages, and limitations.

Technique	*Process Principle*	*Process Parameters*	*Applications*	*Advantages*	*Limitations*
Selective laser melting (SLM)	Employs a high-powered laser to completely melt and amalgamate metallic powder in a layer-by-layer manner within a powder bed.	Laser power, scanning velocity, layer thickness, and hatch spacing influence density, surface quality, and residual stresses.	Aerospace, medical implants, automotive lightweight parts.	High density and mechanical properties; complex geometry capability.	High cost; residual stresses; limited build size; requires support structures.
Selective laser sintering (SLS)	Employs a laser to sinter powdered substances (often polymers or metals) beneath their melting point.	Laser power, scan speed, powder size, bed temperature — affect porosity, surface quality, and accuracy.	Prototyping, tooling, functional polymer or metal parts.	No need for support structures (in some cases); good for polymers.	Lower mechanical strength; rough surface; limited material options.
Electron Beam Melting (EBM)	Uses an electron beam under vacuum to melt metallic powder layer by layer.	Beam current, scan speed, powder size, vacuum level — affect grain size, density, and residual stress.	Aerospace turbine blades, biomedical implants, high-value metal components.	High build rate, good material properties, vacuum environment reduces contamination.	Requires vacuum; limited to conductive materials; rough surface finish.
Direct Energy Deposition (DED)	Utilizes a concentrated energy source (laser, electron beam, or plasma arc) to liquefy powder or wire feedstock introduced through a nozzle.	Laser/e-beam power, feed rate, travel speed — influence bead geometry, dilution, and mechanical properties.	Repair and cladding of aerospace, automotive, and tooling components.	Can process multi-materials; suitable for repair applications; high deposition rate.	Lower resolution; requires post-processing; heat-affected zone issues.
Wire Arc Additive Manufacturing (WAAM)	Utilizes an electric arc as a thermal source to liquefy metal wire, applying it in successive layers.	The arc current, voltage, wire feed speed, and travel speed influence the deposition rate, bead size, and residual stress.	Large structural components in aerospace, marine, and energy industries.	High deposition rate; cost-effective; suitable for large parts.	Poor surface finish; low dimensional accuracy; high residual stresses.

increase laser power to reduce material deposition time and melt it completely. Also, lowering spot diameter increases molten metal energy densities, which promotes adhesion and minimizes porosity. Laser power and scanning velocity favorably affect surface roughness (SR) and elongated grain direction in printed components. Increased scan speed shortens laser energy transmission.

Consequently, when the laser energy is insufficient, melt spreading failure occurs. This suggests that increased cooling rates brought on by high scan speeds cause porosity and inclination to rise. Moreover, the substrate plate and nozzle torch are necessary for mechanical stability and efficient heat dissipation. The availability of powdered materials allows for the easy manufacturing of highly graded and functionalized components. Conversely, the primary benefit of wire form materials in the DED technique are their ease of use and storage, elevated deposition speeds, and high efficiency. Scan speed and laser power have an impact on defect characteristics, the production of intermetallic strengthening phases, the presence of undesired phases, microstructure development, and hence mechanical properties.

Researchers can create components of exceptional quality because of the DED method's inherent rapid heating and cooling speeds. Mechanical and surface quality depend on flow velocity, machine feed rate, and laser power for producing difficult geometric products [86]. DED uses pneumatic or gravity-applied powder feed in an inert gas stream. In powder-based DED, inadequate printing might leave unfused powder, contaminate material, and increase costs. Alloys made of titanium and aluminum are classified as extremely reactive powders that might catch fire. In addition, the repurposing of waste powder in powder-based processes is a challenging and inefficient endeavor. Printing samples with metal wire rather than powder is an alternative technique. DED has different techniques based on their characteristics. Differences fall under material, laser power, powder delivery, inert gas supply, and route planning control. Despite the similarities in features and printing, these technologies have been given a variety of names. Throughout the printing process, the laser is crucial for both melting and hardening the material.

In DED, layer thickness typically varies from 0.3 to 1 mm, however using a micro laser may produce a thickness of 700 to 800 μm. Temperature affects material deposition and component quality in this approach. A limited area is impacted by elevated laser energy density and solidifies rapidly. Due to the fact that it has an effect on the geometry of the part, this procedure makes repeatability more difficult [87]. When the parameters of the process are not optimized, the problem of insufficient repeatability happens. Inadequate melting of the wire or powder is the result of either insufficient laser power or high scan speed. As a consequence of this, the molten layer is insufficient and has not hardened to the appropriate degree. Simultaneously, as the machine liquefies the subsequent layer, it fails to adhere to the preceding layer. As a consequence of this, there is insufficient adhesion and bonding between the layers, which has an effect on the repeatability of the 3D printer. Figure 2 shows DLMD process.

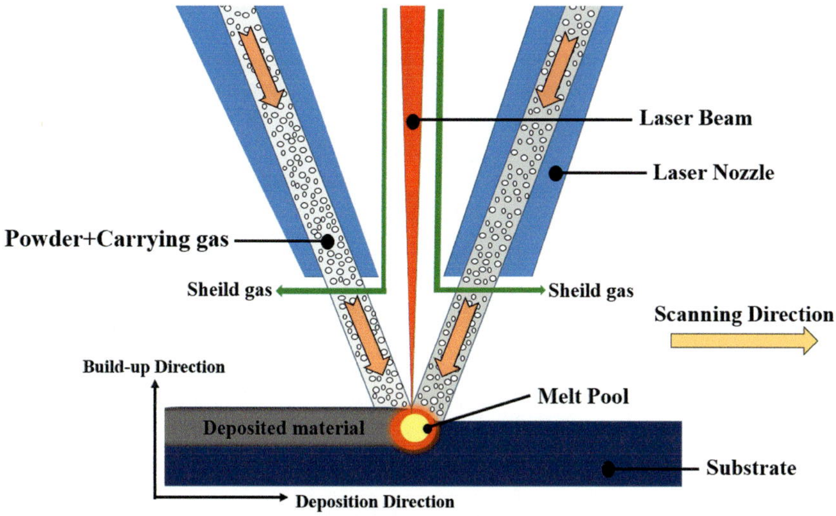

Fig. 2 Schematic diagram of DLMD technique [84].

3.3 Wire Arc-AM (WAAM)

DED technologies like WAAM have been popular in the previous decade. Traditional arc welding underpins WAAM [88, 89]. Electric arc heating is used in the WAAM process to melt the wire, which is then moved to the molten metal pool. The component is then formed layer by layer as it solidifies at the melt pool's edge [90]. The WAAM technology ensures higher manufacturing productivity and cheaper production costs by allowing for the creation of structures that are up to tens of meters in length and the fabrication of completely dense parts that are of substantial size [94, 95]. Additionally, it is useful for the maintenance and repair of components and parts that have been broken.

About 10% more money is spent directly on metal wire used in the WAAM process than on metal powder of the same weight [89]. There are three primary categories that may be applied to the WAAM technique [91, 92]. The category is decided by the type of arc-based welding used to melt the feed wire and create a three-dimensional object or structure. Plasma, gas metal, and gas tungsten arc welding (GTAW) are the three welding processes. A non-consumable tungsten electrode is used in the GTAW-based WAAM process, which results in the formation of an electric arc between the electrode and the substrate. In order to generate heat and achieve the appropriate geometry (breadth, height, cross-section, etc.) and mechanical qualities, a filler wire is injected laterally, melts as a result of the heat generated by the arc, and deposits itself on the specimen. Figure 3 shows the WAAM technique.

The plasma arc welding (PAW) and GTAW methods utilize non-consumable electrodes. PAW exhibits more efficiency than GTAW due to the confinement of the

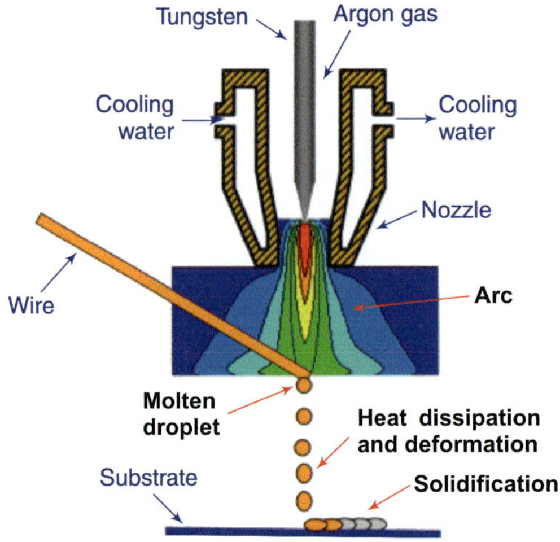

Fig. 3 Schematic of WAAM [106].

plasma arc within a nozzle. PAW possesses a distinct welding torch configuration compared to GTAW. When the water-cooled nozzle and tungsten electrode are connected, PAW generates an arc.

After entering the arcing zone of the torch, the inert gas undergoes ionization and transforms into plasma for the first time. After that, a very small hole is used to divert the plasma jet away from the substrate and toward the substrate. The filler wire causes the filler wire to melt because of the high temperature of the plasma jet. The molten pool is additionally protected from pollutants by shielding gas, which is employed in this process. On the other hand, when looking at PAW in comparison to gas metal arc welding (GMAW) and GTAW, the initial cost is significantly greater. WAAM mostly uses GMAW. This is mostly because of high material deposition rate, enhanced material utilization, and short lead time. GMAW uses wire as a consumable electrode. The wire material is melted by the arc that is created between the wire and specimen, and the molten wire material is subsequently deposited on the specimen.

Metal inert gas (MIG) and metal active gas (MAG) are used in GMAW. When using MIG, an inert gas such as argon or helium is utilized. MIG is utilized for nonferrous metals the majority of the time. CO_2 or mixture of inert gases, more precisely trimix consisting of oxygen, carbon dioxide, and argon, is utilized by MAG for the purpose of shielding. In the realm of ferrous metals, MAG is utilized the majority of the time.

GMAW method's material deposition rate is higher than that of other WAAM approaches. Short-circuiting, globular, spray, and pulsating spray are the four types of metal transfer that are included in gas metal arc welding (GMAW). Each metal transfer method has its own characteristics regarding the kind of electrode, welding current, and equipment [93]. Short-circuiting occurs during the typical short-circuiting transfer mode of the conventional GMAW welding process when the welding electrode comes into touch with the molten pool and submerges, producing a significant current flow. This causes the electrode wire to melt abruptly, which in turn causes droplets to transfer across the molten metal pool's surface tension. After the metal transfer process is completed, the arc ignites again and the high temperature process resumes.

4. Process Parameters of Selective Laser Melting

SLM is especially adept for metal AM, since it can create completely dense, high-strength components with superior mechanical characteristics and intricate geometries. A high-powered laser selectively fuses tiny metal powder layer by layer according to 3D CAD data, enabling precise control over microstructure and material composition [97, 98]. The fast melting and solidification rates attained during SLM result in fine grain structures, hence improving strength and hardness relative to traditionally cast components. SLM accommodates a diverse array of metallic materials, including stainless steel, titanium, aluminum, and nickel-based alloys, rendering it exceptionally adaptable for aerospace, biomedical, and automotive applications. The layer-wise manufacturing technique reduces material waste and facilitates the creation of lightweight lattice structures that are challenging or unattainable by subtractive methods. Furthermore, SLM facilitates design enhancement via topology optimization and the incorporation of internal channels for heat regulation. The technique has exceptional repeatability and customization capabilities, rendering it suitable for low-volume, high-value production. Nevertheless, it necessitates meticulous regulation of parameters like laser power, scanning velocity, and powder quality to guarantee defect-free constructions and optimal material characteristics.

It is necessary for the efficient operation of both automated and manual devices, including SLM, that process parameters be properly defined and maintained [96]. It is essential to make use of the parameters that are employed in SLM machines in order to get the desired results. Laser power, scanning speed, hatch spacing, layer thickness, and powder bed temperature are all factors that can have an impact on the rapid laser melting (SLM) process. Energy input, the rate at which material deposits, and the quality of printed components are all determined by these parameters. Finding the right combination of these components is vital to attaining the necessary surface quality, dimensional accuracy, and structural integrity in a 3D-printed component. The processing parameters of the SLM machine are extremely important for managing and minimizing adverse consequences, such as oxidation throughout the manufacturing process. A variety of additive manufacturing standards have been created in order to guarantee that the SLM process is both consistent and of high quality. One of the most notable standards is ASTM/ISO 52900, which details the processes that must be followed in order to construct three-dimensional items by successively depositing metal and nonmetal. The testing procedures, material specifications, and best practices that are included in these standards are designed to guarantee that the additive manufacturing processes produced are reliable and consistent. It is possible for manufacturers to produce dependable and high-quality outcomes in SLM by following these requirements and carefully managing the process parameters. This will ensure the production of 3D-printed components that are both resilient and useful [99, 100].

The quality and mechanical integrity of created components are greatly influenced by the parameters of additive manufacturing, particularly laser power, scan speed, hatch spacing, and layer thickness. The energy input, which controls the extent of powder melting and subsequent densification, is directly impacted by the laser's strength [102, 102]. Full fusing is ensured by optimal laser power, increasing density and mechanical strength. Overpower can cause thermal distortion, keyhole defects, or vaporization. Together with laser power, scan speed affects energy density; a fast scan speed reduces energy absorption, resulting in partial melting and porosity, whereas a slower scan speed increases energy input, potentially creating residual stress. The overlap between adjacent laser tracks is influenced by the hatch spacing; smaller hatch spacing improves bonding and surface uniformity but raises the risk of overheating, while larger hatch spacing leads to insufficient melting between tracks, reducing density and ultimate tensile strength (UTS). Vertical resolution and melt pool stability are determined by layer thickness; bigger layers speed up build times but weaken bonds, while thinner layers offer better microstructural control and higher density.

There is a notable connection between these process parameters and performance parameters such as density, surface roughness, ultimate tensile strength, and residual stress. While surface roughness is decreased when melting and solidification processes are constant, increased density is frequently associated with optimal energy density, which is achieved by a harmonious combination of laser power and scanning velocity. While inadequate energy produces rough and porous surfaces, excessive energy input exacerbates surface flaws through balling and spattering. Increased density and a refined microstructure enhance UTS; these conditions are facilitated by power and speed combinations that are tuned. On the other hand, abrupt temperature gradients and fast cooling rates cause residual stresses, which may be reduced by adjusting laser settings to promote smooth thermal transitions. Stress relaxation behavior is strongly impacted by changes to hatch spacing and layer thickness, which also affect heat dispersion and solidification dynamics. The overall effectiveness, mechanical dependability, and structural integrity of additively manufactured components are determined by the intricate interplay between process and performance parameters.

5. Process Parameters Optimization of Selective Laser Melting

Metal AM technique SLM uses a high-energy laser beam to melt metal powder in layers in order to construct complex components. Process parameters such as laser power, scanning speed, hatch spacing, and layer thickness influence the

Table 3 Process parameters and performance parameters of selective laser melting process.

Expt Run	*Process Parameters*				*Performance Parameters*			
	Laser Power (W)	*Scan Speed (mm/sec)*	*Hatch Spacing (mm)*	*Layer Thickness (mm)*	*Density (%)*	*Surface roughness (Ra) (μm)*	*UTS (MPa)*	*Residual Stress (Mpa)*
1	200	600	0.08	0.02	95.888	18.02	867.3	122.2
2	200	800	0.12	0.03	99.2	14	874.9	58
3	200	1000	0.16	0.04	96.238	20.44	827	98.4
4	300	600	0.12	0.04	95.288	22.02	818.8	134.2
5	300	800	0.16	0.02	98.886	13.02	894.4	80.2
6	300	1000	0.08	0.03	96.137	19.11	847.8	117.1
7	400	600	0.16	0.03	95.588	20.02	843	146.2
8	400	800	0.08	0.04	93.175	25.54	800.6	177.4
9	400	1000	0.12	0.02	97.051	16.08	877.8	118.8

component quality and mechanical properties. Optimization of above process parameters is important for good surface finish, higher density, precise dimensions, and enhanced mechanical properties. Improper selection of process parameters can result in defects like distortion, porosity, and insufficient melting. By systematically modifying these variables and analyzing their impacts, optimal combinations may be helpful to improve performance. Statistical and multi-criteria decision-making techniques are frequently used to manage the complex and interrelatedness of factors. Therefore, optimizing process factors is crucial for enhancing the efficiency and dependability of components produced using SLM.

Robust multi-criteria decision-making techniques used to effectively optimize SLM process parameters include grey relational analysis (GRA), technique for order preference by similarity to ideal solution (TOPSIS), multi-objective optimization based on ratio analysis (MOORA), and complex proportional assessment (COPRAS). The abovementioned techniques address many performance attributes, including density, surface roughness, and tensile strength, facilitating a balanced optimization approach. In the context of uncertain and limited data circumstances, GRA quantifies the correlations between experimental results and ideal outcomes. TOPSIS provides a rational and cogent framework for making decisions by evaluating parameter combinations based on how close they are to an ideal solution. MOORA enhances optimization through ratio analysis, resulting in computing efficiency and ease of interpretation [104, 105]. Collectively, these methodologies provide the identification of the ideal parameter configuration among the four parameters and three levels, hence enhancing component quality and process consistency.

5.1 Process Parameters Optimization with GRA

A "grey system" is a system that exhibits ambiguity and imperfect understanding. The current state of ambiguity will be addressed by grey systems, which will provide a number of different potential answers. The foundation of the theory, which is known as grey relational analysis, has the potential to be enhanced in order to manage the intricate connections that exist between the prescribed performance indicators [103]. The grey relational grade, often known as the GRG, is an indication of performance characteristics that is utilized in this inquiry. The use of grey relational analysis as a technique for analyzing processes based on a wide variety of performance measures has grown increasingly common over the course of the past several years.

The optimization of a single answer grey relational grade can be used to simplify the hard problem of multiple response optimization in grey relational analysis. The components and their corresponding levels used in the experiment are listed in Table 3. To guarantee consistency across a number of criteria, the normalization process comprises converting data into a dimensionless format. The "larger-the-better" and "smaller-the-better" methods are two common normalization strategies. The aim of this chapter is to limit surface roughness and residual stresses, maximize density, and maximize ultimate tensile strength.

Larger-the-better:

$$_{ij} \quad \frac{y_{ij} \quad \min(y_{ij})}{\max(y_{ij}) \quad \min(y_{ij})} \qquad \text{...(1)}$$

Smaller-the-better:

$$x_{ij} = \frac{\max(y_{ij}) - y_{ij}}{\max(y_{ij}) - \min(y_{ij})} \qquad \text{...(2)}$$

where y_{ij} is the original data.

Table 4 shows normalized density, residual stresses, surface roughness, and ultimate tensile strength. After data pre-processing, a response j from experiment I performed best when its value x_{ij} is equal to or nearer 1 than any other trial. The reference sequence $X0$ is $(x_{01}, x_{02}, x_{0j}, \ldots x_{0n}) = (1, 1, \ldots, 1)$. The purpose of X_{0j} is to discover which experiment's comparability sequence corresponds to the reference sequence. Table 5 shows the deviation sequences.

Table 4 Normalized matrix.

Expt Run	*Density (%)*	*Surface roughness (Ra) (μm)*	*UTS (MPa)*	*Residual Stress (MPa)*
1	0.4503	0.6006	0.7111	0.4623
2	1.0000	0.9217	0.7921	1.0000
3	0.5084	0.4073	0.2814	0.6616

Contd.

Table 4 *Contd.*

Expt Run	*Density (%)*	*Surface roughness (Ra) (μm)*	*UTS (MPa)*	*Residual Stress (MPa)*
4	0.3507	0.2812	0.1940	0.3618
5	0.9479	1.0000	1.0000	0.8141
6	0.4916	0.5136	0.5032	0.5050
7	0.4005	0.4409	0.4520	0.2613
8	0.0000	0.0000	0.0000	0.0000
9	0.6433	0.7556	0.8230	0.4908

Table 5 Deviation sequences.

Expt Run	*Density (%)*	*Surface roughness (Ra) (μm)*	*UTS (MPa)*	*Residual Stress (MPa)*
1	0.5497	0.3994	0.2889	0.5377
2	0.0000	0.0783	0.2079	0.0000
3	0.4916	0.5927	0.7186	0.3384
4	0.6493	0.7188	0.8060	0.6382
5	0.0521	0.0000	0.0000	0.1859
6	0.5084	0.4864	0.4968	0.4950
7	0.5995	0.5591	0.5480	0.7387
8	1.0000	1.0000	1.0000	1.0000
9	0.3567	0.2444	0.1770	0.5092

Next, the grey relational coefficient is utilized to calculate the difference between x_{ij} and x_{0j}. When x_{ij} and x_{0j} are closer together, the grey connection coefficient increases.

$$\Gamma\,(x_{0j} - x_{ij})\ \frac{(\Delta_{\min} + \xi\Delta_{\max})}{(\Delta_{ij} + \xi\Delta_{\max})} \text{ for } i = 1, 2, \ldots\ldots, m \text{ and } j = 1, 2, \ldots\ldots n \quad ...(3)$$

where $\gamma(x_{0j} - x_{ij})$ is the grey relational coefficient between x_{ij} and x_{0j}

$$\Delta_{\min} = \min\{x_{ij}, i = 1, 2, \ldots, m; j = 1, 2\ldots\ldots n\} \quad ...(4)$$

$$\Delta_{\max} = \max\{x_{ij}, i = 1, 2, \ldots, m; j = 1, 2\ldots\ldots n\} \quad ...(5)$$

ξ is the distinguishing coefficient. The value of ξ is assumed as 0.5.

The grey relational grade is calculated by using following equation:

$$\gamma\,(x_0, x_i) = \sum_{j=1}^{n} w_j y(x_{0j}, x_{ij}) \text{ for } i = 1, 2, \ldots.., m \quad ...(6)$$

where $\sum_{j=1}^{n} w_{jj} = 1$

$\gamma(x_0, x_i)$ is the grey relational grade

Table 6 Gray relational coefficients and gray relational grade.

Expt Run	*Density (%)*	*Surface roughness (Ra) (μm)*	*UTS (MPa)*	*Residual Stress (MPa)*	*Grey relational grade*	*Rank*
1	0.4763	0.5560	0.6338	0.4818	0.5370	4
2	1.0000	0.8646	0.7063	1.0000	0.8927	2
3	0.5042	0.4576	0.4103	0.5964	0.4921	6
4	0.4350	0.4102	0.3829	0.4393	0.4169	8
5	0.9056	1.0000	1.0000	0.7289	0.9086	1
6	0.4958	0.5069	0.5016	0.5025	0.5017	5
7	0.4548	0.4721	0.4771	0.4037	0.4519	7
8	0.3333	0.3333	0.3333	0.3333	0.3333	9
9	0.5836	0.6717	0.7386	0.4954	0.6223	3

Table 6 shows that higher grey relationship grades indicate better achievement in numerous categories. Thus, the optimal process parameters are: 300 W laser, 800 mm/sec scan, 0.16 mm hatch spacing, 0.02 mm layer thickness.

5.2 Process Parameters Optimization with TOPSIS

TOPSIS is a multi-criteria decision-making (MCDM) system that ranks options by their proximity to an ideal answer and their distance from a negative-ideal solution.

Step 1: The decision matrix for TOPSIS is shown in Table 3.
Step 2: Normalize the decision matrix.

Normalized decision matrix is found out from the following equation 7 and shown in Table 7.

$$r_{ij} = \frac{x_{ij}}{\sqrt{\sum_{i=1}^{m} x_{ij}^2}} \quad ...(7)$$

Table 7 TOPSIS normalized matrix.

Expt Run	*Density (%)*	*Surface roughness (Ra) (μm)*	*UTS (MPa)*	*Residual Stress (Mpa)*
1	0.3316	0.3151	0.3398	0.3350
2	0.3430	0.2448	0.3428	0.1590

Contd.

Table 7 *Contd.*

Expt Run	*Density (%)*	*Surface roughness (Ra) (μm)*	*UTS (MPa)*	*Residual Stress (Mpa)*
3	0.3328	0.3575	0.3241	0.2697
4	0.3295	0.3851	0.3208	0.3679
5	0.3419	0.2277	0.3505	0.2198
6	0.3324	0.3342	0.3322	0.3210
7	0.3305	0.3501	0.3303	0.4007
8	0.3222	0.4466	0.3137	0.4863
9	0.3356	0.2812	0.3440	0.3256

Step 3: Weighted normalized matrix

Weighted normalized matrix is calculated by giving equal weight (0.25) to each parameter.

Multiply each normalized value by the weight of the corresponding criterion:

$$v_{ij} = w_j \times r_{ij} \qquad \text{...(8)}$$

where w_j = weight of criterion of j

Step 4: Distance from ideal and negative-ideal are calculated from equation 9 and 10.

Distance to ideal

$$D_i^+ = \sqrt{\sum_{j=1}^{n} (v_{ij} - v_j^+)^2} \qquad \text{...(9)}$$

Distance to negative-ideal

$$D_i^- = \sqrt{\sum_{j=1}^{n} (v_{ij} - v_j^-)^2} \qquad \text{...(10)}$$

Step 5: Relative closeness to the ideal solution is calculated from equation 11 and shown in Table 8.

$$C_i = \frac{D_i^-}{D_i^+ + D_i^-} \qquad \text{...(11)}$$

Higher C_i indicates closer to the ideal solution.

Table 8 Normalized weighted matrix, Euclidean distance from ideal best and ideal best, performance score and rank.

Expt Run	*Density (%)*	*Surface roughness (Ra) (μm)*	*UTS (MPa)*	*Residual Stress (Mpa)*	*Euclidean distance from ideal best*	*Euclidean distance from ideal worst*	*Performance Score (Pi)*	*Rank*
1	0.0829	0.0788	0.0850	0.0837	0.0493	0.0506	0.5066	6
2	0.0858	0.0612	0.0857	0.0397	0.0047	0.0965	0.9537	1
3	0.0832	0.0894	0.0810	0.0674	0.0432	0.0587	0.5757	3
4	0.0824	0.0963	0.0802	0.0920	0.0659	0.0335	0.3368	7
5	0.0855	0.0569	0.0876	0.0550	0.0152	0.0868	0.8509	2
6	0.0831	0.0835	0.0831	0.0802	0.0488	0.0503	0.5076	5
7	0.0826	0.0875	0.0826	0.1002	0.068	0.0326	0.3239	8
8	0.0805	0.1117	0.0784	0.1216	0.099	0	0	9
9	0.0839	0.0703	0.0860	0.0814	0.0438	0.0582	0.5706	4

Step 6: Rank the Alternatives based on C_i higher is better and shown in Table 8.

Table 8 shows that the optimum process parameters are 200 W laser power, 800 mm/sec scan speed, 0.12 mm hatching spacing, and 0.03 mm layer thickness.

5.3 Process Parameters Optimization with COPRAS

Process parameters optimization using COPRAS (Complex Proportional Assessment) is a systematic MCDM technique used to identify the most effective combination of process variables. The COPRAS approach was proposed by Kaklauskas, Sarka and Zavadskas [107–110] in 1994. It evaluates alternatives by considering both beneficial and non-beneficial criteria, assigning weights based on their relative importance. COPRAS normalizes the decision matrix and calculates priority values to rank different parameter settings objectively. This method helps in balancing conflicting performance measures such as quality, productivity, and cost. By providing a clear quantitative ranking, COPRAS supports informed and reliable decision-making in manufacturing and engineering processes. The detailed steps are as follows.

5.3.1 Implementation of COPRAS

Step 0: Define Criteria Type

- Beneficial (Maximize): Density (%) and UTS (MPa)
- Non-Beneficial (Minimize): Surface Roughness Ra (μm) and Residual Stress (MPa)

Step 1: Decision Matrix (Performance Parameters) (Table 9).

Table 9 Decision matrix for COPRAS.

Run	*Density*	*Ra*	*UTS*	*Residual Stress*
1	95.888	18.02	867.3	122.2
2	99.200	14.00	874.9	58.0
3	96.238	20.44	827.0	98.4
4	95.288	22.02	818.8	134.2
5	98.886	13.02	894.4	80.2
6	96.137	19.11	847.8	117.1
7	95.588	20.02	843.0	146.2
8	93.175	25.54	800.6	177.4
9	97.051	16.08	877.8	118.8

Step 2: Normalization

$$\text{COPRAS normalization formula: } \frac{w_{ij}}{\sum_{i=1}^{n} x_{ij}} \quad ...(12)$$

Normalized matrix is found from equation 12 and shown in Table 10.

Table 10 Normalized matrix.

Run	*Density*	*Ra*	*UTS*	*Residual Stress*
1	0.1105	0.1071	0.1134	0.1161
2	0.1143	0.0832	0.1143	0.0551
3	0.1109	0.1215	0.1081	0.0935
4	0.1098	0.1309	0.1070	0.1275
5	0.1140	0.0774	0.1169	0.0762
6	0.1108	0.1136	0.1108	0.1113
7	0.1101	0.1190	0.1101	0.1389
8	0.1074	0.1518	0.1046	0.1686
9	0.1119	0.0956	0.1147	0.1129

Step 3: Weight Assignment: Equal weight (0.25) is given to each parameter.

Step 4: Weighted normalized matrix is found from equation 13 and represented in Table 11

$$n_{ij} = W_j * n_{ij} \quad ...(13)$$

Table 11 Weighted normalized matrix, beneficial (S^+) and non-beneficial (S^-) parameters, relative significance (Qi), utility degree (Ui), and rank.

Run	*Density*	*Ra*	*UTS*	*Residual Stress*	S^+	S^-	*Qi*	*Ui (%)*	*Rank*
1	0.0276	0.0268	0.0284	0.0290	0.0560	0.0558	0.675	63.9	5
2	0.0286	0.0208	0.0286	0.0138	0.0572	0.0346	**1.057**	**100**	**1**
3	0.0277	0.0304	0.0270	0.0234	0.0547	0.0538	0.697	65.9	4
4	0.0275	0.0327	0.0268	0.0319	0.0543	0.0646	0.588	55.6	8
5	0.0285	0.0193	0.0292	0.0191	0.0577	0.0384	0.958	90.6	2
6	0.0277	0.0284	0.0277	0.0278	0.0554	0.0562	0.671	63.5	6
7	0.0275	0.0298	0.0275	0.0347	0.0550	0.0645	0.590	55.8	7
8	0.0268	0.0379	0.0261	0.0422	0.0529	0.0801	0.484	45.8	9
9	0.0280	0.0239	0.0287	0.0282	0.0567	0.0521	0.721	68.2	3

Step 5: Calculate S^+ and S^-: Beneficial (S^+) and non-beneficial (S^-) parameters are found out form equation 14 and 15 and shown in Table 11.

Beneficial (Density, UTS)

$$B_i = \sum_{j=1}^{k} N_{ij} \qquad ...(14)$$

Non-Beneficial (Ra, Residual Stress)

$$C_i = \sum_{j=k+1}^{m} N_{ij} \qquad ...(15)$$

Step 6: Relative significance (Q_i): Relative Significance is calculated from equation 16

and shown in Table 11.

$$Q_i = B_i + \frac{\min(C_i) * \sum_{i=1}^{n} C_i}{C_i * \sum_{i=1}^{n} \left(\frac{\min(c_i)}{c_i} \right)} \qquad ...(16)$$

Step 7: Utility degree (U_i): Utility degree is evaluated from equation 17 and represented in Table 11.

$$UD_i = \frac{Q_i}{\max(Q_i)} * 100 \qquad ...(17)$$

Step 8: Rank: Rank for COPRAS is shown in Table 11.

Final Result: From Table 11 it can be seen that optimum combination of process parameters for best performance outcomes is for experimental run 2. Optimum process parameters are laser power 200 W, scanning speed 800 mm/sec, hatch spacing 0.12 mm and layer thickness 0.03 mm.

5.4 Multi-Objective Optimization Ratio Analysis (MOORA)

Multi objective optimization ratio analysis (MOORA) is a decision-making technique used to evaluate and rank alternatives when multiple, often conflicting, criteria are involved. An objective (non-subjective) technique has been introduced by Brauers et al. [111]. It works by normalizing the performance values of each alternative to make criteria comparable, regardless of their units of measurement. MOORA then separates beneficial criteria (to be maximized) from non-beneficial criteria (to be minimized) and combines them into a single assessment score. This structured approach helps reduce bias and improves clarity in complex decision scenarios. The method is valued for its simplicity, computational efficiency, and strong theoretical foundation. As a result, MOORA is widely applied in fields such as engineering design, supply chain management, and resource allocation.

5.4.1 Step-by-Step MOORA Implementation

Step 1: Identification of alternatives and criteria: Beneficial criteria are density and UTS while non-beneficial criteria are surface roughness and residual stresses. Beneficial criteria are maximized and non-beneficial criteria are minimized.

Step 2: Decision Matrix: Decision matrix for MOORA is shown in Table 12.

Table 12 Decision matrix for MOORA.

Run	*Density*	*Ra*	*UTS*	*Residual Stress*
1	95.888	18.02	867.3	122.2
2	99.200	14.00	874.9	58.0
3	96.238	20.44	827.0	98.4
4	95.288	22.02	818.8	134.2
5	98.886	13.02	894.4	80.2
6	96.137	19.11	847.8	117.1
7	95.588	20.02	843.0	146.2
8	93.175	25.54	800.6	177.4
9	97.051	16.08	877.8	118.8

Step 3: Normalization of decision matrix: Normalized decision matrix is calculated from equation 18 and shown in Table 13.

MOORA uses **vector normalization:** $X_{ij} = \frac{X_{ij}}{\sqrt{\sum_{i=1}^{m}(X_{ij})^2}}$ $(j = 1 \ldots n)$ ...(18)

Table 13 Normalized decision matrix.

Run	*Density*	*Ra*	*UTS*	*Residual Stress*
1	0.3302	0.3057	0.3396	0.3316
2	0.3417	0.2376	0.3424	0.1574
3	0.3314	0.3467	0.3237	0.2671
4	0.3283	0.3736	0.3205	0.3643
5	0.3407	0.2209	0.3500	0.2176
6	0.3310	0.3243	0.3319	0.3178
7	0.3292	0.3397	0.3297	0.3967
8	0.3209	0.4333	0.3133	0.4815
9	0.3343	0.2728	0.3435	0.3224

Step 4: Weight assignment: Weighted normalized matrix is found out by giving the equal weight (0.25) to each parameter. Normalized matrix is calculated from equation 19 and represented in Table 14.

Step 5: Weighted Normalized Matrix

$$y_{ij} = x_i * w_j \quad ...(19)$$

Table 14 Weighted normalized matrix, MOORA assessment value (Zi) and rank.

Run	*Density*	*Ra*	*UTS*	*Residual Stress*	*Zi*	*Rank*
1	0.0826	0.0764	0.0849	0.0829	0.0082	5
2	0.0854	0.0594	0.0856	0.0394	**0.0722**	**1**
3	0.0828	0.0867	0.0809	0.0668	0.0102	4
4	0.0821	0.0934	0.0801	0.0911	−0.0223	8
5	0.0852	0.0552	0.0875	0.0544	0.0631	**2**
6	0.0827	0.0811	0.0830	0.0795	0.0051	6
7	0.0823	0.0849	0.0824	0.0992	−0.0194	7
8	0.0802	0.1083	0.0783	0.1204	−0.0702	9
9	0.0836	0.0682	0.0859	0.0806	0.0207	**3**

Step 6: MOORA Assessment Value (Zi): Assessment value is calculated from equation 20 and shown in Table 14.

$$Z_i = (\text{Density} + \text{UTS}) - (\text{Ra} + \text{Residual Stress}) \quad ...(20)$$

Step 7: Ranking of Alternatives: Finally, MOORA rank is found out and shown in Table 14. From Table 14 it can be seen that optimum combination of process parameters for best performance outcomes is for experimental run 2. Optimum process parameters are laser power 200 W, scanning speed 800 mm/sec, hatch spacing 0.12 mm, and layer thickness 0.03 mm.

Table 15 Rank as per different optimization techniques.

Expt run	*Laser Power (W)*	*Scan Speed (mm/sec)*	*Hatch Spacing (mm)*	*Layer Thickness (mm)*	*Laser Power (W)*	*GRA*	*TOPSIS*	*COPRAS*	*MOORA*
1	200	600	0.08	0.02	200	4	6	5	5
2	**200**	**800**	**0.12**	**0.03**	**200**	**2**	**1**	**1**	**1**
3	200	1000	0.16	0.04	200	6	3	4	4
4	300	600	0.12	0.04	300	8	7	8	8
5	**300**	**800**	**0.16**	**0.02**	**300**	**1**	**2**	**2**	**2**
6	300	1000	0.08	0.03	300	5	5	6	6
7	400	600	0.16	0.03	400	7	8	7	7
8	400	800	0.08	0.04	400	9	9	9	9
9	400	1000	0.12	0.02	400	3	4	3	3

From Table 15, it is observed that according to the TOPSIS, COPRAS, and MOORA optimization techniques, experimental run 2 yields the optimum process parameters and provides the best results. The second-best performance is obtained in experimental run 5. However, based on the GRA technique, experimental run 5 offers the optimum process parameters, while experimental run 2 is identified as the second-best option.

6. Conclusion

- It was concluded that powder-bed fusion techniques (SLM, SLS, EBM) are best suited for producing highly complex and high-precision components, especially for aerospace and biomedical applications, but they come with high costs, size limitations, and residual stress or surface finish challenges. Directed energy and wire-based processes (DED, WAAM) excel in large-scale manufacturing, repair, and high deposition rates, offering better cost efficiency and scalability, though at the expense of lower resolution, surface quality, and dimensional accuracy.
- MCDM techniques, namely GRA, TOPSIS, COPRAS, and MOORA, were effectively used to optimize SLM process parameters considering multiple performance criteria. Strong consistency among TOPSIS, COPRAS,

and MOORA methods are observed which give same optimized process parameters. The close agreement among TOPSIS, COPRAS, and MOORA confirms the robustness and reliability of these methods for SLM process parameter optimization.

- As per TOPSIS, COPRAS, and MOORA optimization techniques, the optimized process parameters are 200 W laser power, 800 mm/s scan speed, 0.12 mm hatch spacing, and 0.03 mm layer thickness for best performance parameters. The corresponding optimized performance parameters are density 99.2%, surface roughness 14 μm, UTS 874.9 MPa, and residual stress 58 MPa.
- As per GRA optimization technique, the optimized process parameters are 300 W laser power, 800 mm/s scan speed, 0.16 mm hatch spacing, and 0.02 mm layer thickness for best performance parameters. The corresponding optimized performance parameters are density 98.886%, surface roughness 13.02 μm, UTS 894.4 MPa, and residual stress 80.2 MPa. The difference observed in GRA and other techniques ranking highlights the importance of using different MCDM techniques to get diverse weighting and relational assessment perspectives.
- The comparative optimization findings indicate that a multi-criteria decision-making framework offers a thorough and dependable strategy for selecting optimal SLM process parameters, enhancing part quality and process efficiency in metal additive manufacturing.

References

[1] Vafadar, Ana, Ferdinando Guzzomi, Alexander Rassau and Kevin Hayward. "Advances in metal additive manufacturing: a review of common processes, industrial applications, and current challenges." *Applied Sciences* 11, no. 3 (2021): 1213.

[2] Citarella, Roberto and Venanzio Giannella. "Additive manufacturing in industry." *Applied Sciences* 11, no. 2 (2021): 840.

[3] Pesode, Pralhad and Shivprakash Barve. "Additive manufacturing of magnesium alloys and its biocompatibility." *Bioprinting* 36 (2023): e00318.

[4] Wong, Kaufui V. and Aldo Hernandez. "A review of additive manufacturing." *International Scholarly Research Notices* 2012, no. 1 (2012): 208760.

[5] Rejeski, David, Fu Zhao and Yong Huang. "Research needs and recommendations on environmental implications of additive manufacturing." *Additive Manufacturing* 19 (2018): 21–28.

[6] Sala, Dariusz and Maria Richert. "Perspectives of Additive Manufacturing in 5.0 Industry." Materials 18, no. 2 (2025): 429.

[7] Pesode, Pralhad and Shivprakash Barve. "Additive manufacturing of metallic biomaterials: sustainability aspect, opportunity and challenges." *Journal of Industrial and Production Engineering* 40, no. 6 (2023): 464–505.

[8] Joshua, Rajan John Nekin, Sakthivel Aravind Raj, Mohamed Thariq Hameed Sultan, Andrzej Łukaszewicz, Jerzy Józwik, Zbigniew Oksiuta, Krzysztof Dziedzic, Arkadiusz Tofil and Farah Syazwani Shahar. "Powder bed fusion 3D printing in precision manufacturing for biomedical applications: a comprehensive review." *Materials* 17, no. 3 (2024): 769.

[9] Khaimovich, Alexander, Andrey Balyakin, Maxim Oleynik, Artem Meshkov and Vitaly Smelov. "Optimization of process parameters for powder bed fusion additive manufacturing using a linear programming method: a conceptual framework." *Metals* 12, no. 11 (2022): 1976.

[10] Tusher, Md Mehide Hasan and Ayhan Ince. "Comprehensive review of fabrication process parameters influencing defect formation in laser powder bed fused (L-PBF) Al-Si alloys." *Materials & Design* (2025): 114374.

[11] Maurya, Anoop Kumar, Pasupuleti Lakshmi Narayana, Jong-Taek Yeom, Jae-Keun Hong and Nagireddy Gari Subba Reddy. "Artificial neural network modeling of Ti-6Al-4V alloys to correlate their microstructure and mechanical properties." *Materials* 18, no. 5 (2025): 1099.

[12] Fan, Zhechao and Hongwei Feng. "Study on selective laser melting and heat treatment of Ti-6Al-4V alloy." *Results in Physics* 10 (2018): 660–664.

[13] Reda, Reham, Adel Nofal and Abdel-Hamid Hussein. "Effect of single and duplex stage heat treatment on the microstructure and mechanical properties of cast Ti–6Al–4V alloy." Metallography, *Microstructure and Analysis* 2, no. 6 (2013): 388–393.

[14] Herzog, Dirk, Vanessa Seyda, Eric Wycisk and Claus Emmelmann. "Additive manufacturing of metals." *Acta Materialia* 117 (2016): 371–392.

[15] Bartkowiak, Konrad, Sven Ullrich, Thomas Frick and Michael Schmidt. "New developments of laser processing aluminium alloys via additive manufacturing technique." *Physics procedia* 12 (2011): 393–401.

[16] Bartkowiak, Konrad, Sven Ullrich, Thomas Frick and Michael Schmidt. "New developments of laser processing aluminium alloys via additive manufacturing technique." *Physics procedia* 12 (2011): 393–401.

[17] Schmidtke, Katja, Frank Palm, A. Hawkins and Claus Emmelmann. "Process and mechanical properties: applicability of a scandium modified Al-alloy for laser additive manufacturing." *Physics Procedia* 12 (2011): 369–374.

[18] Fulcher, Benjamin A., David K. Leigh and Trevor J. Watt. "Comparison of AlSi10Mg and Al 6061 processed through DMLS." (2014).

[19] Louvis, Eleftherios, Peter Fox and Christopher J. Sutcliffe. "Selective laser melting of aluminium components." *Journal of Materials Processing Technology* 211, no. 2 (2011): 275–284.

[20] Bajaj, Priyanshu, Avinash Hariharan, Anoop Kini, Philipp Kürnsteiner, Dierk Raabe and Eric Aimé Jägle. "Steels in additive manufacturing: A review of their microstructure and properties." *Materials Science and Engineering*: A 772 (2020): 138633.

[21] Abdelaal, Ahmed F., Avijit Chakrobarty, Md Nazmus Sakib, Arnab Mustafi Arka and Ezazul Haque Sabuz. "Porosity, residual stress, wear properties and impact toughness of additively manufactured low-alloy steel: A review." *Next Materials* 9 (2025): 101288.

[22] Yao, Ju, Qiyang Tan, Jeffrey Venezuela, Andrej Atrens and Ming-Xing Zhang. "Additive manufacturing of high-strength low-alloy AISI 4340 steel with an optimal strength-ductility-toughness trade-off." *Additive Manufacturing* 94 (2024): 104496.

[23] You, Hang, Guang Fu, Shangkun Jin, Shengliang Luo, Kaifei Zhang, Zhihao Ren, Taiqian Mo and Zhengwen Zhang. "Advances in laser powder bed fusion of hard-to-weld superalloys: a comprehensive review." *Materials & Design* (2025): 114932.

[24] Ling, Li-Shi-Bao, Zheng Yin, Zhi Hu, Jin-Hui Liang, Zhi-Yong Wang, Jun Wang and Bao-De Sun. "Effects of the γ''-Ni3Nb phase on mechanical properties of Inconel 718 Superalloys with different heat treatments." *Materials* 13, no. 1 (2019): 151.

[25] Sun, Hongyuan, Dexin Ma, Yunxing Zhao, Jianhui Wei, Xiaoyi Gong and Zhongyuan Sun. "Dendrite Structure Refinement and Mechanical Property Improvement of a Single-Crystal Superalloy." Metals 15, no. 3 (2025): 295.

[26] del Bosque, Antonio, Pablo Fernández-Arias and Diego Vergara. "Advances in the Additive Manufacturing of Superalloys." *Journal of Manufacturing and Materials Processing* (2025).

[27] Pröbstle, M., S. Neumeier, P. Feldner, R. Rettig, H. E. Helmer, R. F. Singer and M. Göken. "Improved creep strength of nickel-base superalloys by optimized γ/γ' partitioning behavior of solid solution strengthening elements." *Materials Science and Engineering*: A 676 (2016): 411–420.

[28] Fleischmann, Ernst, Michael K. Miller, Ernst Affeldt and Uwe Glatzel. "Quantitative experimental determination of the solid solution hardening potential of rhenium, tungsten and molybdenum in single-crystal nickel-based superalloys." *Acta Materialia* 87 (2015): 350–356.
[29] Štamborská, Michaela, Juraj Lapin and Kateryna Kamyshnykova. "Preparation, microstructure and mechanical behaviour of Ni3Al-based superalloy reinforced with carbide particles." Intermetallics 149 (2022): 107667.
[30] Cartón-Cordero, Marta, Mónica Campos, Lisa P. Freund, Markus Kolb, Steffen Neumeier, Mathias Göken and Jose M. Torralba. "Microstructure and compression strength of Co-based superalloys hardened by γ' and carbide precipitation." *Materials Science and Engineering*: A 734 (2018): 437–444.
[31] Li, Shuai, Qingsong Wei, Yusheng Shi, Zicheng Zhu and Danqing Zhang. "Microstructure characteristics of Inconel 625 superalloy manufactured by selective laser melting." *Journal of Materials Science & Technology* 31, no. 9 (2015): 946–952.
[32] Jafari, Davoud, Tom HJ Vaneker and Ian Gibson. "Wire and arc additive manufacturing: Opportunities and challenges to control the quality and accuracy of manufactured parts." Materials & Design 202 (2021): 109471.
[33] Huang, Hui, Ninshu Ma, Jian Chen, Zhili Feng and Hidekazu Murakawa. "Toward large-scale simulation of residual stress and distortion in wire and arc additive manufacturing." *Additive manufacturing* 34 (2020): 101248.
[34] Chauvet, Edouard, Paraskevas Kontis, Eric A. Jägle, Baptiste Gault, Dierk Raabe, Catherine Tassin, Jean-Jacques Blandin et al. "Hot cracking mechanism affecting a non-weldable Ni-based superalloy produced by selective electron Beam Melting." *Acta Materialia* 142 (2018): 82–94.
[35] O'Donnell, Katie, Maria J. Quintana and Peter C. Collins. "Understanding the Effect of Electron Beam Melting Scanning Strategies on the Aluminum Content and Materials State of Single Ti-6Al-4V Feedstock." Materials 16, no. 19 (2023): 6366.
[36] Svetlizky, David, Mitun Das, Baolong Zheng, Alexandra L. Vyatskikh, Susmita Bose, Amit Bandyopadhyay, Julie M. Schoenung, Enrique J. Lavernia and Noam Eliaz. "Directed energy deposition (DED) additive manufacturing: Physical characteristics, defects, challenges and applications." *Materials Today* 49 (2021): 271–295.
[37] Imran, Muhammad Mu'az, Azam Che Idris, Liyanage Chandratilak De Silva, Yun-Bae Kim and Pg Emeroylariffion Abas. "Advancements in 3D printing: directed energy deposition techniques, defect analysis and quality monitoring." *Technologies* 12, no. 6 (2024): 86.
[38] Wang, Zhen, Jiang Wang, Shurong Xu, Bo Liu, Qingxuan Sui, Fengjun Zhao, Le Gong and Jun Liu. "Influence of powder characteristics on microstructure and mechanical properties of Inconel 718 superalloy manufactured by direct energy deposition." *Applied Surface Science* 583 (2022): 152545.
[39] Dejene, Naol Dessalegn and Hirpa G. Lemu. "Current status and challenges of powder bed fusion-based metal additive manufacturing: literature review." Metals 13, no. 2 (2023): 424.
[40] Liu, Jikai, Andrew T. Gaynor, Shikui Chen, Zhan Kang, Krishnan Suresh, Akihiro Takezawa, Lei Li et al. "Current and future trends in topology optimization for additive manufacturing." Structural and multidisciplinary optimization 57, no. 6 (2018): 2457–2483.
[41] Wang, Xun, Lianyong Xu, Lei Zhao, Wenjing Ren, Qiwen Li and Yongdian Han. "Machine learning method for estimating the defect-related mechanical properties of additive manufactured alloys." *Engineering Fracture Mechanics* 291 (2023): 109559.
[42] Nikam, Sumedh, Shailendra P. Shisode and Pralhad Pesode. "A Review on Magnesium Matrix Composites for Biomedical Applications: Materials, Fabrication Techniques, Coatings, and Applications." *Regenerative Engineering and Translational Medicine* (2025): 1–33.
[43] Pesode, Pralhad, Shivprakash Barve and Shailendra Dayane. "Antibacterial coating on magnesium alloys by MAO for biomedical applications." *Research on Biomedical Engineering* 40, no. 2 (2024): 409–433.

[44] Pesode, Pralhad, Shivprakash Barve, Sagar V. Wankhede and Manoj Mugale. "Manufacturing methods of Mg alloys for biomedical applications." In Magnesium Alloys for Biomedical Applications, pp. 20–61. CRC Press, 2024.

[45] Wankhede, Sagar V., Akbar Ahmad, Pralhad Pesode and Manoj Mugale. "Reuse, remanufacturing, and recycling of mg alloys." In Magnesium Alloys for Biomedical Applications, pp. 189–206. CRC Press, 2024.

[46] Manjhi, Shambhu Kumar, Prithivirajan Sekar, Srikanth Bontha and A. S. S. Balan. "Additive manufacturing of magnesium alloys: characterization and post-processing." *International Journal of Lightweight Materials and Manufacture* 7, no. 1 (2024): 184–213.

[47] Davis, Bruce. "The application of magnesium alloys in aircraft interiors—changing the rules." In Magnesium Technology 2015, pp. 5-5. Cham: Springer International Publishing, 2015.

[48] Pesode, Pralhad, Shivprakash Barve, Yogesh Mane, Shailendra Dayane, Snehal Kolekar and Kahtan A. Mohammed. "Recent advances on biocompatible coating on magnesium alloys by micro arc oxidation technique." *Key Engineering Materials* 944 (2023): 117–134.

[49] Wang, Gerry Gang and J. P. Weiler. "Recent developments in high-pressure die-cast magnesium alloys for automotive and future applications." *Journal of Magnesium and Alloys* 11, no. 1 (2023): 78–87.

[50] Hirsch, Juergen and Talal Al-Samman. "Superior light metals by texture engineering: Optimized aluminum and magnesium alloys for automotive applications." *Acta Materialia* 61, no. 3 (2013): 818–843.

[51] Chen, Jiayu and Bin Chen. "Progress in additive manufacturing of magnesium alloys: a review." Materials 17, no. 15 (2024): 3851.

[52] Pesode, Pralhad and Shivprakash Barve. "Additive manufacturing of metallic biomaterials and its biocompatibility." Materials Today: Proceedings (2022).

[53] Gneiger, Stefan, Johannes A. Österreicher, Aurel R. Arnoldt, Alois Birgmann and Martin Fehlbier. "Development of a high strength magnesium alloy for wire arc additive manufacturing." Metals 10, no. 6 (2020): 778.

[54] Takagi, Hisataka, Hiroyuki Sasahara, Takeyuki Abe, Hiroki Sannomiya, Shinichiro Nishiyama, Shuichiro Ohta and Kunimitsu Nakamura. "Material-property evaluation of magnesium alloys fabricated using wire-and-arc-based additive manufacturing." *Additive Manufacturing* 24 (2018): 498–507.

[55] Armstrong, Mark, Hamid Mehrabi and Nida Naveed. "An overview of modern metal additive manufacturing technology." *Journal of Manufacturing Processes* 84 (2022): 1001–1029.

[56] Rouf, Saquib, Abrar Malik, Navdeep Singh, Ankush Raina, Nida Naveed, Md Irfanul Haque Siddiqui and Mir Irfan Ul Haq. "Additive manufacturing technologies: Industrial and medical applications." *Sustainable Operations and Computers* 3 (2022): 258–274.

[57] Gibson, Ian, David Rosen and Brent Stucker. "Directed energy deposition processes." In Additive manufacturing technologies: 3D printing, rapid prototyping, and direct digital manufacturing, pp. 245–268. New York, NY: Springer New York, 2015.

[58] Mostafaei, Amir, Amy M. Elliott, John E. Barnes, Fangzhou Li, Wenda Tan, Corson L. Cramer, Peeyush Nandwana and Markus Chmielus. "Binder jet 3D printing—Process parameters, materials, properties, modeling, and challenges." Progress in Materials Science 119 (2021): 100707.

[59] Gibson, Ian, David Rosen, Brent Stucker, Mahyar Khorasani, David Rosen, Brent Stucker and Mahyar Khorasani. Additive manufacturing technologies. Vol. 17. Cham, Switzerland: Springer, 2021.

[60] Joshua, Rajan John Nekin, Sakthivel Aravind Raj, Mohamed Thariq Hameed Sultan, Andrzej Łukaszewicz, Jerzy Józwik, Zbigniew Oksiuta, Krzysztof Dziedzic, Arkadiusz Tofil and Farah Syazwani Shahar. "Powder bed fusion 3D printing in precision manufacturing for biomedical applications: a comprehensive review." Materials 17, no. 3 (2024): 769.

[61] Tebianian, Mohaddeseh, Sara Aghaie, Nazanin Sadat Razavi Jafari, Seyed Reza Elmi Hosseini, António B. Pereira, Fábio AO Fernandes, Mojtaba Farbakhti, Chao Chen and Yuanming Huo. "A review of the metal additive manufacturing processes." Materials 16, no. 24 (2023): 7514.

[62] Gao, Bingwei, Hongjian Zhao, Liqing Peng and Zhixin Sun. "A review of research progress in selective laser melting (SLM)." Micromachines 14, no. 1 (2022): 57.

[63] Srivastava, Manu, Sandeep Rathee, Vivek Patel, Atul Kumar and Praveennath G. Koppad. "A review of various materials for additive manufacturing: Recent trends and processing issues." *Journal of Materials Research and Technology* 21 (2022): 2612–2641.

[64] Konda Gokuldoss, Prashanth, Sri Kolla and Jürgen Eckert. "Additive manufacturing processes: Selective laser melting, electron beam melting and binder jetting—Selection guidelines." materials 10, no. 6 (2017): 672.

[65] Nandhakumar, R. and K. Venkatesan. "A process parameters review on selective laser melting-based additive manufacturing of single and multi-material: Microstructure, physical properties, tribological and surface roughness." *Materials Today Communications* 35 (2023): 105538.

[66] Gupta, Khanish and Kusum Meena. "Artificial bone scaffolds and bone joints by additive manufacturing: A review." *Bioprinting* 31 (2023): e00268.

[67] Periyasamy, Manikandan and N. Srinivasan. "A Review on Auxetic Structures: Applications, Properties, Classifications, Finite Element Analysis, Manufacturing Methods and Selective Laser Melting Process Parameters." *Results in Engineering* (2025): 107660.

[68] Santos, Luis S., Satyandra K. Gupta and Hugh A. Bruck. "Simulation of buckling of internal features during selective laser sintering of metals." *Additive Manufacturing* 23 (2018): 235–245.

[69] Kundakcioglu, Erdem, Ismail Lazoglu and Suraj Rawal. "Transient thermal modeling of laser-based additive manufacturing for 3D freeform structures." *The International Journal of Advanced Manufacturing Technology* 85, no. 1 (2016): 493–501.

[70] Megahed, Mustafa, Hans-Wilfried Mindt, Narcisse N'Dri, Hongzhi Duan and Olivier Desmaison. "Metal additive-manufacturing process and residual stress modeling." *Integrating Materials and Manufacturing Innovation* 5, no. 1 (2016): 61–93.

[71] Steuben, John C., Athanasios P. Iliopoulos and John G. Michopoulos. "Discrete element modeling of particle-based additive manufacturing processes." *Computer Methods in Applied Mechanics and Engineering* 305 (2016): 537–561.

[72] Shen, Fei, Shangqin Yuan, Chee Kai Chua and Kun Zhou. "Development of process efficiency maps for selective laser sintering of polymeric composite powders: Modeling and experimental testing." *Journal of Materials Processing Technology* 254 (2018): 52–59.

[73] Peyre, Patrice, Yann Rouchausse, Denis Defauchy and Gilles Régnier. "Experimental and numerical analysis of the selective laser sintering (SLS) of PA12 and PEKK semi-crystalline polymers." *Journal of Materials Processing Technology* 225 (2015): 326–336.

[74] Salonitis, Konstantinos, Laurent D'Alvise, Babis Schoinochoritis and Dimitrios Chantzis. "Additive manufacturing and post-processing simulation: laser cladding followed by high speed machining." *The International Journal of Advanced Manufacturing Technology* 85, no. 9 (2016): 2401–2411.

[75] Kim, Mingyu, Minseok Kang and Hyun Chung. "Simplified welding distortion analysis for fillet welding using composite shell elements." *International Journal of Naval Architecture and Ocean Engineering* 7, no. 3 (2015): 452–465.

[76] Uçak, Necati, Adem Çiçek and Kubilay Aslantas. "Machinability of 3D printed metallic materials fabricated by selective laser melting and electron beam melting: A review." *Journal of Manufacturing Processes* 80 (2022): 414–457.

[77] Körner, Carolin. "Additive manufacturing of metallic components by selective electron beam melting—a review." *International Materials Reviews* 61, no. 5 (2016): 361–377.

[78] Rafi, H.K., N.V. Karthik, Haijun Gong, Thomas L. Starr and Brent E. Stucker. "Microstructures and mechanical properties of Ti6Al4V parts fabricated by selective laser melting and electron

beam melting." *Journal of materials engineering and performance* 22, no. 12 (2013): 3872–3883.

[79] Dezaki, Mohammadreza Lalegani, Ahmad Serjouei, Ali Zolfagharian, Mohammad Fotouhi, Mahmoud Moradi, M.K.A. Ariffin and Mahdi Bodaghi. "A review on additive/subtractive hybrid manufacturing of directed energy deposition (DED) process." *Advanced powder materials 1*, no. 4 (2022): 100054.

[80] Bajaj, Priyanshu, Avinash Hariharan, Anoop Kini, Philipp Kürnsteiner, Dierk Raabe and Eric Aimé Jägle. "Steels in additive manufacturing: A review of their microstructure and properties." *Materials Science and Engineering*: A 772 (2020): 138633.

[81] Jinoop, Arackal Narayanan, C.P. Paul and K.S. Bindra. "Laser-assisted directed energy deposition of nickel super alloys: a review." Proceedings of the Institution of Mechanical Engineers, Part L: *Journal of Materials: Design and Applications* 233, no. 11 (2019): 2376–2400.

[82] Yi, Hui-Jun, Jin-Woo Kim, Young-Lak Kim and Sangyong Shin. "Effects of cooling rate on the microstructure and tensile properties of wire-arc additive manufactured Ti–6Al–4V alloy." *Metals and Materials International* 26, no. 8 (2020): 1235–1246.

[83] Saboori, Abdollah, Donato Gallo, Sara Biamino, Paolo Fino and Mariangela Lombardi. "An overview of additive manufacturing of titanium components by directed energy deposition: microstructure and mechanical properties." *Applied Sciences* 7, no. 9 (2017): 883.

[84] Moradi, Mahmoud, Arman Hasani, Zeynab Pourmand and Jonathan Lawrence. "Direct laser metal deposition additive manufacturing of Inconel 718 superalloy: Statistical modelling and optimization by design of experiments." *Optics & Laser Technology* 144 (2021): 107380.

[85] Yan, Chenxiao, Hongbiao Han, Jiayang Hu, Peng Zhang and Jun Li. "The Influence of Process Parameters on the Morphology of Laser Wire Additive Manufacturing." *International Journal of Optics* 2025, no. 1 (2025): 3085857.

[86] Moradi, Mahmoud, Ali Ashoori and Arman Hasani. "Additive manufacturing of stellite 6 superalloy by direct laser metal deposition–Part 1: Effects of laser power and focal plane position." *Optics & Laser Technology* 131 (2020): 106328.

[87] Wang, Xinlin, Dewei Deng, Yingbin Hu, Fuda Ning, Hui Wang, Weilong Cong and Hongchao Zhang. "Overhang structure and accuracy in laser engineered net shaping of Fe-Cr steel." *Optics & Laser Technology* 106 (2018): 357–365.

[88] Shah, Abid, Rezo Aliyev, Henning Zeidler and Stefan Krinke. "A review of the recent developments and challenges in wire arc additive manufacturing (WAAM) process." *Journal of Manufacturing and Materials Processing* 7, no. 3 (2023): 97.

[89] Frazier, William E. "Metal additive manufacturing: a review." *Journal of Materials Engineering and performance* 23, no. 6 (2014): 1917–1928.

[90] Ahn, Dong-Gyu. "Direct metal additive manufacturing processes and their sustainable applications for green technology: A review." *International Journal of Precision Engineering and Manufacturing-Green Technology* 3, no. 4 (2016): 381–395.

[91] Ahn, Dong-Gyu. "Direct metal additive manufacturing processes and their sustainable applications for green technology: A review." *International Journal of Precision Engineering and Manufacturing-Green Technology* 3, no. 4 (2016): 381–395.

[92] Xiong, Jun and Guangjun Zhang. "Adaptive control of deposited height in GMAW-based layer additive manufacturing." *Journal of Materials Processing Technology* 214, no. 4 (2014): 962–968.

[93] Tawfik, M.M., M.M. Nemat-Alla and M.M. Dewidar. "Enhancing the properties of aluminum alloys fabricated using wire+ arc additive manufacturing technique-a review." *Journal of materials research and technology* 13 (2021): 754–768.

[94] Treutler, Kai and Volker Wesling. "The current state of research of wire arc additive manufacturing (WAAM): a review." *Applied Sciences* 11, no. 18 (2021): 8619.

[95] Rodríguez-González, Paula, Elisa María Ruiz-Navas and Elena Gordo. "Wire arc additive manufacturing (WAAM) for aluminum-lithium alloys: a review." *Materials* 16, no. 4 (2023): 1375.

[96] Soni, Neetesh, Gilda Renna and Paola Leo. "Advancements in metal processing additive technologies: selective laser melting (SLM)." Metals 14, no. 9 (2024): 1081.

[97] Kim, Moo-Sun. "Effects of processing parameters of selective laser melting process on thermal conductivity of alsi10mg alloy." *Materials* 14, no. 9 (2021): 2410.

[98] Wysocki, Bartłomiej, Joanna Idaszek, Joanna Zdunek, Krzysztof Rożniatowski, Marcin Pisarek, Akiko Yamamoto and Wojciech Święszkowski. "The influence of selective laser melting (SLM) process parameters on in-vitro cell response." *International journal of molecular sciences* 19, no. 6 (2018): 1619.

[99] Kaufmann, Niko, Muhammad Imran, Tim Marten Wischeropp, Claus Emmelmann, Shafaqat Siddique and Frank Walther. "Influence of process parameters on the quality of aluminium alloy EN AW 7075 using selective laser melting (SLM)." *Physics Procedia* 83 (2016): 918–926.

[100] Delgado, Jordi, Joaquim Ciurana and Ciro A. Rodríguez. "Influence of process parameters on part quality and mechanical properties for DMLS and SLM with iron-based materials." *The International Journal of Advanced Manufacturing Technology* 60, no. 5 (2012): 601–610.

[101] Zhang, Bi, Yongtao Li and Qian Bai. "Defect formation mechanisms in selective laser melting: a review." *Chinese Journal of Mechanical Engineering* 30, no. 3 (2017): 515–527.

[102] Xiang, Hongliang, Guanglei Chen, Wei Zhao and Chaochao Wu. "Densification behavior and build quality of duplex stainless steel fabricated by laser powder bed fusion." *Metals* 13, no. 4 (2023): 741.

[103] Pesode, Pralhad and Shivprakash Barve. "A grey relational analysis of the micro arc oxidation process parameters and their effects on Ti-6Al-7Nb coating performance and corrosion resistance for biomedical uses." *Engineering Research Express* 6, no. 4 (2024): 045512.

[104] Wankhede, Sagar, Pralhad Pesode, Sanjay Gaikwad, Sumod Pawar and Amar Chipade. "Implementing combinative distance base assessment (CODAS) for selection of natural fibre for long lasting composites." *In Materials Science Forum*, vol. 1081, pp. 41–48. Trans Tech Publications Ltd, 2023.

[105] Wankhede, Sagar, Pralhad Pesode, Sumod Pawar and Rayan Lobo. "Comparison study of GRA, COPRAS and MOORA for ranking of phase change material for cooling system." *Materials Today: Proceedings* (2023).

[106] Li, Yan, Chen Su and Jianjun Zhu. "Comprehensive review of wire arc additive manufacturing: Hardware system, physical process, monitoring, property characterization, application and future prospects." *Results in Engineering* 13 (2022): 100330.

[107] Podvezko, V., The comparative analysis of MCDA methods SAW and COPRAS. Inzinerine *Ekonomika-Engineering Economics*, 22(2), (2011); 134–146.

[108] Zagorskas, J., Burinskiene, M., Zavadskas, E. and Turskis, Z., Urbanistic assessment of city compactness on the basis of GIS applying the COPRAS method. Ekologija, 53, (2007); 55–63.

[109] Zavadskas, E.K., Kaklauskas, A., Peldschus, F. and Turskis, Z., Multi-attribute assessment of road design solutions by using the COPRAS method. *Baltic Journal of Road & Bridge Engineering*, 2(4), (2007); 193–203.

[110] Zavadskas, E.K., Kaklauskas, A., Turskis, Z. and Tamosaitiene, J., Selection of the effective dwelling house walls by applying attributes values determined at intervals. *Journal of Civil Engineering and Management*, 14(2), (2008); 85–93.

[111] Brauers, W.K., Zavadskas, E.K., The MOORA method and its application to privatization in a transition economy, Control and Cybernetics, 35, (2006); 445–469.

Index

About the Editors

Vijaykumar S. Jatti is a Professor in the Symbiosis Skills and Professional University, Kiwale, Pune, Maharashtra, India. He earned his Ph.D. in Mechanical Engineering from Symbiosis International (Deemed University). He also holds an M.Tech in Mechanical Engineering from National Institute of Technology Calicut and a B.E. in Mechanical Engineering from Visvesvaraya Technological University. He brings nearly 15 years of combined teaching, research, and industrial experience. His scholarly contributions include over 100 publications indexed in major databases such as Web of Science and Scopus, appearing in reputed journals and conference proceedings. He has 10 published patents to his credit and has authored 12 book chapters along with 11 books with leading publishers. He actively contributes to the academic community as a reviewer for international journals and conferences. He has successfully completed a funded research project and has guided several Ph.D. scholars, while also serving on doctoral committees. In addition, he has delivered invited talks at numerous academic and industrial forums. His contributions to academia and research have been recognized with 16 awards and appreciations.

A. Saiyathibrahim is currently affiliated with the Department of Mechanical Engineering, Saveetha School of Engineering, Saveetha Institute of Medical and Technical Sciences, Chennai, Tamil Nadu, India. He holds a Ph.D. in Mechanical Engineering from Anna University, with a specialization in functionally graded materials. His research interests encompass advanced machining, centrifugal casting, functionally graded materials, and additive manufacturing. He has contributed extensively to the field through numerous book chapters, research articles in reputed scientific journals, and patented innovations. With over thirteen years of academic experience, he is also an active member of several professional bodies, including ISME, IAAM, IAAC, and ISDS.

For Product Safety Concerns and Information please contact our
EU representative GPSR@taylorandfrancis.com Taylor & Francis
Verlag GmbH, Kaufingerstraße 24, 80331 München, Germany